Jakub Yaghob (Ed.)

ITAT 2015: Information Technologies – Applications and Theory

Proceedings of the 15th conference ITAT 2015

Slovenský Raj, Slovakia, September 17–21, 2015

ITAT 2015: Information Technologies – Applications and Theory
Proceedings of the 15th conference ITAT 2015
Hotel Čingov, Slovenský Raj, Slovakia, September 17-21, 2015
Jakub Yaghob (Ed.)
Cover design: Róbert Novotný
Cover photo: Zdeněk Svoboda

Publisher: CreateSpace Independent Publishing Platform, 2015
ISBN 978-1515120650

Also published online by CEUR Workshop Proceedings vol. 1422
http://ceur-ws.org/Vol-1422/
ISSN 1613-0073

These proceedings contain papers from the conference ITAT 2015. All authors agreed to publish their papers in these proceedings. All papers were reviewed by at least two anonymous referees.

http://www.itat.cz/

Introduction

This volume contains papers from the main track and associated workshops of the 15th ITAT conference. The conference was held in Hotel Čingov, Slovenský Raj, Slovakia on September 17–21, 2015.

ITAT is a computer science conference with the primary goal of presenting new results of young researchers and doctoral students from Slovakia and the Czech Republic. The conference serves as a platform for exchange of information within the community, and also provides opportunities for informal meetings of the participants in a mountainous regions of the Czech Republic and Slovakia.

The traditional topics of the conference include software engineering, data processing and knowledge representation, information security, theoretical foundations of computer science, computational intelligence, parallel and distributed computing, natural language processing, and computer science education.

The conference accepts papers describing original previously unpublished results, significant work-in-progress reports, as well as reviews of special topics of interest to the conference audience.

The conference program this year included the main track of 9 contributed papers, two invited lectures, and two workshops. Two specialized workshops were held as a part of the conference:

- Slovenskočeský NLP workshop — SloNLP
  (organized by Petra Barančíková and Rudolf Rosa)

- Computational Intelligence and Data Mining — WCIDM
  (organized by Martin Holeňa)

Overall, 31 papers were submitted to all conference tracks. These proceedings present in the first part 9 papers of the main track which were selected by the program committee based on at least two reviews by the program committee members. The second part contains 6 papers of SloNLP and the last part presents 13 papers of the WCIDM. All workshop papers were anonymously reviewed and selected on the base of at least two reviews by their corresponding program comittees.

These proceedings also contain abstract of two invited lectures by Zbyněk Falt (Google Zürich) and Ivan Zelinka (VŠB-Technical University of Ostrava).

I would like to thank all program committee members, conference organizers, invited speakers, and authors of the papers for helping to create an exciting scientic program for ITAT 2015. Special thanks deserves H. Bílková for preparing the conference proceedings.

Jakub Yaghob
Charles University in Prague
Chair of the Program Committee

Steering Committee of ITAT 2015

Peter Vojtáš, Charles University in Prague (chair)
Tomáš Horváth, Pavol Jozef Šafárik University in Košice
Filip Zavoral, Charles University in Prague
Martin Holeňa, Academy of Sciences of the Czech Republic
Tomáš Vinař, Comenius University in Bratislava

Program Committee of ITAT 2015

Jakub Yaghob, Charles University in Prague (chair)
David Bednárek, Charles University in Prague
Mária Bieliková, Slovak University of Technology in Bratislava
Broňa Brejová, Comenius University in Bratislava
Marek Ciglan, Slovak Academy of Sciences
Jiří Dokulil, Charles University in Prague
Tomáš Holan, Charles University in Prague
Martin Holeňa, Academy of Sciences of the Czech Republic
Tomáš Horváth, Pavol Jozef Šafárik University in Košice
Daniela Chudá, Slovak University of Technology in Bratislava
Jozef Jirásek, Pavol Jozef Šafárik University in Košice
Jana Katreniaková, Comenius University in Bratislava
Rastislav Kráľovič, Comenius University in Bratislava
Michal Krátký, VŠB-Technical University of Ostrava
Martin Kruliš, Charles University in Prague
Věra Kůrková, Academy of Sciences of the Czech Republic
Markéta Lopatková, Charles University in Prague
Dana Pardubská, Comenius University in Bratislava
Štefan Pero, Pavol Jozef Šafárik University in Košice
Tomáš Plachetka, Comenius University in Bratislava
Martin Plátek, Charles University in Prague
Jaroslav Pokorný, Charles University in Prague
Karel Richta, Charles University in Prague
Gabriel Semanišin, Pavol Jozef Šafárik in Košice
Roman Špánek, Academy of Sciences of the Czech Republic
Ondrej Šuch, Matej Bel University Banská Bystrica
Tomáš Vinař, Comenius University in Bratislava
Filip Zavoral, Charles University in Prague

Organizing Committee

Peter Gurský, Pavol Jozef Šafárik University in Košice (chair)
Július Malčovský, Pavol Jozef Šafárik University in Košice

Contents

Building a Computer System for the World's Information

Zbyněk Falt

Google Zürich, Switzerland

Abstract: Google's mission is to organize the world's information and make it universally accessible and useful. Obviously, Google has to deal with many problems to successfully fulfill such mission. The talk will address some of these problems with particular emphasis on the issues regarding efficient and scalable information storage, processing, and querying.

The first part of the talk will be focused on the computing platforms used at Google and its hardware design philosophy. We will also explain why it could be beneficial to build warehouse-scale computer facilities using commodity components instead of enterprise components, which are more reliable.

The second part will focus on the software infrastructure. We will describe the architecture of Google filesystem, BigTable data storage, Chubby lock service, and the MapReduce framework. Finally, we will demonstrate how are these technologies combined when solving real-life problems.

Zbyněk Falt (born in 1985) graduated at the Charles University in Prague in 2010. He continued his studies at the Charles University and obtained Ph.D. degree in Software Systems in 2014. His research area was parallel computing and its optimization for modern hardware. Since 2014 he works in Google Zürich as a software engineer in the Search Infrastructure team.

ITAT

Evolutionary Algorithms — Selected Topics

Ivan Zelinka

Department of Computer Science, Faculty of Electrical Engineering and Computer Science VŠB-TUO

Abstract: This keynote is focused on mutual intersection of interesting fields of research: bio-inspired algorithms, deterministic chaos and complex systems.

The first part will discuss main principles of bio-inspired methods, its historical background and its use on various examples including real world ones. Examples include plasma reactor control, optimal signal routing in the network of portable meteorological stations, complex system design as antenna design, nonlinear system and controllers design and more. Also its use on deterministic chaos control with focusing on simple chaotic systems (logistic, Hennon, . . .) as well as CML systems exhibiting spatiotemporal chaos will be mentioned and explained.

The second part will discuss use of deterministic chaos instead of pseudo-random number generators inside evolutionary algorithms with application on well known evolutionary algorithms (differential evolution, PSO, SOMA, genetic algorithms, . . .) and test functions. Mutual comparison will be presented, based on our research. Also will be discussed question whether evolutionary dynamics really need pseudo- random numbers.

At the end will be mentioned a novel approach joining evolutionary dynamics, complex networks and CML systems exhibiting chaotic behavior. Reported methodology and results are based on actual state of art (that is a part of this tutorial) as well as on our own research.

Ivan Zelinka (born in 1965, ivanzelinka.eu) is currently associated with the Technical University of Ostrava (VSB-TU), Faculty of Electrical Engineering and Computer Science. He graduated consequently at the Technical University in Brno (1995 – MSc.), UTB in Zlin (2001 - Ph.D.) and again at Technical University in Brno (2004 – Assoc. Prof.) and VSB-TU (2010 – Professor).

Prof. Zelinka is responsible supervisor of several grant researches of Czech grant agency GAČR as for example Unconventional Control of Complex Systems, Security of Mobile Devices and Communication (bilateral project between Czech and Vietnam) and co-supervisor of grant FRVŠ – Laboratory of parallel computing amongst the others. He was also working on numerous grants and two EU projects as member of team (FP5 – RESTORM) and supervisor (FP7 – PROMOEVO) of the Czech team. He is also head of research team NAVY http://navy.cs.vsb.cz/.

Prof. Zelinka was awarded by Siemens Award for his Ph.D. thesis, as well as by journal Software news for his book about artificial intelligence. He is a member of the British Computer Society, Machine Intelligence Research Labs (MIR Labs – http://www.mirlabs.org/czech.php), IEEE (committee of Czech section of Computational Intelligence), a few international program committees of various conferences, and three international journals. He is also the founder and editor-in-chief of a new book series entitled Emergence, Complexity and Computation (Springer series 10624, see also www.ecc-book.eu).

J. Yaghob (Ed.): ITAT 2015 pp. 3–8
Charles University in Prague, Prague, 2015

ITAT

Stability of Extents in One-Sided Fuzzy Concept Lattices*

Ľubomír Antoni, Stanislav Krajči, and Ondrej Krídlo

Institute of Computer Science, Faculty of Science,
Pavol Jozef Šafárik University in Košice, Jesenná 5, 040 01 Košice
lubomir.antoni@student.upjs.sk, stanislav.krajci@upjs.sk, ondrej.kridlo@upjs.sk

Abstract: The efficient selection of relevant extents is an important issue for investigation in formal concept analysis. The notion of stability has been adopted for this reasoning. We present three different methods for evaluation of stability and we summarize the comparative remarks.

1 Introduction

The efficient selection of relevant formal concepts is an interesting and important issue for investigation and several studies have focused on this scalability question in formal concept analysis. The stability index [27] represents the proportion of subsets of attributes of a given concept whose closure is equal to the extent of this concept (in an extensional formulation). A high stability index signalizes that extent does not disappear if the extent of some of its attributes is modified. It helps to isolate concepts that appear because of noisy objects in [22] and the complete restoring of the original concept lattice is possible with combination of two other indices. The phenomenon of the basic level of concepts is advocated to select important formal concepts in [11]. Five quantitative approaches on the basic level of concepts and their metrics are comparatively analyzed in [12]. The approaches on selecting of the formal concepts and simplifying the concept lattices are examined in [15], as well.

In this paper, we present three methods which concern the selection of the relevant formal concepts from the set of all one-sided fuzzy formal concepts. We recall the modified Rice-Siff algorithm, extend the results on the quality subset measure and propose a new index for the stability of one-sided fuzzy formal concepts taking into account the probabilistic aspects in the fuzzy formal contexts. We would like to emphasize that the best results one can obtain by the combination of various methods.

2 Classical approach

A central role in this section will be played by the notions of a formal context (Fig. 1), a polar (Fig. 2), a formal concept and a concept lattice (Fig. 3). We recall the definitions and we refer to [18] for more details.

Definition 1. *Let B and A be the nonempty sets and let $R \subseteq B \times A$ be a relation between B and A. A triple $\langle B, A, R \rangle$ is called a formal context, the elements of set B are called objects, the elements of set A are called attributes and the relation R is called incidence relation.*

	i	ii	iii
a	×		
b	×	×	
c		×	

Figure 1: A formal context $\langle \{a,b,c\}, \{i,ii,iii\}, R \rangle$

Definition 2. *Let $\langle B, A, R \rangle$ be a formal context and $X \in \mathscr{P}(B)$, $Y \in \mathscr{P}(A)$. Then the maps $\nearrow: \mathscr{P}(B) \to \mathscr{P}(A)$ and $\swarrow: \mathscr{P}(A) \to \mathscr{P}(B)$ defined by*

$$\nearrow (X) = X^{\nearrow} = \{y \in A : (\forall x \in X)\langle x, y \rangle \in R\}$$

and

$$\swarrow (Y) = Y^{\swarrow} = \{x \in B : (\forall y \in Y)\langle x, y \rangle \in R\}$$

are called concept-forming operators (also called derivation operators or polars) of a given formal context.

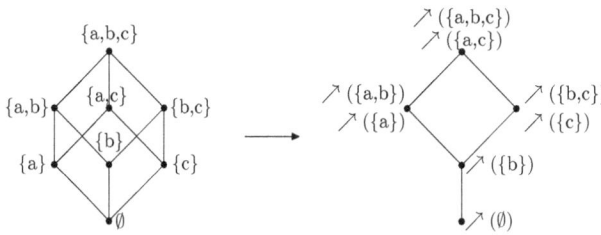

Figure 2: Polar $\nearrow$ of $(\mathscr{P}(B), \subseteq)$

Definition 3. *Let $\langle B, A, R \rangle$ be a formal context, $\nearrow$ and $\swarrow$ are concept-forming operators and $X \in \mathscr{P}(B), Y \in \mathscr{P}(A)$. A pair $\langle X, Y \rangle$ such that $X^{\nearrow} = Y$ and $Y^{\swarrow} = X$ is called a formal concept of a given formal context. The set X is called extent of a formal concept and the set Y is called intent of a formal concept. The set of all formal concepts of a formal context $\langle B, A, R \rangle$ is a set*

$$\mathscr{C}(B, A, R) = \{\langle X, Y \rangle \in \mathscr{P}(B) \times \mathscr{P}(A) : X^{\nearrow} = Y, \ Y^{\swarrow} = X\}.$$

*This work was supported by the Scientific Grant Agency of the Ministry of Education, Science, Research and Sport of the Slovak Republic under contract VEGA 1/0073/15.

Definition 4. *Let* $\langle X_1, Y_1 \rangle, \langle X_2, Y_2 \rangle \in C(B,A,R)$ *be two formal concepts of a formal context* $\langle B,A,R \rangle$. *Let* $\preceq$ *be a partial order in which* $\langle X_1, Y_1 \rangle \preceq \langle X_2, Y_2 \rangle$ *if and only if* $X_1 \subseteq X_2$. *A partially ordered set* $(\mathscr{C}(B,A,R), \preceq)$ *is called a concept lattice of a given context and is denoted by* $\mathrm{CL}(B,A,R)$.

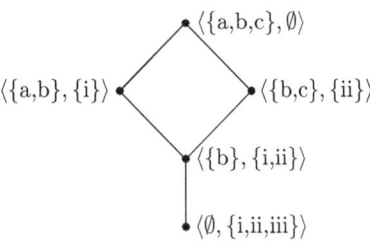

Figure 3: A concept lattice of $\langle \{\text{a,b,c}\}, \{\text{i,ii,iii}\}, R \rangle$

3   One-sided fuzzy approach

The statements that people use to communicate facts about the world are usually not bivalent. The truth of such statements is a matter of degree, rather than being only true or false. Fuzzy logic and fuzzy set theory are frameworks which extend formal concept analysis in various independent ways [5, 6, 9, 23]. Here, we recall the basic definitions of fuzzy formal context. The structures of partially ordered set, complete lattice or residuated lattice are applied here to represent data. The last one allows to speed up the computing.

Definition 5. *Consider two nonempty sets B a A, a set of truth degrees T and a mapping R such that* $R : B \times A \longrightarrow T$. *Then the triple* $\langle B,A,R \rangle$ *is called a (T)-fuzzy formal context, the elements of the sets B and A are called objects and attributes, respectively. The mapping R is a fuzzy incidence relation.*

In the definition of (T)-fuzzy formal context, we often take the interval $T = [0,1]$, because it is a scale of truth degrees commonly used in many applications. For such replacement, the terminology of $[0,1]$-fuzzy formal context has been adopted. Analogously, one can define the (more general) notion of L-fuzzy formal context, or P-fuzzy formal context, having replaced the interval $[0,1]$ by the algebraic structures of complete residuated lattice L, partially ordered set P or other plausible scale of truth degrees. Several extensions were advocated by the authors to provide the knowledge extraction from (T)-fuzzy formal contexts, whereby the set of truth degrees $T \in \{L; P; [0,1]; \{0,0.5,1\}; \{a_1, \ldots, a_n\}; \ldots\}$ is frequently selected.

The one-sided approach [23] one can represent by the concept-forming operators in a non-symmetric way. For $[0,1]$-fuzzy formal context and for every crisp subset of

R	a_1	a_2	a_3
b_1	1	0.9	0.8
b_2	0.8	0.7	0.7
b_3	0.3	0.3	0.3
b_4	0.8	0.6	0.9

Figure 4: Example of $[0,1]$-fuzzy formal context

objects X, the first function assigns the specific truth degree of the attribute (each object from X has this attribute at least in this specific truth degree):

Definition 6. *Let* $X \subseteq B$ *and* $\uparrow : \mathscr{P}(B) \longrightarrow [0,1]^A$. *Then* $\uparrow$ *is a mapping that assigns to every crisp set X of objects a fuzzy membership function* $X^\uparrow$ *of attributes, such that a value in a point* $a \in A$ *is:*

$$X^\uparrow(a) = \inf\{R(b,a) : b \in X\}. \tag{1}$$

Conversely, for each fuzzy membership function of attributes, the second concept-forming operator assigns the specific crisp set of objects (each included object has all attributes at least in a truth degree given by this fuzzy membership function):

Definition 7. *Let* $f : A \to [0,1]$ *and* $\downarrow : [0,1]^A \longrightarrow \mathscr{P}(B)$. *Then* $\downarrow$ *is a mapping that assigns to every fuzzy membership function f of attributes a crisp set* $\downarrow (f)$ *of objects, such that:*

$$f^\downarrow = \{b \in B : (\forall a \in A) R(b,a) \geq f(a)\}. \tag{2}$$

Lemma 1. *The pair* $\langle \uparrow, \downarrow \rangle$ *forms a Galois connection.*

Proof. Take $X, X_1, X_2 \subseteq B$ and $f, f_1, f_2 \in [0,1]^A$. The inequality $f_1 \leq f_2$ expresses that $f_1(a) \leq f_2(a)$ for all $a \in A$. From Eq. (1) and (2), it holds that

- $X_1 \subseteq X_2$ implies that $X_1^\uparrow \geq X_2^\uparrow$,

- $f_1 \leq f_2$ implies that $f_1^\downarrow \supseteq f_2^\downarrow$,

- $X \subseteq X^{\uparrow\downarrow}$,

- $f \leq f^{\downarrow\uparrow}$,

which are the assumptions on the pair of mappings to be a Galois connection. $\qquad\square$

In addition, the composition of Eq. (1) and (2) allows us to define the notion of one-sided fuzzy concept.

Definition 8. *Let* $X \subseteq B$ *and* $f \in [0,1]^A$. *The pair* $\langle X, f \rangle$ *is called a one-sided fuzzy concept, if* $X^\uparrow = f$ *and* $f^\downarrow = X$. *The crisp set of objects X is called the extent and the fuzzy membership function* $X^\uparrow$ *is called the intent of one-sided fuzzy concept.*

The set of all one-sided fuzzy concepts ordered by inclusion of extents forms a complete lattice, called one-sided fuzzy concept lattice, as introduced in [23]. This construction is a generalization of classical approach from [18]. The one-sided fuzzy concept lattices for a fuzzy formal context from Fig. 4 is illustrated in Fig. 5.

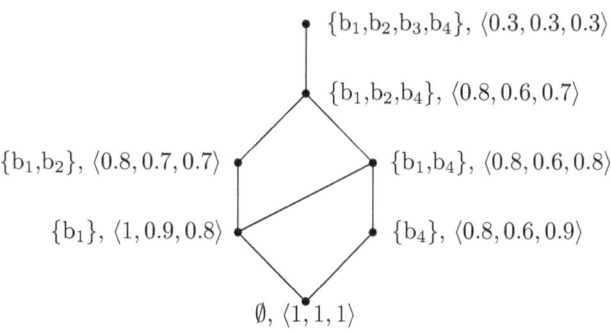

$\{b_1, b_2, b_3, b_4\}, \langle 0.3, 0.3, 0.3 \rangle$

$\{b_1, b_2, b_4\}, \langle 0.8, 0.6, 0.7 \rangle$

$\{b_1, b_2\}, \langle 0.8, 0.7, 0.7 \rangle$ $\{b_1, b_4\}, \langle 0.8, 0.6, 0.8 \rangle$

$\{b_1\}, \langle 1, 0.9, 0.8 \rangle$ $\{b_4\}, \langle 0.8, 0.6, 0.9 \rangle$

$\emptyset, \langle 1, 1, 1 \rangle$

Figure 5: One-sided fuzzy concept lattice for $[0,1]$-fuzzy formal context from Fig. 4

3.1 Modified Rice-Siff algorithm

In effort to reduce the number of one-sided fuzzy formal concepts, the Rice-Siff algorithm was modified and applied in [23, 25]. The method focuses on the distance function and its metric properties. The distance function $\rho : \mathscr{P}(B) \times \mathscr{P}(B) \to \mathbb{R}$ is defined for $X_1, X_2 \subseteq B$ by:

$$\rho(X_1, X_2) = 1 - \frac{\sum_{a \in A} \min\{\uparrow(X_1)(a), \uparrow(X_2)(a)\}}{\sum_{a \in A} \max\{\uparrow(X_1)(a), \uparrow(X_2)(a)\}}.$$

This function is a metric on the set of all extents. The function is a cornerstone of Alg. 1.

Algorithm 1. (Modified Rice-Siff algorithm)
input: $\langle B, A, R \rangle$
  $\mathscr{C} \leftarrow \mathscr{D} \leftarrow \{\{b\}^{\uparrow\downarrow} : b \in B\};$
  while $(|\mathscr{D} > 1|)$ *do* $\{$
    $m \leftarrow \min\{\rho(X_1, X_2) : X_1, X_2 \in \mathscr{D}, X_1 \neq X_2\}$
    $\Psi \leftarrow \{\langle X_1, X_2 \rangle \in \mathscr{D} \times \mathscr{D} : \rho(X_1, X_2) = m\}$
    $\mathscr{V} \leftarrow \{X \in \mathscr{D} : (\exists Y \in \mathscr{D})\langle X, Y \rangle \in \Psi\}$
    $\mathscr{N} \leftarrow \{(X_1 \cup X_2)^{\uparrow\downarrow} : \langle X_1, X_2 \rangle \in \Psi\}$
    $\mathscr{D} \leftarrow (\mathscr{D} \setminus \mathscr{V}) \cup \mathscr{N}$
    $\mathscr{C} \leftarrow \mathscr{C} \cup \mathscr{N}$
  $\}$
output: $\mathscr{C}$

Notice that set $\mathscr{D}$ is changed in each loop by excluding elements of $\mathscr{V}$ and joining a member of $\mathscr{N}$ in each loop. It assures that set $\mathscr{D}$ is still decreasing. More particular, two clusters with minimal distance are joined in each step of algorithm and the closure of their union is returned as the output. Such closures are gathered in a tree-based structure on the subset hierarchy with the cluster of all objects in the root. The zero iterations gather the closures of singletons, therefore the value of minimal distance function is not computed in the zero step. The more detailed properties of this clustering method with the special defined metric are described in [23, 24, 25].

3.2 Subset quality measure

Snášel et al. in [38] reflect the transformation of the original $[0,1]$-fuzzy formal context to the sequence of classical

formal contexts (from Definition 1) using the binary relations called α-cuts for $\alpha \in [0,1]$. The core of our novel modification in this approach (i. e. lower cuts and interval cuts) follows and it can be fruitfully applied for real data.

Definition 9. *Let* $\langle B, A, R \rangle$ *be* $[0,1]$-*fuzzy formal context and let* $\alpha \in [0,1]$. *Then the binary relation* $R_\alpha \subseteq B \times A$ *is called*

- *the upper* α-*cut if* $\langle b, a \rangle \in R_\alpha$ *is equivalent to* $R(b,a) \geq \alpha$,

- *the lower* α-*cut if* $\langle b, a \rangle \in R_\alpha$ *is equivalent to* $R(b,a) \leq \alpha$.

The binary relation $R_{\alpha\beta} \subseteq B \times A$ *is called*

- *the interval* $\alpha\beta$-*cut if* $\langle b, a \rangle \in R_{\alpha\beta}$ *is equivalent to* $R(b,a) \in [\alpha, \beta]$.

It can be seen that the triple $\langle B, A, R_\alpha \rangle$ for every $\alpha \in [0,1]$ forms the formal context given by Definition 1. For each formal context, one can build the corresponding concept lattice $\mathrm{CL}(\langle B, A, R_\alpha \rangle)$ by Definition 4. With respect to the division of the interval $[0,1]$ into n parts, we can define the subset quality measure as follows.

Definition 10. *Let* $X \subseteq B$ *and* R_α *be the upper* α-*cuts for* $\alpha \in [0,1]$. *Then the upper quality measure of the subset* X *is the value*

$$q_{\mathrm{upp}}(X,n) = \frac{\left| \left\{ p \in \{0,1,\ldots,n\} : (\exists Y \subseteq A)\, \langle X, Y \rangle \in \mathrm{CL}\left(B, A, R_{\frac{p}{n}}\right) \right\} \right|}{n+1},$$

whereby $n+1$ *is the count of different values of* α *which divides interval* $[0,1]$ *into* n *partitions.*

The formula of the lower quality measure $q_{\mathrm{low}}(X,n)$ of the subset X one can build analogously. However, the slight modification is naturally needed if we consider the interval $\alpha\beta$-cuts.

Definition 11. *Let* $X \subseteq B$ *and* $R_{\alpha\beta}$ *be the interval* $\alpha\beta$-*cuts for* $\alpha, \beta \in [0,1]$. *Then the interval quality measure of the subset* X *is the value*

$$q_{\mathrm{int}}(X,n) = \frac{\left| \left\{ \langle p, r \rangle \in J : (\exists Y \subseteq A)\, \langle X, Y \rangle \in \mathrm{CL}\left(B, A, R_{\frac{p}{n}\frac{r}{n}}\right) \right\} \right|}{|J|},$$

whereby $J = \{\langle p, r \rangle \in \{0,1,\ldots,n\} \times \{0,1,\ldots,n\} \wedge p < r\}$ *and* $n+1$ *is the count of different values of* α *and simultaneously* β *which divides interval* $[0,1]$ *into* n *partitions.*

In this paper, we omit the definitions of α-concepts, since more details about the properties of these structures can be found in [26, 2]. Moreover, the reduction of concepts from generalized one-sided concept lattices based on the method of upper α-cuts is introduced in the recently published book chapter of Butka et al. [14]. A method for α, β-cut of bipolar fuzzy formal contexts with illustrative examples is proposed in [36, 37].

3.3 Gaussian probabilistic index

In our recent work [3], the notions of $[0,1]$-fuzzy formal contexts and random variables are connected in effort to define the randomized formal contexts and to explore the stability of extents of one-sided fuzzy formal concepts.

We will consider the sample space Ω as a set of all possible finite or infinite outcomes of a random study. An event T is an arbitrary subset of Ω. The probability function p on a finite $(\{\omega_1,\dots,\omega_n\})$ or infinite (e. g. interval of real numbers) sample space Ω assigns to each event $T \subseteq \Omega$ a number $p(T) \in [0,1]$ such that $p(\Omega) = 1$ and $p(T_1 \cup T_2 \cup \dots) = p(T_1) + p(T_2) + \dots$ for $T_1, T_2, \dots$ which are disjoint. From $T \cup T^c = \Omega$, we deduce that $p(T^c) = 1 - p(T)$. Events $T_1, T_2, \dots T_m$ are called independent if $p(T_1 \cap T_2 \cap \dots \cap T_m) = \prod_{i=1}^{m} p(T_i)$.

Definition 12. *Let $\langle B,A,R \rangle$ be $[0,1]$-fuzzy formal context. For $i \in \{1,\dots,n\}$, consider the system of $[0,1]$-fuzzy formal contexts $\langle B,A,R_i \rangle$ such that*

$$R_i(b,a) = \min\Big\{1, \max\big\{0, R(b,a) + \varepsilon_{b,a,i}\big\}\Big\},$$

whereby $\varepsilon_{b,a,i}$ is a normally distributed value of a random variable $\mathscr{E}_{b,a}$ with the mean 0 and variance σ^2, i. e. $\mathscr{E}_{b,a} \sim \mathrm{N}(0,\sigma^2)$, for all $b \in B, a \in A$.

Let $X \subseteq B$. The Gaussian probability index gpi : $\mathscr{P}(B) \times \mathbb{R}^+ \to [0,1]$ *is the function given by*

$$\mathrm{gpi}(X,\sigma) = p(X \text{ is an extent of } \langle B,A,R_i \rangle)$$

for an arbitrary subset of objects X, an arbitrary standard deviation σ and mean 0. The [0,1]-fuzzy formal context $\langle B,A,R_i \rangle$ will be called the randomized (fuzzy) formal context for each $i \in \{1,\dots,n\}$.

R	a_1	a_2
b_1	0.8	0.1
b_2	0.5	0.7
b_3	0.1	0.9

→

R_1	a_1	a_2
b_1	0.87	0.07
b_2	0.46	0.72
b_3	0	0.93

R_2	a_1	a_2
b_1	1	0.11
b_2	0.53	0.82
b_3	0.08	0.78

…

R_n	a_1	a_2
b_1	0.78	0
b_2	0.51	0.66
b_3	0.15	1

Figure 6: Example of randomized formal contexts

The values of Gaussian probability index express the probability of X being the extent of the arbitrary randomized formal context by supposing the standard deviation σ in the values of the incidence relation R_i in comparison with the original incidence relation R. Alternatively, the values of the Gaussian probability index one can compute by the following construction. Consider the randomized formal contexts $\langle B,A,R_1 \rangle, \langle B,A,R_2 \rangle \dots, \langle B,A,R_n \rangle$ for a large positive integer n (see Fig. 6). Then by the classical definition of probabilistic function p one can write

$$\mathrm{gpi}(X,\sigma) = \frac{|i, i \in \{1,2,\dots,n\} : X \text{ is an extent of } \langle B,A,R_i \rangle|}{n}. \quad (3)$$

The computation of Eq. (3) is described by Alg. 2.

Algorithm 2. (Algorithm of Gaussian probabilistic index)
input: $\langle B,A,R \rangle$, X, σ, n

```
k ← 0;
for i := 1 to n do
  {
  for all b ∈ B do
    for all a ∈ A do
      {
      ε_{b,a,i} ← Random.nextGaussian() * σ;
      R_i(b,a) ← min{1, max{0, R(b,a) + ε_{b,a,i}}};
      }
  if (X is an extent of ⟨B,A,R_i⟩) then
    k ← k + 1;
  }
gpi(X,σ) ← k/n;
```

output: $\mathrm{gpi}(X,\sigma)$

In effort to express the values of Gaussian probabilistic index directly from the input $[0,1]$-fuzzy formal context, we explore the probabilistic aspects of randomized formal contexts including the boundary test conditions in [3].

Theorem 1. *Let $X \subseteq B$ and let $\langle B,A,R_i \rangle$ be a randomized formal context for some $i \in \{1,\dots,n\}$, i. e.*

$$R_i(b,a) = \min\Big\{1, \max\big\{0, R(b,a) + \varepsilon_{b,a,i}\big\}\Big\}$$

for the $[0,1]$-fuzzy formal context $\langle B,A,R \rangle$ and normally distributed value $\varepsilon_{b,a,i}$ of random variable $\mathscr{E}_{b,a} \sim \mathrm{N}(0,\sigma^2)$ for all $b \in B, a \in A$. Then the value of Gaussian probabilistic index for the subset $X \subseteq B$ and standard deviation σ is given by

$$\mathrm{gpi}(X,\sigma) = p\Big(\bigcap_{o \in B \setminus X} \Big(\bigcap_{a \in A} \Big(\Big(\bigcap_{x \in X} T_x\Big)^c\Big)^c\Big)\Big),$$

where T_x represents the event

$$\mathscr{E}_{o,a} - \mathscr{E}_{x,a} < R(x,a) - R(o,a) \quad \wedge$$
$$\mathscr{E}_{o,a} < 1 - R(o,a) \quad \wedge$$
$$\mathscr{E}_{x,a} > -R(o,a).$$

For more details, see the results from [3]. Here, we emphasize that the set of pairs $\{\langle X, \mathrm{gpi}(X,\sigma) \rangle : X \subseteq B\}$ for some σ can be ordered by the second coordinate, which gives the opportunity to use the Gaussian probabilistic index to select the relevant one-sided formal concepts in the applications.

3.4 Comparative remarks

The relationship between the Gaussian probabilistic index and the methods from Subsection 3.1 and 3.2 is now briefly outlined:

- every cluster $\mathscr{N}$ obtained by modified Rice-Siff algorithm is the extent of one-sided fuzzy formal concept of the input formal context (because we have that $\mathscr{N} = \{(X_1 \cup X_2)^{\uparrow\downarrow} : \langle X_1, X_2 \rangle \in \Psi\}$),

- modified Rice-Siff algorithm represents the crisp index for selection of one-sided concepts, the Gaussian probabilistic index is a fuzzy index,

- the subset quality measure and the Gaussian probabilistic index can be applied also for the subsets which are not the extents of the one-sided formal concepts of the input [0,1]-fuzzy formal context,

- the clusters obtained by modified Rice-Siff algorithm have mostly the higher $\text{gpi}(X, \sigma)$ as the other extents of one-sided formal concepts, some exceptions exist,

- the Gaussian probabilistic index gpi works with data tables (relations) which need not to be ordinally equivalent. The relationship between the ordinally equivalent relations were explored by Bělohlávek [7],

- we conclude that it is important to understand the advantages of the available methods and to apply them separately or in their mutual combination.

The comparative example on the modified Rice-Siff algorithm and the Gaussian probabilistic index can be found in [3] including the interpretation and explanations.

4 Applications and future work

An extensive overview of papers which apply formal concept analysis in various domains including software mining, web analytics, medicine, biology and chemistry data is provided in [32]. Particularly, we mention the conceptual difficulties in the education of mathematics [34], the techniques for analyzing and improving integrated care pathways [33] or evaluation of questionnaires [10, 8]. In [20], formal concept analysis is applied as a tool for image processing and detection of inaccuracies. Recently, the morphological image and signal processing from the viewpoint of fuzzy formal concept analysis was presented in [1]. The main results offer the possibility to interpret the binary images as the classical formal concepts and open digital signals as fuzzy formal concepts.

Regarding one-sided fuzzy approach from Section 3, a set of representative symptoms for the disease are investigated in [21]. Furthermore, the application of fuzzy concepts clustering in the domain of text documents [13, 35] or attribute characterizations of cars in generalized one--sided concept lattices [19] are the subjects of study.

In our future work, our aim is to extend the results presented in [23, 24, 25] and to verify the methods in the applications from the educational area or in the area of social networks. More particular, for a given set of students from a longitudinal survey about the relationships between the students in the secondary school classes, we can compute

a) the clusters of students sensed similar by their school-mates (by modified Rice-Siff algorithm from Subsection 3.1),

b) the clusters of more popular students or less popular students (by upper or lower cuts of subset quality measure from Subsection 3.2),

c) the stable clusters of students sensed similar due to random fluctuation of data (by Gaussian probabilistic index from Subsection 3.3).

The another possibility is to consider a set of students and their scores of the tests from different subjects (see Fig. 7). Take for example student b_2 and find the students with better results as b_2 in all subjects. From Section 3 we have that $\{b_2\}^{\uparrow\downarrow} = \{b_1, b_2\}$. Will it be valid after the repeated exams? We suppose that student b_3 will not be better than b_1 or b_2. However, how about student b_4? What is the probability of that some other student will join the group $\{b_1, b_2\}$ in other testing?

R	a_1	a_2	a_3
b_1	1	0.9	0.8
b_2	0.8	0.7	0.7
b_3	0.3	0.3	0.3
b_4	0.8	0.6	0.9

Figure 7: Students and their scores

We can answer these question by Gaussian probabilistic index presented in Subsection 3.3. The Gaussian normal distribution one can replace by real observations of teachers who can estimate the standard deviations for each individual student. We can suppose that one of the students will obtain roughly 90% in the most of exams, but once a time it can happened that he/she will pass 70% for different reasons, otherwise will reach 98%.

In another way, the paper [31] compares several collaborative-filtering techniques on a dataset from courses with only a few of students. The random Galois lattices [16], the randomized formal contexts of a discrete random variable, a generalized probability framework [17] and stability in a multi-adjoint framework [28, 29, 30] or heterogeneous framework [4] will be the point of interest in our future work, as well.

References

[1] Alcalde, C., Burusco, A., Fuentes-González, R.: Application of the L-fuzzy concept analysis in the morphological image and signal processing. Ann. Math. Artif. Intell. **72** (2014) 115–128

[2] Antoni, L., Krajči, S.: Quality measure of fuzzy formal concepts. In: Abstracts of 11th International Conference on Fuzzy Set Theory and Applications FSTA (2012), p. 18

[3] Antoni, L., Krajči, S., Krídlo, O.: Randomized fuzzy formal contexts and relevance of one-sided concepts. In: Baixeries, J., Sacarea, Ch. (eds.) ICFCA 2015. LNCS (LNAI) **9113**, Springer, Heidelberg (2015) 183–199

[4] Antoni, L., Krajči, S., Krídlo, O., Macek, B., Pisková, L.: On heterogeneous formal contexts. Fuzzy Sets Syst. **234** (2014) 22–33

[5] Bělohlávek, R.: Lattices generated by binary fuzzy relations. Tatra Mt. Math. Publ. **16** (1999) 11–19

[6] Bělohlávek, R.: Concept Lattices and Order in Fuzzy Logic. Ann. Pure Appl. Logic **128** (2004) 277–298

[7] Bělohlávek, R.: Ordinally equivalent data: A measurement-theoretic look at formal concept analysis of fuzzy attributes. Int. J. Approx. Reason. **54**(9) (2013) 1496–1506

[8] Bělohlávek, R., Sigmund, E., Zacpal, J.: Evaluation of IPAQ questionnaires supported by formal concept analysis. Inf. Sci. **181**(10) (2011) 1774–1786

[9] Bělohlávek, R., Sklenář, V., Zacpal, J.: Crisply generated fuzzy concepts. In: Ganter, B., Godin, R. (eds.) ICFCA 2005. LNCS (LNAI) **3403**, Springer, Heidelberg (2005) 269–284

[10] Bělohlávek, R., Sklenář, V., Zacpal, J., Sigmund, E.: Evaluation of questionnaires by means of formal concept analysis. In: Diatta, J., Eklund, P., Liquiére, M. (eds.) Proceedings of the Fifth International Conference on Concept Lattices and Their Applications CLA (2007) 100–111

[11] Bělohlávek, R., Trnečka, M.: Basic level of concepts in formal concept analysis. In: Domenach, F., Ignatov, D., Poelmans, J. (eds.) ICFCA 2012. LNCS (LNAI) **7278**, Springer, Heidelberg (2012) 28–44

[12] Bělohlávek, R., Trnečka, M.: Basic Level in Formal Concept Analysis: Interesting Concepts and Psychological Ramifications. In: Rossi, F. (eds.) IJCAI 2013, AAAI Press (2013) 1233–1239

[13] Butka P.: Aplikácia zhlukovania fuzzy konceptov v doméne textových dokumentov. In: Vojtáš, P. (eds), ITAT, Information technologies – Applications and Theory (2005) 31–40

[14] Butka, P., Pócs, J., Pócsová, J.: Reduction of concepts from generalized one-sided concept lattice based on subsets quality measure. In: Zgrzywa, A., Choros, A., Sieminski, A. (eds.) New Research in Multimedia and Internet Systems. Advances in Intelligent Systems and Computing **314**, Springer (2015) 101–111

[15] Dias, S. M., Vieira, N. J.: Concept lattices reduction: Definition, analysis and classification. In: Expert Syst. Appl. **42**(20) (2015) 7084–7097

[16] Emillion, R., Lévy, G.: Size of random Galois lattices and number of closed frequent itemsets. Discrete Appl. Math. **157** (2009) 2945–2957

[17] Frič, R., Papčo, M.: A categorical approach to probability theory. Stud. Logica **94**(2) (2010) 215–230

[18] Ganter, B., Wille, R.: Formal Concept Analysis: Mathematical Foundation. Springer, Heidelberg (1999)

[19] Halaš, R., Pócs, J.: Generalized one-sided concept lattices with attribute preferences. Inf. Sci. **303** (2015) 50–60

[20] Horák, Z., Kudělka, M.: Snášel, V.: FCA as a tool for inaccuracy detection in content-based image analysis. In: IEEE International Conference on Granular Computing (2010) 223–228

[21] Kiseliová, T., Krajči, S.: Generation of representative symptoms based on fuzzy concept lattices. In: Reusch, B. (eds.) Computational Intelligence, Theory and Applications. Advances in Soft Computing **33**, Springer (2005) 349–354

[22] Klimushkin, M., Obiedkov, S., Roth, C.: Approaches to the selection of relevant concepts in the case of noisy data. In:

Kwuida, L., Sertkaya, B. (eds.) ICFCA 2010. LNCS **5986**, Springer, Heidelberg (2010) 255–266

[23] Krajči, S.: Cluster based efficient generation of fuzzy concepts. Neural Netw. World **13**(5) (2003) 521–530

[24] Krajči, S.: Social Network and Formal Concept Analysis. In: Pedrycz, W., Chen, S. M. (eds.) Social Networks: A framework of Computational Intelligence. Studies in Computational Intelligence **526**, Springer (2014) 41–62

[25] Krajči, S., Krajčiová, J.: Social network and one-sided fuzzy concept lattices. In: Spyropoulos, C. (eds.) Fuzz-IEEE 2007. Proceedings of the IEEE International Conference on Fuzzy Systems, IEEE Press (2007) 1–6

[26] Krídlo, O., Krajči, S.: Proto-fuzzy concepts, their retrieval and usage. In: Bělohlávek, R., Kuznetsov, S. (eds.) Proceedings of the Sixth International Conference on Concept Lattices and Their Applications CLA (2008) 83–95

[27] Kuznetsov, S.O.: On stability of a formal concept. Ann. Math. Artif. Intell. **49** (2007) 101–115

[28] Medina, J., Ojeda-Aciego, M., Ruiz-Calviño, J.: Formal concept analysis via multi-adjoint concept lattices. Fuzzy Sets Syst. **160**(2) (2009) 130–144

[29] Medina, J., Ojeda-Aciego, M., Valverde, A., Vojtáš, P.: Towards biresiduated multi-adjoint logic programming. In: Conejo, R., Urretavizcaya, M., Pérez-de-la-Cruz, J.-L. (eds.) CAEPIA-TTIA 2003. LNCS (LNAI) **3040**, Springer, Heidelberg (2004) 608–617

[30] Medina, J., Ojeda-Aciego, M., Vojtáš, P.: Similarity-based unification: a multi-adjoint approach. Fuzzy Sets Syst. **146** (2004) 43–62

[31] Pero, Š., Horváth, T.: Comparison of Collaborative-Filtering Techniques for Small-Scale Student Performance Prediction Task. In: Sobh, T., Elleithy, K. (eds.) Innovations and Advances in Computing, Informatics, Systems Sciences, Networking and Engineering: Lecture Notes in Electrical Engineering **313**, Springer (2015) 111–116

[32] Poelmans, J., Ignatov, D. I., Kuznetsov, S. O., Dedene, G.: Formal concept analysis in knowledge processing: A survey on applications. Expert Syst. Appl. **40**(16) (2013) 6538–6560

[33] Poelmans, J., Dedene, G., Verheyden, G., Van der Mussele, H., Viaene, S., Peters, E, 2010. Combining business process and data discovery techniques for analyzing and improving integrated care pathways. In: Perner, P. (eds.) ICDM 2010, Lecture Notes in Computer Science **6171** (2010) 505–517

[34] Priss, U., Riegler, P., Jensen, N.: Using FCA for Modelling Conceptual Difficulties in Learning Processes. In: Domenach, F., Ignatov, D.I., Poelmans, J. (eds.) Contributions to the 10th International Conference ICFCA (2012) 161–173

[35] Sarnovský M., Butka P., Tkáč E.: Distribuovaná tvorba hierarchie konceptov z textových dokumentov v gridovom prostredí. In: Vojtáš, P. (eds), ITAT, Information technologies – Applications and Theory (2007) 97–102

[36] Singh, P.K., Kumar, Ch.A: Bipolar fuzzy graph representation of concept lattice. Inf. Sci. **288** (2014) 437–448

[37] Singh, P.K., Kumar, Ch.A: A note on bipolar fuzzy graph representation of concept lattice. Int. J. Comp. Sci. Math. **5**(4) (2014) 381–393

[38] Snášel, V., Ďuráková, D., Krajči, S., Vojtáš, P.: Merging concept lattices of α-cuts of fuzzy contexts. Contrib. Gen. Algebra **14** (2004) 155–166

J. Yaghob (Ed.): ITAT 2015 pp. 9–16
Charles University in Prague, Prague, 2015

Hybrid Flow Graphs: Towards the Transformation of Sequential Code into Parallel Pipeline Networks

Michal Brabec, David Bednárek

Department of Software Engineering
Faculty of Mathematics and Physics, Charles University Prague
{brabec,bednarek}@ksi.mff.cuni.cz

Abstract: Transforming procedural code for execution by specialized parallel platforms requires a model of computation sufficiently close to both the sequential programming languages and the target parallel environment. In this paper, we present Hybrid Flow Graphs, encompassing both control flow and data flow in a unified, pipeline based model of computation. Besides the definition of the Hybrid Flow Graph, we introduce a formal framework based on graph rewriting, used for the specification of Hybrid Flow Graph semantics as well as for the proofs of correctness of the associated code transformations. As a formalism particularly close to pipeline-based runtime environments which include many modern database engines, the Hybrid Flow Graphs may become a powerful means for the automatic parallelization of sequential code under these environments.

Keywords: compiler, graph, optimization, parallelism

1 Introduction

Parallelization of procedural code has been studied for several decades, resulting in a number of mature implementations, usually tightly coupled with compilers of FORTRAN or C. In most cases, the parallelization techniques are targeted at multi-threaded environment, sometimes augmented with task-based parallelism. This shared-memory approach is perfectly suited for numeric applications; however, there are niches of computing which may benefit from a different run-time paradigm. In particular, database systems make use of execution plans which explicitly denote the data flows between operators; similar graph-based representation is often used in streaming systems. Besides other advantages, the explicitly denoted data flow allows easier distribution of the computation across different nodes, compared to the complexity of potentially random access in the shared-memory model of computation.

Nevertheless, database and streaming systems lack the generality of procedural programming languages. To extend the applicability of such runtime systems towards general programming, a system capable of compiling (a subset of) a well-known procedural language is required. This paper demonstrates an important step towards such a system – an intermediate representation capable of describing procedural code using the language of data-flow oriented graphs.

Our approach is focused at target environments in which an application is presented in two layers – the declarative upper layer represented as a graph of operators, the lower layer consisting of a procedural implementation of individual operators. In such a setting, a language front-end transforms the application source code into an intermediate representation; then, a strategic phase decides where to put the boundary between the declarative upper layer and the procedural lower layer. Finally, the upper layer is converted into the final graph while procedural code (in source, bytecode, or binary form) is generated from the lower layer.

Realization of this idea involves many particular problems; in this paper, we deal with the intermediate representation which serves as the backbone connecting the front-end, strategy, and generator phases. Since the upper layer will finally be represented by a graph, it seems reasonable that also the intermediate representation be graph based. However, such a graph shall also be able to completely represent the lower, procedural part of the application. Although the theory of compiling uses graphs in many applications including the representation of the code, the control-flow and data-flow aspects of the code are usually approached differently. In our approach, the control flow and the data flow is represented by the same mechanism because the strategic phase must be able to make the cuts inside both the control flow and the data flow.

In this paper, we present *Hybrid Flow Graphs* (HFG) as an intermediate representation of procedural code, capable of representing control-flow and data-flow in the same layer. We will demonstrate two forms of the HFG, the *Sequential HFG* and the *General HFG*. The former form has its execution constrained to sequential, equivalent to the execution of the procedural code which it was generated from. The latter form exhibits independent execution of individual operations, showing available parallelism. Of course, making operations independent requires careful dependency analysis – we assume that the required points-to/alias/dependency analyses were made before the conversion to HFG, using some of the algorithms known from the theory of compiler construction.

We will describe the operational semantics of both the sequential and the independent forms of HFG and we will also demonstrate the conversion between them. We will use graph rewriting as the vehicle for the description of both the semantics and conversions. The use of graph rewriting systems allows for some generality of the approach, both in the description and in the implementation. In particular, the set of operations provided by the intermediate code may be defined almost arbitrarily as long as their semantics is depicted by graph rewriting rules. On the other hand, the set of control-flow operations is fixed since it forms the non-trivial part of the transformation from sequential code.

In this paper, we will use the set of types and operators borrowed from the C# language, more exactly, from its standard bytecode representation called CIL [1]. This includes some artifacts specific to the CIL, namely the stack and the associated operations. Since the Sequential HFG mimics the behavior of the CIL abstract machine, the stack operations are preserved in the HFG. However, they must disappear in the transformation to the General HFG; otherwise, the presence of stack operations would prohibit almost any parallelism. Since the removal of stack operations is a non-trivial sub-problem, we include the corresponding (CIL-specific) algorithm in this paper, as well as the description of the preceding conversion of CIL into the sequential HFG.

The rest of the paper is organized as follows: We review the related work in Section 2. The hybrid flow graph and its semantics is defined in Section 3. We define the sequential hybrid flow graph in Section 4. In Section 5, we present an algorithm producing a sequential HFG from a source code. Section 6 presents the algorithms that transform a general HFG to a sequential HFG.

2   Related Work

The flow graph described in this paper is similar but not identical to other modeling languages, like Petri nets [2] or Kahn process networks [3]. The main difference is that the flow graph was designed for automatic generation from the source code, where the other languages are generally used to model the application prior to implementation [4] or to verify a finished system [5]. The flow graph is similar to the graph rewriting system [6], which can be used to design and analyze applications, but it is not convenient for execution. There are frameworks that generate GRS from procedural code like Java [7], though the produced graphs are difficult to optimize. The flow graph has similar traits to frameworks that allow applications to be generated from graphs, like UML diagrams [8] [9], but we concentrate both on graph extraction and execution.

The flow graph is closely related to graphs used in compilers, mainly the dependence and control flow graphs [10], where the flow graph merges the information from both. The construction of the flow graph and its subsequent optimization relies on compiler techniques, mainly *points-to* analysis [11], *dependence* testing [12] and *control-flow* analysis [13]. In compilers, graphs resulting from these techniques are typically used as additional annotation over intermediate code.

The flow graph is not only a compiler data representation, it is a processing model as well, similar to KPN graphs [14]. It can be used as a source code for specialized processing environments, where frameworks for pipeline parallelism are the best target, since these frameworks use similar models for applications [15]. One such a system is the Bobox framework [16], where the flow graph can be used to generate the execution plan similarly to the way Bobox is used to execute SPARQL queries [17].

2.1   Graph Rewriting

A *graph rewriting system* (GRS) is a set of rules R transforming one graph to another. The rewriting systems are very similar to grammars where each rule has a left side that has to be matched to the graph and a right side that replaces it.

The GRS are very simple and they are best explained by an example. Figure 2 shows the CIL based rewriting system, where each rule emulates a single CIL instruction.

We use extended rewriting rules where $*$ represents any node (wildcard) and we add Greek letters to identify unique wildcard nodes where necessary. Variable names (x or y) represent only nodes with data. Expressions are used to simplify data manipulation. This is necessary so we do not have to create a rule for each combination of instruction or inputs.

3   Hybrid Flow Graph

A *hybrid flow graph* (HFG) is an annotated directed graph, where nodes together with their labels represent operations and edges represent queues for transferring data. The direction of the edge indicates the direction of data flow. Figure 1 shows a general hybrid flow graph for a simple function (see Listing 1 for source code).

```
void SimpleLoop(int a) {
for(int i=a; i<5; )
{
i = call(i);
}
}
```

Listing 1 Simple function containing a single loop

A hybrid flow graph is a general concept that can be used regardless of operations O or data types T. A *platform specific flow graph model* is defined by specifying O and T based on the target platform. The platform specific definition is $HFG\_MODEL_{platform} = (O \cup \{special\ operations\}, T)$, where O and T are platform specific and *special operations* are platform independent (see Section 3.1). The hybrid flow graph semantics (explained in Section 3.2) are also platform specific.

As our research is focused on C#, we use CIL [1] instructions to specify node operations. The most common

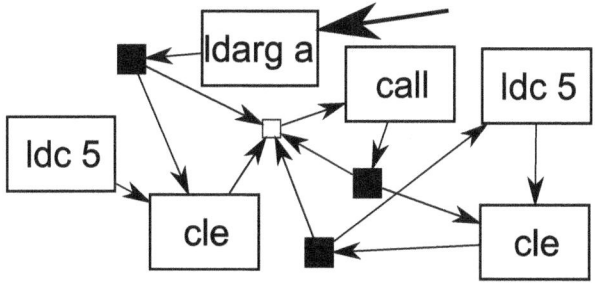

Figure 1: The HFG representation of the C# code from Listing 1

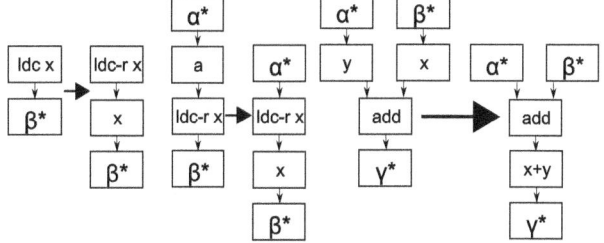

Figure 2: Semantics of selected basic operations

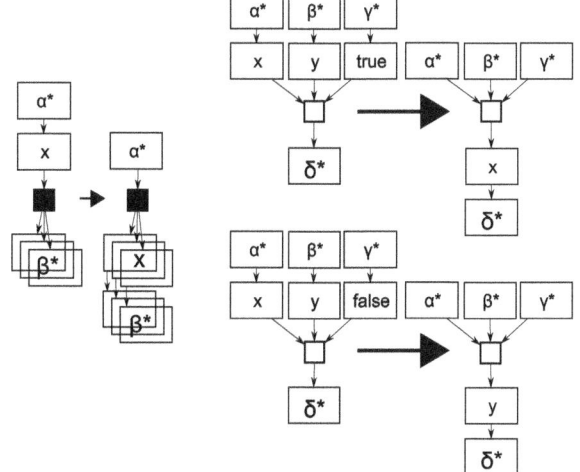

Figure 3: Semantics of selected special operations

instruction in our examples are: *ldl x* (load variable or constant), *stl x* (write to a variable), *cle* (compare less or equal), *add* etc. (mathematical operations) [1]. We omit data types for the edges, because they are not important for the graph construction.

3.1 Representation of Control Flow

Control flow operations are transformed into platform independent *special operations* which interact with the dataflow carried through the platform specific *basic operations*. Graphical notation of special operations is in Figure 3, where a broadcast is a black square and a merge is a white square.

A *broadcast node* has a single input and a variable number of outputs. It represents an operation that creates a copy of the input for each output.

A *merge node* has a single output and three inputs. It represents an operation that accepts data from two sources and passes them to a single operation based on a condition, it is used to merge data flow after a conditional branch.

A *loop merge node* is a special version of the merge node with two inputs for conditional branches, it is used to merge data in loops. The input is split into two pairs, where each contains one data input and one condition input. The node first reads data from the first pair and then it loops, while the second pair has a positive value in the condition input.

3.2 Semantics of Hybrid Flow Graphs

The *hybrid flow graph semantics* define the way nodes process data and communicate. Basically, nodes represent operations or data and edges represent unbounded queues (FIFO) for data transfer.

Semantics are platform specific, similar to the hybrid flow graph model. We define the *hybrid flow graph semantics* using the *graph rewriting systems* according to the platform specific operations logic.

We define the HFG semantics for CIL using graph rewriting. Figure 2 shows the example of definition for the basic instructions. Instruction load constant (*ldc*) has a special behavior, it produces a single constant and then

it repeats while it has an input, we specify this behavior by changing its label to *ldc − r*. The definition of special operations is illustrated in Figure 3, the figure contains the definition of broadcast and merge. We use special graphical notation for special operations, since they are platform independent (broadcast is black square, merge is a white square).

We can use the graph rewriting rules defined in Figure 2 and 3 to perform the computation of a hybrid flow graph. The computation of a flow graph with a for-loop is shown in Figure 4. Each part of Figure 4 contains the HFG state after application of a single rule and the computation starts after application the *ldc*5 rule.

4 Sequential HFG

In this section, we define the semantics of a sequential hybrid flow graph using the graph transformation systems. A *sequential hybrid flow graph* is a special HFG that uses a program counter a memory nodes instead of special operations defined in Section 3.1. It has a strict control flow that closely follows that of the source program. It is useless for the parallel pipelines or other applications, but it allows us to create a general HFG from the source code.

We use special operations for variables and the stack (CIL virtual machine is stack based [1]), these operations

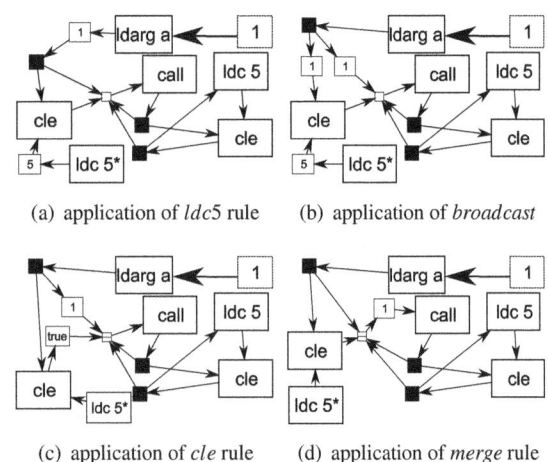

(a) application of *ldc5* rule   (b) application of *broadcast*

(c) application of *cle* rule   (d) application of *merge* rule

Figure 4: Computation of a hybrid flow graph

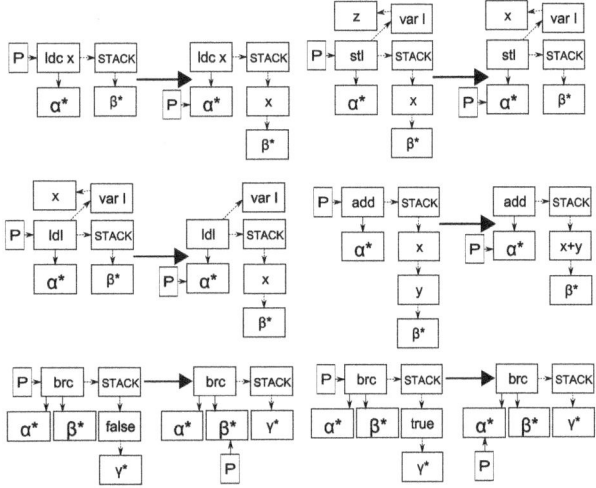

Figure 5: Semantics of selected CIL sequential HFG operations

are used to model the data flow. Plus we add a special node P (program counter) to model the control flow.

4.1 Semantics of a CIL Sequential HFG

The *sequential hybrid flow graph semantics* is defined using the graph rewriting, same as any other hybrid flow graph. The transformation rules for the CIL based sequential hybrid flow graph are defined in Figure 5, we present only the rules necessary for our examples.

The GRS rules defined in Figure 5 are completely based on the CIL instruction definition [1]. The control flow is controlled by the node P, which follows the edges between instructions. The instruction edges are constructed according to the following rules: 1) The standard instructions follow one another 2) The branches have two outgoing edges, one following the jump target and the other leading to the next instruction.

The data management is handled by the stack and variable operations, where each instruction modifies the

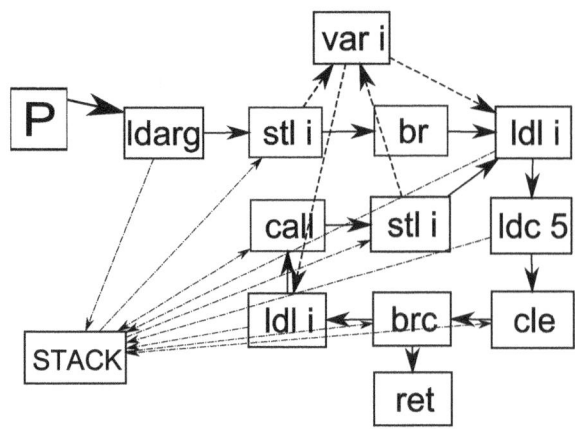

Figure 6: Sequential HFG created from code in Listing 1

proper number of stack levels and variables, based on the instruction definition.

The actual computation is the same as in the case of the .NET virtual machine, the control flow follows the same path, variables and stack contain the same value at after every evaluated instruction. Graphically, the computation is similar to Figure 4.

5 Transformation of CIL to Sequential HFG

In this section, we explain how to create a sequential hybrid flow graph from a CIL code. We present an algorithm that produces a sequential HFG that follows the CIL computation step by step, including control and data flow. It allows us to define the advanced transformation algorithm in Section 6.

The basic transformation of a CIL code to a sequential HFG requires two steps: 1) create nodes based on the instructions 2) create edges based on the control and data flow.

The entire transformation is in Algorithm 1, function v maps nodes to operations. The node P is basically a program counter, it points to the actual instruction and it is used to handle the control flow. The input sets I, M, P_i are obtained from the source code. Sr_i and Sw_i are based directly on CIL specification [1]. The transformation is best presented on an example, Listing 1 contains CIL source code and Figure 6 contains the resulting graph.

6 Transformation of Sequential HFG to General HFG

In this section, we present an algorithm that transforms a procedural code to a hybrid flow graph. We create the CIL sequential HFG (Section 5) and convert it to a general HFG that requires neither the program counter P, nor the

Algorithm 1 Transformation of CIL to Sequential HFG

Require: I – set of instruction addresses
  M – set containing all variables
  C_i, P_i – code and parameter of the instruction i
  Sr_i – number of stack values read by instruction i
  Sw_i – number of stack values written by instruction i

Ensure: V – nodes of the flow graph
  $v(V_i)$ – label of the node V_i
  E – edges of the flow graph

1: $V := \left\{ V^{P}, V^{STACK} \right\}$
2: $V := V \cup \left\{ V_i^{INSTR} : i \in I \right\}$
3: $v(V_i^{INSTR}) := \langle C_i \, P_i \rangle$ for each $i \in I$
4: $V := V \cup \left\{ V_m^{VAR} : m \in M \right\}$
5: $v(V_m^{VAR}) := \langle \text{var } m \rangle$ for each $m \in M$
6: $E := \emptyset$
7: **for all** $i \in I$ **do**
8:    **if** $i = branch$ **then**
9:       $E := E \cup \left\{ (V_i^{INSTR}, V_{P_i}^{INSTR}) \right\}$
10:    **end if**
11:    $E := E \cup \left\{ (V_i^{INSTR}, V_{i+1}^{INSTR}) \right\}$
12: **end for**
13: **for all** $i \in I$ **do**
14:    **if** $i = stl$ **then**
15:       $E := E \cup \left\{ (V_i^{INSTR}, V_{P_i}^{VAR}) \right\}$
16:    **end if**
17:    **if** $i = ldl$ **then**
18:       $E := E \cup \left\{ (V_{P_i}^{VAR}, V_i^{INSTR}) \right\}$
19:    **end if**
20:    **if** $Sw_i > 0$ **then**
21:       $E := E \cup \left\{ (V_i^{INSTR}, V^{STACK}) \right\}$
22:    **end if**
23:    **if** $Sr_i > 0$ **then**
24:       $E := E \cup \left\{ (V^{STACK}, V_i^{INSTR}) \right\}$
25:    **end if**
26: **end for**

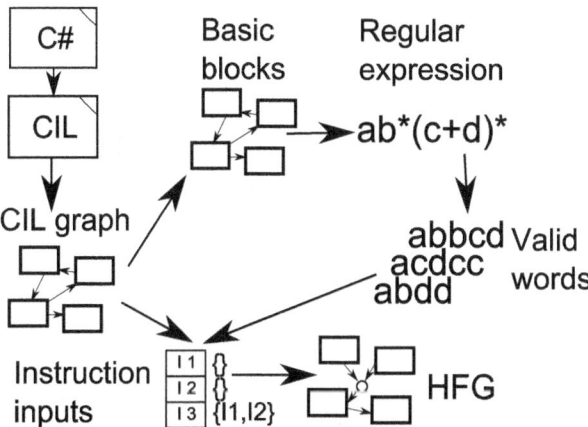

Figure 7: Transformation from CIL to a general HFG

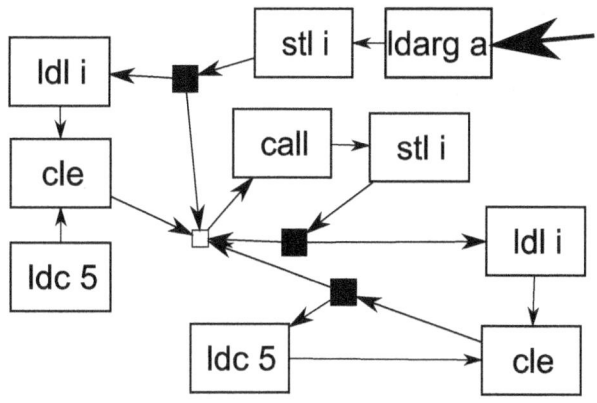

Figure 8: General HFG of the SimpleLoop function before empty operation elimination

memory nodes. The input for the algorithm is an intermediate code; in our case CIL.

The basic idea of the algorithm is that every CIL instruction is transformed to a node. The data and control flow is then modeled using the sequential HFG presented in Section 5. We emulate computation on the graph and note what data can reach which instruction, by what path. Finally, we create edges according to the collected information and we introduce special nodes where necessary.

6.1 Flow Graph Construction

To construct a complete flow graph, we create nodes based on instructions and use a modified sequential HFG to create edges and special nodes. We use basic blocks to model all valid execution paths which we use to create all the necessary edges. Algorithm 2 describes the entire process, the other algorithms are discussed in the following sections. The result of Algorithm 2, applied to the code in Listing 1, is the HFG in Figure 1 and Figure 8 shows the graph before NOP elimination.

The first step of the algorithm is to create basic nodes according to the instructions of the source code (lines 1 to 3 in Algorithm 2).

To crate the edges, we generate necessary execution paths using the *CIL path generator* (Section 6.4) and analyze all the possible inputs by the *CIL reachability* algorithm (Section 6.3). Basically, we emulate computations for every valid execution path recording the possible inputs, and the paths leading to them, for every instruction.

First, we convert the basic blocks to a regular expression (function *to_regular_expression*), then we generate all the valid paths using the *CIL path generator*. Next, we have to update the G_{CIL} so it contains the correct number of iterations (function *update*) for every loop (this due to the IDs assigned to branches as shown in Figure 9). Finally, we use the *CIL reachability* to create sets $S_{[0:N]}$ containing all the possible inputs for each instruction (lines 4 to 10). $S_{[0:N]}$ represents sets S_0 to S_N that contain inputs for instructions indexed 0 through N.

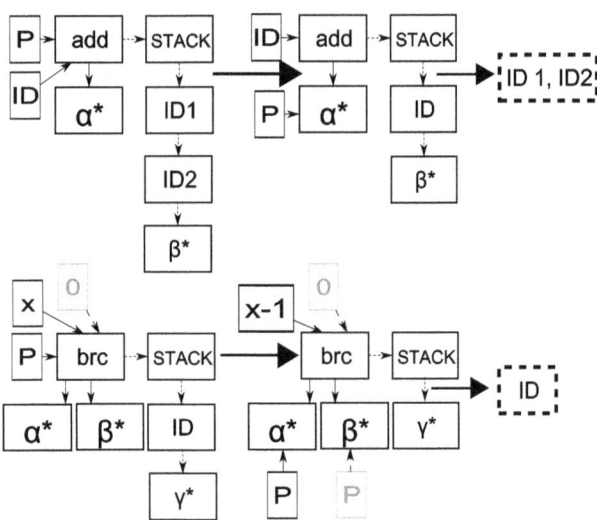

Figure 9: Symbolic semantics of selected Sequential HFG operations

We can create edges, once the input sets $S_{[0:N]}$ are ready. We start by discarding duplicate sources, because each instruction can produce only one value, which can reach the instruction i through multiple paths (different iterations in-between), we can keep any instance as they are equivalent (lines 12 to 16). Then we check the number of inputs left, if there is the same number of sources as the instruction i has inputs then we simply create edges (lines 17 to 21).

If there is more sources than i has inputs then we create merge nodes to select the correct input value and we connect them to a condition that decides what input is used. We iterate over the sources and we compare the attached paths, the paths always start the same (i.e. the first instruction) and they continue until there is a conditional branch that changes the execution order for one of them. We create a *merge* node, connect both sources to it and we connect the branch condition to the condition input. This way, we reduce the number of incoming edges, until it matches the number of inputs.

The last step of Algorithm 2 is the elimination of empty operations that no longer have any purpose. This means mainly loading and storing local variables (*ldl* and *stl*). *Eliminate NOP* removes any nodes that do not change the data.

6.2 Symbolic Semantics of Sequential HFG

We use modified semantics for the CIL sequential HFG defined (Section 5). The modified rules work with instruction identifiers instead of actual values. This way, we can analyze which instructions consume the value produced other instructions. The rule modification is shown in Figure 9.

We assign each instruction an identifier equal to its index ($\forall i \in I : id(i) = i$). The identifiers of branches decide how many times the branch is *true* before it is *false*, we

---

Algorithm 2 Hybrid flow graph construction

Require: I – set of instructions
$\quad$ B – graph of basic blocks
$\quad$ G_{CIL} – modified CIL graph
$\quad$ R_{CIL} – modified rules for CIL graph
Ensure: N – nodes of the flow graph
$\quad$ E – edges of the flow graph
1: $N := \{N_i : i \in I \land v(N_i) = O_i\}$
2: $regex := to\_regular\_expression(B)$
3: $S_{[0:N]} := \emptyset$
4: **for all** $W \in CIL\_path\_generator(regex, G_{CIL})$ **do**
5: $\quad$ **for all** $s_i \in CIL\_reachability(I, update(G_{CIL}), R_{CIL})$
$\quad\quad$ **do**
6: $\quad\quad$ $S_i := S_i \cup s_i$
7: $\quad$ **end for**
8: **end for**
9: **for all** $s_i \in S_{[0:N]}$ **do**
10: $\quad$ **for all** $p \in s_i$ **do**
11: $\quad\quad$ **while** $\exists q \in s_i : q.first = p.first \land q \neq p$ **do**
12: $\quad\quad\quad$ $s_i := s_i \setminus q$
13: $\quad\quad$ **end while**
14: $\quad$ **end for**
15: $\quad$ **if** $|s_i| = input\_count(i)$ **then**
16: $\quad\quad$ **for all** $p \in s_i$ **do**
17: $\quad\quad\quad$ $E := E \cup \{(p, i)\}$
18: $\quad\quad$ **end for**
19: $\quad$ **else**
20: $\quad\quad$ **while** $|s_i| > input\_count(i)$ **do**
21: $\quad\quad\quad$ $a := s_i[0]$
22: $\quad\quad\quad$ $b := s_i[1]$
23: $\quad\quad\quad$ $V := V \cup \{merge\}$
24: $\quad\quad\quad$ $E := E \cup \{a.first, merge\}$
25: $\quad\quad\quad$ $E := E \cup \{b.first, merge\}$
26: $\quad\quad\quad$ $E := E \cup \{merge, i\}$
27: $\quad\quad\quad$ **while** $a.second[pos] = b.second[pos]$ **do**
28: $\quad\quad\quad\quad$ $pos := pos + 1$
29: $\quad\quad\quad$ **end while**
30: $\quad\quad\quad$ $E := E \cup \{pos, merge\}$
31: $\quad\quad$ **end while**
32: $\quad$ **end if**
33: **end for**
34: $eiminate_N OP(V, E)$

---

use this to drive the control flow (explained in the next section). Last modification is that the rules output identifiers of consumed values, this is indicated by the bold dashed boxes on the right side in Figure 9.

6.3 CIL Reachability

We use the modified HFG semantics to locate all possible inputs for each instruction i, where an input is an instructions producing a value consumed by i. We do this by emulating CIL computation and storing the inputs for each instruction, implemented by Algorithm 3.

Algorithm 3 CIL reachability

Require: I – set of instructions
    G_{CIL} – modified CIL graph
    R_{CIL} – modified rules for CIL graph
Ensure: $S_{[0:N]}$ – sets of sources for each instruction
1: $path := ()$ – empty path
2: **while** $\exists r \in R_{CIL} : applicable(G_{CIL}, r)$ **do**
3:     $i := program\_counter(G_{CIL})$ – current instruction
4:     $consumed := apply(G_{CIL}, r)$ – apply rule
5:     **for all** $id \in consumed$ **do**
6:        $S_i := S_i \cup \{(id, path)\}$
7:     **end for**
8:     $path := path \cdot i$ – update execution path
9: **end while**

Algorithm 3 uses several simple functions: *applicable* tells whether the rule can be matched on the graph, *apply* applies the rule and returns the input IDs (dashed boxes in Figure 9), *program_counter* returns the instruction connected to P.

6.4 CIL Path Generator

We have to examine all valid execution paths to make sure that the HFG is complete. We do this by creating a *regular expression* representing the control flow and then generating all valid words (paths).

We create basic block graph based on the source code, this is a very simple algorithm explained in [10]. The basic block graph can be interpreted as a *finite state machine*, the source code is finite, producing only a finite number of states. We disregard the actual values produced in the blocks that define the actual control flow. Instead, we treat the basic block graph as a finite state machine that accepts a set of words where each word defines a valid order of basic blocks. Therefore the execution paths can be represented by a *regular expression* [18]. Based on the basic blocks, we create a regular expression modeling all the valid execution paths.

Algorithm 4 CIL path generator

Require: R – regular expression
Ensure: W – set of words
1: **for all** $l \in [1 : 3 * length(R)]$ **do**
2:     **for all** $w \in \{v : v = e^* \wedge e \in \Sigma \wedge |v| = l\}$ **do**
3:     **if** $valid(w, R)$ **then**
4:        $W := W \cup \{w\}$
5:     **end if**
6:     **end for**
7: **end for**

Algorithm 4 generates all valid execution paths, where the function *valid* simply checks if the word is valid according to the regular expression and Σ is the alphabet. We limit the word length, thus limiting the loop iteration

count, because multiple loop iterations do not add new information (see Section 6.5 for more details).

6.5 Algorithm Correctness

In the following sections, we discuss the reasoning behind the hybrid flow graph construction presented in Section 6, mainly why it is equivalent to the input source code. Here, we present only the main ideas of a proof, because a complete proof would require too much space.

We have to focus mainly on three parts of the transformation algorithm: 1) the CIL sequential hybrid flow graph is equivalent to the CIL source code 2) the analyzed execution paths contain all the possible inputs 3) the merge nodes created according to the execution paths are placed correctly.

6.6 CIL HFG Equivalence To Code

We prove the equivalence of the CIL code and the CIL hybrid graph (Section 5) by making sure that they have the same control and data flow. We can do this, because the CIL graph contains a node for program counter and nodes for variables and stack.

We base the proof on the fact that the HFG operations behave the same way as the instructions. The program counter is passed from one instruction to another in the same way as in CIL – control flow is the same. The instructions store data using special operations (nodes) for stack and variables in the same way they do in memory – data flow is equivalent. We omit the full proof, because it would be too long and technical.

6.7 Flow Graph Logic

We cannot use the same approach for a general hybrid flow graph as for CIL sequential HFG, because a general HFG does not model the platform specific memory features and control logic. Instead, we focus on the transformation of a CIL sequential HFG to a general HFG. We prove their equivalence by tracing all valid execution paths.

Lemma 1 (Necessary execution paths). *Let W we a set of words produced by the CIL path generator algorithm (Algorithm 4) and $E := \{e* : e \in \Sigma\}$ set of all words, then $\bigcup_{w \in W} reachable(w) = \bigcup_{e \in E} reachable(e)$, where the function reachable represents the CIL reachability algorithm.*

Proof. The lemma says that the paths generated by Algorithm 4 produce all possible inputs for all instructions. This is true because there is only a limited number of paths that can produce new inputs.

The basic block graph can be viewed as a finite state machine. The regular language accepted by the state machine can be infinite, but only due to the *Pumping lemma* [19], which means infinite loop iterations.

The important effect of loops is that the instructions use values produced outside the loop in the first iteration and then they consume values produced in the loop. The only possible exception is a conditional branch in the loop, but we check the paths with both branches taken in the first iteration. □

Lemma 2 (Data flow path merging)**.** *Let pairs* $(S_{[1:N]}, P_{[1:N]})$ *be sources for instruction* i, *then* $input\_count(i) = N \implies (S_{[1:N]}, i) \in E(HFG)$ *or* $input\_count(i) < N \implies \exists m \in V(HFG) \wedge (S_{[1:N]}, m) \in E(HFG)$

Proof. The lemma says that either all the sources were used to create appropriate edges or that a merge nodes were introduced if there is too many sources.

The first implication is true, because Algorithm 2 creates an edge for every source (lines 17 to 20).

The second implication is correct, because the algorithm creates merge nodes when there is too many sources (line 25). The merge is created based on the execution path stored for each input, it is connected to the last instruction of the prefix shared by both paths (lines 29 to 32). This means that the merge is driven by the conditional branch that changed the execution path and decided what source would be used. □

7   Conclusions

We designed the flow graph to represent a procedural code along with important information about its structure and behavior. We defined the HFG semantics using graph rewriting, which is basically a grammar on graphs. We defined semantics for both the sequential and general HFG using the formalism, thus making the sequential HFG a special version of the general HFG. We designed an algorithm that transforms a sequential source code (C# compiled to CIL) to a general HFG using mainly the HFG semantics and graph algorithms. Therefore, the algorithm can be modified for other platforms.

We have implemented a prototype of the general HFG that uses the defined semantics and is capable of completely parallel computation. The prototype serves as the proof of concept, showing that the HFG is defined correctly. It is implemented using the Bobox framework [16].

Acknowledgments

This paper was partially supported by the Grant Agency of Charles University (GAUK) project 122214 and by the Czech Science Foundation (GACR) project P103/13/08195.

References

[1] "TG3. Common Language Infrastructure (CLI). Standard ECMA-335, June 2005."

[2] Peterson, J. L.: Petri nets. ACM Comput. Surv., **9 (3)** (Sep. 1977), 223–252, [Online]. Available: http://doi.acm.org/10.1145/356698.356702

[3] Gilles, K.: The semantics of a simple language for parallel programming. In Information Processing: Proceedings of the IFIP Congress **74** (1974), 471–475

[4] Josephs, M. B.: Models for data-flow sequential processes. In: Communicating Sequential Processes. The First 25 Years, Springer, 2005, 85–97

[5] Ezpeleta, J., Colom, J. M., Martinez, J.: A Petri net based deadlock prevention policy for flexible manufacturing systems. Robotics and Automation, IEEE Transactions on **11 (2)** (1995), 173–184

[6] Ehrig, H., Ehrig, K., Prange, U., Taentzer, G.: Graph transformation systems. Fundamentals of Algebraic Graph Transformation (2006), 37–71

[7] Corradini, A., Dotti, F. L., Foss, L., Ribeiro, L.: Translating java code to graph transformation systems. In: Graph Transformations, Springer, 2004, 383–398

[8] Geiger, L., Zündorf, A.: Graph based debugging with fujaba. Electr. Notes Theor. Comput. Sci. **72 (2)** (2002), 112

[9] Balasubramanian, D., Narayanan, A., van Buskirk, C., Karsai, G.: The graph rewriting and transformation language: Great. Electronic Communications of the EASST **1** (2007)

[10] Allen, R., Kennedy, K.: Optimizing compilers for modern architectures. Morgan Kaufmann San Francisco, 2002

[11] Sridharan, M., Bodík, R.: Refinement-based context-sensitive points-to analysis for Java. ACM SIGPLAN Notices **41 (6)** (2006), 387–400

[12] Muchnick, S. S.: Advanced compiler design implementation. Morgan Kaufmann Publishers, 1997

[13] Reps, T.: Program analysis via graph reachability. Information and software technology **40 (11)** (1998), 701–726

[14] Geilen, M., Basten, T.: Requirements on the execution of Kahn process networks. In: Programming languages and systems, Springer, 2003, 319–334

[15] Navarro, A., Asenjo, R., Tabik, S., Cascaval, C.: Analytical modeling of pipeline parallelism. In: Parallel Architectures and Compilation Techniques, 2009, PACT'09, 18th International Conference on, IEEE, 2009, 281–290.

[16] Falt, Z., Kruliš, M., Bednárek, D., Yaghob, J., Zavoral, F.: Locality aware task scheduling in parallel data stream processing. In: Intelligent Distributed Computing VIII, ser. Studies in Computational Intelligence, D. Camacho, L. Braubach, S. Venticinque, and C. Badica, Eds., Springer International Publishing, 2015, **570**, 331–342

[17] Falt, Z., Čermák, M., Dokulil, J., Zavoral, F.: Parallel SPARQL query processing using Bobox. International Journal On Advances in Intelligent Systems **5 (3,4)** (2012), 302–314

[18] McNaughton, R., Yamada, H.: Regular expressions and state graphs for automata, 1960.

[19] Jaffe, J.: A necessary and sufficient pumping lemma for regular languages. ACM SIGACT News **10 (2)** (1978), 48–49

J. Yaghob (Ed.): ITAT 2015 pp. 17–22
Charles University in Prague, Prague, 2015

Assessing Applicability of Power-Efficient Embedded Devices for Micro-Cloud Computing

Martin Kruliš, Petr Stefan, Jakub Yaghob, and Filip Zavoral

Parallel Architectures/Algorithms/Applications Research Group
Faculty of Mathematics and Physics, Charles University in Prague
Malostranské nám. 25, Prague, Czech Republic
{krulis,yaghob,zavoral}@ksi.mff.cuni.cz, ptr.stef@gmail.com

Abstract: Distributed computing and cloud phenomenon have become an intensively studied topic in the past decade. These technologies have been enveloped with attractive business models, where the customer pays only for the resources or services which have been actually utilized. Even though this popularity lead to rapid development of distributed algorithms, virtualization platforms, and various cloud services, many issues are still waiting to be solved. One of these issues is the question of power efficiency. In this paper, we investigate possibilities of applying single-board computers as platform for distributed systems and cloud computing. These small devices (such as Raspberry Pi) are quite power efficient and relatively cheap, so they may reduce the overall cost for cloud services. Furthermore, they may be employed to create small clusters that could replace traditional enterprise servers and achieve lower cost and better robustness for some tasks.

Keywords: reliability, distributed systems, cloud computing, micro cloud, power efficiency, Raspberry PI

1 Introduction

Distributed computing has been an intensively studied topic since the dawn of computer science. The idea of utilizing multiple ordinary devices instead of a single powerful one brought many advantages, such as much easier scaling, possibly higher robustness, or better utilization of spare hardware. On the other hand, distributed computing is encumbered with many challenges that include the question of efficiency, communication and synchronization overhead, or the necessity of handling failures of individual nodes.

In combination with modern technologies and hardware virtualization, the distributed computing lead to the inception of the cloud phenomenon, where large complex systems are presented to users not in a form o a distributed system, but as virtual hardware, programming platform, or even specialized services. In this form, the user is completely shielded from tedious details of system design. Furthermore, the concept of cloud allows much more effi cient allocation of hardware resources, from which benefits both the cloud providers (since they have less hardware to buy and maintain) and cloud customers (who pay only for resources they really utilize).

With the growth of the cloud infrastructure, the power efficiency become a more and more important problem. Despite the fact that the cloud services utilize underlying hardware more efficiently than it could have been used by individual users, the pressure to reduce power consumption of this infrastructure is raising steadily. One of the possibilities is to utilize more efficient hardware that requires less power to perform the same task. A quite promising platform are the ARM CPUs which are currently utilized in mainstream mobile and other handheld devices. However, the majority of enterprise servers and professional solutions use CPUs based on x86 architecture which have more computational power. Nevertheless, these solutions are considered less efficient at least for some tasks.

Some specialized problems cannot utilize cloud solutions for various reasons such as security or domain-specific constraints, hence they must be hosted on privatized clusters. Beside the power efficiency issues, small clusters may benefit from small ARM-based devices in other ways. For instance, utilizing many single-board computers instead of a few enterprise server may be cheaper. Furthermore, using many devices allows more fine-grained performance scaling.

In this paper, we study issues of power efficiency in distributed systems, clouds, and micro-cloud solutions. We have selected the Raspberry Pi single-board computer as a representative of power efficient hardware based on ARM platform. We have tested performance of this device using our own application benchmark and compare the results with a commodity desktop PC and an enterprise server to determine the power-to-performance ratio and relative applicability for various problems. Even though the results are only approximate, the Raspberry Pi seems to be a viable candidate for green micro-cloud solutions.

The paper is organized as follows. More detailed overview of distributed systems and cloud solutions is provided in Section 2. Section 3 revises related work on micro-cloud systems. In Section 4, we present details about our tested platform – the Raspberry Pi device. Section 5 summarizes our empirical evaluation, Section 6 outlines possible applicability of these technologies, and Section 7 concludes the paper.

2 Distributed Applications and Cloud Solutions

Among the most important computing technologies that are in use nowadays are Distributed Systems and Cloud Computing Systems. Distributed system [1] is a collection of computers that work together and appear as one large computer. These computers cooperate to solve usually complex tasks; they are mutually interconnected to provide a massive computing power.

The basic advantages of distributed systems are:

- High performance

- Transparency

- Resource sharing

- Reliability and availability

- Incremental extensibility

On the other hand, the disadvantages that we may face in distributed systems are complexity, software development difficulties, networking problems, and security issues.

Contemporary cloud solutions has evolved from the earlier distributed systems. Cloud computing (despite the term has no exact definition) can be considered as a specialized form of distributed computing where virtualized resources are available as a service over the internet. These services usually include infrastructure, platform, applications, storage space and many other vendor-specific modules, libraries and frameworks. The users pay only for the services or resources they actually use. The underlying resources, such as storage, processors, memory, are completely abstracted from the consumer. The vendor of the cloud service is responsible for the reliability, performance, scalability and security of the service.

Cloud computing has many benefits, but cases exist where some data cannot be moved to the cloud for various reasons. In some cases, data may be generated at rates that are too big to move or at rates that exceed transfer capacity, for example in surveillance, operations in remote areas, and telemetry applications. In other cases, security concerns or regulatory compliance requirements might limit the use of the cloud.

Green computing [2] [3] refers to the environmentally responsible use of computers and any other technology related resources. Green computing includes the implementation of best practices, such as energy efficiency central processing units (CPUs), peripherals and servers [4]. Green Cloud is a computing facility that is entirely built, managed and operated on green computing principles. It provides the same features and capabilities of a typical cloud solution but uses less energy and space, and its design and operation are environmentally friendly.

3 Micro-Cloud Solutions

The recent introduction of the Raspberry Pi, a low-cost, low-power single-board computer, has made the construction of miniature green cloud systems more affordable.

Glasgow Raspberry Pi Cloud [5] is a model of a micro-cloud solution composed of clusters of Raspberry Pi devices. The PiCloud emulates every layer of a cloud stack, ranging from resource virtualisation to network behaviour, providing a full-featured cloud computing research and educational environment.

Iridis-pi [6] cluster consists of 64 Raspberry Pi Model B nodes each equipped with a 700 MHz ARM processor, 256 Mbit of RAM and a 16 GiB SD card for local storage. The cluster has a number of advantages that are typical for micro-clouds, such as low total power consumption, easy portability due to its small size and weight, and passive, ambient cooling. These attributes make Iridis-Pi ideally suited to educational applications, where it provides a low-cost starting point to inspire and enable students to understand principles of high-performance computing.

Sher.ly [7] builds a network-attached storage (NAS) device, the Sherlybox, that comes with its own peer-to-peer virtual private network and file server. The Sherylbox is built around the Raspberry Pi Model B computer. It comes with 512 MB of RAM, two USB 2.0 ports, 802.11n Wi-FI, and a 100mb Ethernet port. Instead of just the naked board, the Sherylbox comes with a case, a 4GB eMMC flash drive, and an optional 1 TB hard-drive. The company claims that with external USB drives, it can support up to 127 USB drives.

Tonido [8] offer a compelling alternative to public cloud file services allowing consumers to leverage their existing computers or IT infrastructures to keep control over their own data. It is available for a wide list of operating systems running on different hardware including Raspberry Pi using Raspbian or Raspbmc OS. Nimbus [9] is another example of a micro-cloud solution.

Although all of the abovementioned solutions are intended especially to personal or educational use (and a majority of scientific papers expect such use-cases), we claim that, under certain conditions, there may exist a wider range of possible applications. Some of them are discussed in Section 6.

4 Single-board Computers

Single-board computers constitute a special brand of computational devices which aim for compactness and power efficiency. These devices have various applications in robotics, intelligent household devices, smart monitoring stations, and many other domains. Even though their performance cannot compete with mainstream desktop PC and servers, they may achieve better power to performance and power to cost ratios. In this section, we present a few examples of compact single-board devices and revise the properties of Raspberry Pi device, which was selected as a representative for our research.

4.1 Computer Examples

Arandale Board [10] is a single-board computer powered by Samsung Exynos 5, which is an ARM CPU. The board is equipped with 2GB of RAM and various common peripherals such as USB 3.0, WiFi, GPS module, or interface for LCD display. The board is mainly designed for tablets and embedded computers; however, it also provide adequate performance for cloud computing. On the other hand, most of its peripherals are undesired for a solely computational solution and they may increase the overall cost.

AMD presented a *Gizmo 2* board [11], which is also a single-board computer that is compatible with x86 architecture. It comprises specialized double core APU (clocked at 1 GHz), which is a single chip that integrates power efficient CPU and Radeon GPU, and 1 GB of DDR3 RAM. The board is designed to provide all-in-one PC solution, so it is equipped with traditional interfaces such as USB, Gigabit Ethernet, or HDMI. The integrated GPU may provide excellent performance (with respect to power consumption); however, the price of the board is rather high in comparison to similar devices.

Intel entered the domain of power efficient single-board devices with *Galileo* development board [12]. It is equipped with Intel Quark X1000 CPU, which is a single-core Pentium-based 32-bit processor clocked at 400 MHz, and 256 MB of DRAM. The board is compatible with Arduino [13] device specification, which allows it to share peripherals and extensions designed for this platform.

Another similar platform is *Intel Edison*. It also contains Intel Quark CPU, but the Edison platform aims mainly at wearable devices and extensive miniaturization.

The *Parallela* board [14] is a relatively novel accomplishment in the field of efficient parallel hardware. Unlike many other devices, Parallela was designed by a small company Adapteva. It is equipped with ARM Cortex-9 CPU, FPGA, and a Epiphany coprocessor. The coprocessor is perhaps the most intriguing part of this hardware, since it is a specialized power-efficient parallel processing unit which organizes the cores in a 2D grid. This device may be the most promising alternative for a Raspberry Pi in the terms of power efficiency and total performance. On the other hand, Parallela is approximately $3\times$ more expensive than Raspberry Pi.

4.2 Raspberry Pi

Raspberry Pi [15] is one of the first low-cost devices that is capable of running a traditional operating system (in this case Linux), so it can be used as a modest desktop PC. It was originally created as a cheap platform that would allow children to learn basics of programming, but it was quickly adopted for various applications, such as embedded devices, simple audio and video players, etc.

At present, there are several configurations available (models A, B, and B+) and a new version called Raspberry Pi 2 was introduced to the market. In this work, we present (and measure) the properties of Raspberry Pi model B+, which is the newest revision of the original Raspberry Pi (before its second version was released).

The device is powered by Broadcom BCM2835 CPU, which is an ARMv6 processor clocked at 700 MHz. The graphics is rendered by VideoCore IV GPU clocked at 250 MHz. The GPU is capable of decoding a full-HD video in real time; however, there is currently no API (such as OpenCL) provided for computations. The system holds 512 MB of DDR2 RAM, which is shared both by CPU and GPU. Persistent memory is not integrated on the board, but it contains an interface for memory cards. We have used commodity 32 GB Kingston MicroSDHC card (class 10) as the persistent data storage.

Raspberry Pi has many external interfaces. Beside traditional USB or HDMI connector, it also holds custom GPIO port or I2C bus, which make the device suitable as a high-level controller for many electronic devices. The most important interface for our intentions is the 100 Mb Ethernet. Unfortunately, the Ethernet interface is internally connected via USB 2.0 bus. This bridged solution does not reduce the overall throughput, but slightly increases the communication latency.

The system is designed mainly for Linux operating system, but it can accomodate virtually any system that can run on ARM CPU (e.g., RiscOS). For the convenience of the users, the community has prepared modified distribution of Debian Linux called Raspbian and some other distributions based on Ubuntu or Fedora are also available. We have used the Raspbian in our experiments, since it is the recommended system.

5 Experimental Results

We have subjected Raspberry Pi to a custom set of performance tests to assess its applicability for distributed computing and cloud applications. The performance results are compared with results from a desktop PC and commodity server in the perspective of the power consumption. Let us emphasize that the results measured for Raspberry Pi and for full-sized computers are not directly comparable and provide only approximate comparison since our measurements of power consumption does not use same methodology and our benchmark is only single-threaded.

5.1 Experimental Setup

The parameters of Raspberry Pi are detailed in Section 4.2. The referential desktop PC is equipped with Intel Core i7 870 CPU, which has four physical (8 logical) cores clocked at 2,93 GHz, and 16 GB DDR3-1600 RAM. The persistent storage is represented by two 100 GB SSD disks connected in RAID 1.

The referential server is Dell PowerEdge M910. It is 4-way cache-coherent NUMA[1] system, where each node has 8 physical (16 logical) cores clocked at 2 GHz. Each node manages 32 GB RAM – i.e., the whole system comprises 64 logical cores and 128 GB of internal memory. The server was connected to Infortrend ESDS 3060 disk array comprising two 400 GB SSD disks and 14 magnetic disks of 4 TB each. Both desktop PC and server are running Red Hat Enterprise Linux 7 as an operating system.

To asses the performance, we measure the real execution time of prepared tests. All tests are executed on the same data inputs and the size of the input is selected so that the test takes reasonable time on Raspberry Pi and at least a few seconds on desktop PC and server. Each test was repeated $10\times$ and the average time is presented as the final result. The values were processed by statistical methods to remove outliers (times tainted with errors of measurement).

The power consumption of the Raspberry Pi was determined by KCX-017 device, which measures voltage and current on an USB power cord, since Raspberry Pi is powered via USB. The power consumption is equal to voltage times current ($P = UI$) and we employ additional correction factor of $1/0.8$, which simulates loss on power source with efficiency of 80%. The power consumption was between 1.2 W (idle device) to 1.7 W (performing cryptographical tests).

The power of our server was measured on its power controller embedded in server chassis. We also include estimated partial consumption of the chassis itself and additional equipment (such as cooling infrastructure), hence we will operate with aggregated approximate consumption of 500 W. The power consumption of the desktop PC was calculated from the component specifications since we were not able to measure this value with reasonable effort. For our purposes and intentions, we will operate with the value 250 W.

5.2 Tests

The performance experiments were design to test various aspects of the device. Since we are trying to determine applicability of Raspberry Pi as a platform for distributed system and cloud infrastructure, we have selected algorithms that cover many different domains:

- *aes* – The Rijndael (Advanced Encryption Standard) algorithm [16] for symmetric cryptography.

- *scrypt* – Computing scrypt [17] hash function.

- *sha256* – Computing SHA256 hash function.

- *dijkstra* – Finding shortest path in a sparse graph using Dijkstra algorithm [18] with regular heaps.

- *hash* – Simulation of database hash-join operation using integer keys.

- *merge* – Simulation of database merge-join on sorted data streams using integer keys.

- *levenshtein* – Wagner-Fischer dynamic programming algorithm [19] that computes Levenshtein edit distance

- *multiply* – Naïve ($O(N^3)$) algorithm for matrix multiplication on float numbers.

- *strassen* – Strassen algorithm for matrix multiplication on float numbers.

- *quicksort* – Quicksort [20] in memory sorting algorithm implemented in C++ `std::sort` routine applied on integers.

- *zlib* – DEFLATE [21] compression algorithm implemented in Zlib.

Beside these application tests, we have performed additional tests designed to determine the speed of internal memory, effectivity of its CPU caches, and performance of the persient storage (i.e., the SD flash card). However, we do not present detailed results of all these tests for the sake of the scope.

5.3 Results

The application benchmark results are presented in Figure 1. The results depict computational power efficiency normalized relatively to Raspberry Pi (individually for each algorithm) – i.e., higher value means greater power consumption with respect to computational performance. Hence, we can directly determine, which platform is better and which is worse for a particular problem. Let us note that we have adjusted the results so that they take the multi-core and multi-processor nature of the desktop PC and the server, since our benchmark is only single threaded. The performance of the full-sized computers were multiplied by the number of their physical cores.

The results indicate that Raspberry Pi is quite efficient for memory-intensive tasks. For some tests (especially database merge joins), the Raspberry Pi even outperforms both desktop PC and server. On the other hand, number crunching operations (such as the matrix multiplication on float numbers) are more suitable for x86 architecture, since it may employ SIMD instructions. We have performed additional synthetic memory-oriented experiments and they have confirmed this observation.

In addition to application tests, we have measured performance of the persistent storage. The throughput of individual operations is presented in Table 1. Let us emphasize that the Raspberry Pi has only a commodity SD card, while the server uses enterprise disk array.

---

[1]Nonuniform Memory Architecture

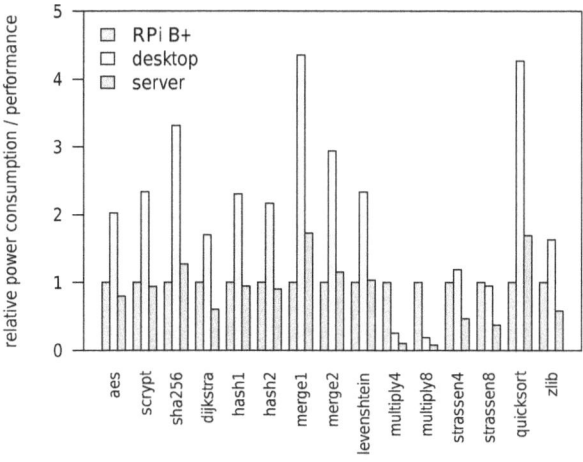

Figure 1: Relative efficiency of application tests

	Rasbperry Pi	desktop PC	server
rand. read	0.7	9.3	5.3
seq. read	17.3	171.9	404.5
seq. write	1.7	58.1	86.9

Table 1: Persistent storage performance (MB/s)

The results indicate that the performance of the Raspberry Pi is approximately 10-100× worse than the performance of other two platforms. On the other hand, if the data are distributed evenly among the devices, each Raspberry Pi has to handle two orders of magnitude smaller amount of data, so the performance is comparable. Furthermore, the devices may also utilize external disk array connected via 100 Mbit ethernet, which should provide data transfers around 5 MB/s.

6 Applicability

In this section, we would like to outline possible applicability of single-board devices for various problems. Besides the obvious cost issue, the presented solutions are expected to take advantage of two greatest benefits over traditional servers or desktop PCs:

- increased robustness

- and better heat dissipation.

The robustness is one of the expected properties of many distributed systems. However, when one server fails, the total drop of performance could be significant, especially in case of smaller and mid-sized clusters. When small devices such as Raspberry Pi are used, the failure of a single device is hardly noticeable on the overall performance and the faulty hardware could be replaced more quickly. Furthermore, small devices permit more fine-grained redundancy in the system.

The heat dissipation presents a challenging problem for modern servers as most powerful x86 processors easily produce over a hundred watts of thermal power. Hence, the servers, their chassis, and the server racks employ sophisticated cooling mechanism to drive the undesired heat out off the server room. In case of smaller devices, the produced heat has much lower watt per area ratio, so it is much easier to cool these devices.

6.1 Replacing Tradional Servers

A direct applicability of a Raspberry Pi cluster could be to replace traditional enterprise servers. Based on the scale, this solution could work for a small cluster within one server room or as a large distributed system that provides cloud services. In any case, the main advantage of such solution is the more evenly distributed heat output. Hence, the system does not to have a server room with powerful cooling system.

It may even be considered to place most of the hardware outside of a server room and integrate the single board computers into the infrastructure of a building or into regular rooms (offices, etc.). The Raspberry Pi does not require a cooling fan, hence such solution would not increase background noise inside the building. Furthermore, the heat produced by the devices may be used as part of internal heating system and the I/O ports (USB or GPIO) could be used to operate building sensors.

6.2 Outdoor Micro-Clouds

The compactness and low consumption of single-board computers may be utilized in many applications which could be characterized as *outside the server room* projects. Such projects would include robotics, autonomous vehicles and aircraft, probes and intelligent exploration devices, etc. A micro-cloud solution could increase robustness of these devices, which could be important since their hardware is subjected to much harsh physical conditions than hardware located in a server room or in an office.

Let us use an autonomous car (which is a domain that spawned an intensive research in the past few years) as an example of such outdoor device that required nontrivial computational power. A cluster of single-board computers may provide much scalable hardware for navigation computations. For instance, when the car is driving on a straight road in an unpopulated area, it requires much less computational power to track and analyse surrounding environment. Hence, it may shut down most of the devices in the cluster to save energy. On the other hand, when driving inside a city, it may turn on the whole cluster to get necessary computational power. Finally, the decentralized nature of the hardware may provide enough computational power even in extreme cases, such as when part of the vehicle is compromised in a car crash.

7 Conclusions

In this paper, we have addressed the issue of power efficiency in cloud systems. Many systems would benefit greatly from a hardware that provide less computational power, but which is more power efficient and has lower initial and maintenance costs. We have designed an application benchmark for small devices that tests various known algorithms. The benchmark was applied on the Raspberry Pi, which is one of the first single-board computers. The Raspberry Pi is very power efficient and cost around $30, which makes it a good candidate to be a worker in a green cluster or a micro cloud. The benchmark results indicate that current version of Raspberry Pi is competitive with desktop PC as well as an enterprise server in tasks that can be ideally distributed.

In our future work, we would like to test other similar devices, especially the second version Raspberry Pi and the Parallela board with Epiphany coprocessor. Furthermore, we are planning to build a small cluster from these devices to measure the total consumption more precisely and to determine the communication overhead of various distributed algorithms.

Acknowledgment

This paper was supported by Czech Science Foundation (GACR) projects P103/13/08195 and P103/14/14292P and by SVV-2015-260222.

References

[1] Tanenbaum, A. S., Van Steen, M.: Distributed systems: principles and paradigms (2nd Ed.). Prentice-Hall, 2006

[2] Priya, B., Pilli, E., Joshi, R.: A survey on energy and power consumption models for greener cloud. In: Advance Computing Conference (IACC), 2013 IEEE 3rd International, 2013, 76–82

[3] Kumar, S., Buyya, R.: Green cloud computing and environmental sustainability. In: Harnessing Green It, John Wiley and Sons, Ltd, UK, 2012, 315–339

[4] Yanovskaya, O., Yanovsky, M., Kharchenko, V.: The concept of green cloud infrastructure based on distributed computing and hardware accelerator within fpga as a service. In: Design Test Symposium (EWDTS), 2014 East-West, 2014, 1–4

[5] Tso, F. P., White, D. R., Jouet, S., Singer, J., Pezaros, D. P.: The glasgow raspberry pi cloud: A scale model for cloud computing infrastructures. In: Distributed Computing Systems Workshops (ICDCSW), 2013 IEEE 33rd International Conference on, IEEE, 2013, 108–112

[6] Cox, S. J., Cox, J. T., Boardman, R. P., Johnston, S. J., Scott, M., O'Brien, N. S.: Iridis-pi: a low-cost, compact demonstration cluster. In: Cluster Computing 17 (2014), 349–358

[7] sher.ly, "Private Cloud Solution for Sensitive Data Sharing with Secure Access Control." [Online]. Available: http://sher.ly/

[8] Tonido, "Turn your Raspberry PI into your personal cloud." [Online]. Available: http://www.tonido.com

[9] Cloudnimbus, "Nimbus - Personal Cloud for Raspberry Pi." [Online]. Available: www.cloudnimbus.org

[10] "Arandale board." [Online]. Available: http://www.arndaleboard.org/

[11] Gizmosphere, "Amd gizmo 2." [Online]. Available: http://www.gizmosphere.org/products/gizmo-2/

[12] Intel, "Galileo Gen 2 Development Board, url=http://www.intel.com/content/www/us/en/do-it-yourself/galileo-maker-quark-board.html,."

[13] M. Banzi, D. Cuartielles, T. Igoe, G. Martino, and D. Mellis, "Arduino uno." [Online]. Available: http://www.arduino.cc/

[14] Adapteva, "Parallela." [Online]. Available: https://www.parallella.org/

[15] Raspberry Pi Foundation, "Raspberry Pi single-board computer." [Online]. Available: https://www.raspberrypi.org/

[16] Gladman, B.: A specification for rijndael, the aes algorithm. at fp. gladman. plus. com/cryptography_technology/rijndael/aes. spec 311 (2001), 18–19

[17] Percival, C.: Stronger key derivation via sequential memory-hard functions. Proceedings BSD Canada, 2009

[18] Dijkstra, E. W.: A note on two problems in connexion with graphs. Numerische Mathematik 1 (1) (1959), 269–271

[19] Wagner, R. A., Fischer, M. J.: The string-to-string correction problem. Journal of the ACM (JACM) 21 (1) (1974), 168–173

[20] Hoare, C. A.: Quicksort. The Computer Journal 5 (1) (1962), 10–16

[21] Deutsch, L. P.: Deflate compressed data format specification version 1.3, 1996

J. Yaghob (Ed.): ITAT 2015 pp. 23–29
Charles University in Prague, Prague, 2015

Free or Fixed Word Order: What Can Treebanks Reveal?

Vladislav Kuboň and Markéta Lopatková

Charles University in Prague, Faculty of Mathematics and Physics, Institute of Formal and Applied Linguistics
Malostranské nám. 25, Prague 1, 118 00, Czech Republic
{lopatkova,vk}@ufal.mff.cuni.cz

Abstract: The paper describes an ongoing experiment consisting in the attempt to quantify word-order properties of three Indo-European languages (Czech, English and German). The statistics are collected from the syntactically annotated treebanks available for all three languages. The treebanks are searched by means of a universal query tool PML-TQ. The search concentrates on the mutual order of a verb and its complements (subject, object(s)) and the statistics are calculated for all permutations of the three elements. The results for all three languages are compared and a measure expressing the degree of word order freedom is suggested in the final section of the paper.

This study constitutes a motivation for formal modeling of natural language processing methods.

1 Introduction

General linguistics, see esp. [1, 2] studies natural languages from the point of view of similarities and differences in their syntactic structure, their development and historical changes, as well as from the point of view of language functions. It studies mutual influence of particular groups of features and, on the basis of similarities of language phenomena it introduces the so called language typology [3, 4]. The freedom or, on the other hand, strictness of the word order definitely belongs among the most important phenomena. General linguistics, for example, studies whether and how a particular language handles the order of words in sentences – whether the word is determined primarily by syntactic categories (e.g., a noun or a pronoun, without any additional morphological signs, located on the first sentential position represents a subject in English), or whether syntactic categories are primarily determined by other means than by the word order (for example, in Slavic languages, the subject tends to be a noun in the nominative case, regardless of its position in the sentence).

Particular natural languages cannot be, of course, strictly characterized by a single feature (for example word order), they are typically categorized into individual language types by a mixture of characteristic features. If we concentrate on word order, we study the prevalent order of the verb and its main complements – indo-european languages are thus characterized as SVO (SVO reflecting the order Subject, Verb, Object) languages. English and other languages with a fixed word order typically follow this order of words in declarative sentences; although Czech,

Russian and other Slavic languages are the so-called languages with a high degree of word order freedom, they still stick to the same order of word in a typical (unmarked) sentence. As for the VSO-type languages, their representatives can be found among semitic (Arabic, classical Hebrew) or Celtic languages, while (some) Amazonian languages belong to the OSV type. These characteristics, which are traditionally mentioned in classical textbooks of general linguistics [5], have been specified on the basis of excerptions and careful examination by many linguists.

Today, when we have at our disposal a wide range of linguistic data resources for tens of languages, we can easily confirm (or enhance by quantitative clues) their conclusions. This paper represents one of the steps in this direction.

The Institute of Formal and Applied Linguistics at the Charles University in Prague, has established a repository for linguistic data and resources LINDAT/CLARIN[1]. This repository enables experiments with syntactically annotated corpora, so called treebanks, for several tens of languages. Wherever it is possible due to the license agreements, the corpora are trasformed into a common format, which enables – after a very short period of getting acquainted with each particular treebank – a comfortable search and analysis of the data from a particular language. The HamleDT[2] (HArmonized Multi-LanguagE Dependency Treebank) project has already managed to transform more than 30 treebanks from all over the world [6] into a common format.

In this pilot study we concentrate on three Indo-european languages which substantially differ by the degree of word freedom – Czech, German and English. We investigate their typological properties on the basis of the Prague Dependency Treebank [7], the English part of the Prague Czech-English Dependency Treebank[8] and the German treebank TIGER [9] by means of the interface of PML-TQ Tree Query [10], which enables the access to the treebanks from the HamleDT.[3]

2 Setup of the Experiment

The analysis of syntactic properties of natural languages constitutes one of our long term goals. The phenomenon of word order has been in a center of our investigations

---

[1]https://lindat.mff.cuni.cz/cs/
[2]http://ufal.mff.cuni.cz/hamledt
[3]https://lindat.mff.cuni.cz/services/pmltq/

for a very long time. Our previous investigations concentrated both on studying individual properties of languages with higher degree of word-order freedom (as, e.g., non-projective constructions (long-distance dependencies) [11] as well as on the endeavor to find some general measures enabling to more precisely characterize concrete natural languages with regard to the degree of their word-order freedom (see, e.g. [12]).

The experiment presented in this paper continues in the same direction. It is driven by the endeavor to find an objective way how to compare natural languages from the point of view of the degree of their word-order freedom. While the previous experiments concentrated on more formal approach, this one builds upon a thorough analysis of available data resources. Let us briefly introduce them in the subsequent subsections.

When investigating syntactic properties of natural languages, it is very often the case that the discussion concentrates on individual phenomena, their properties and their influence on the order of words. The mere presence of some phenomenon (or its more detailed properties) is, of course, important and definitely influences the degree of word-order freedom but this kind of investigation cannot be complete without stating also the quantitative properties of the given phenomenon. A linguistically interesting, but marginal phenomenon does not tell us so much as a basic phenomenon occurring relatively frequently. This observation constitutes the basis of our current experiment. In order to capture the quantitative characteristic of a natural language, let us take a representative sample of its syntactically annotated data and let us calculate the distribution of individual types of word order for the three main syntactic components – subject, predicate and object. It is obvious that the more free is the word order of a given language, the more equally they are going to be distributed.

2.1 Available Treebanks

The extensive quantitative analysis of the same linguistic phenomenon for different languages would not be feasible without a common platform which makes it possible to compare various data resources from the same point of view. Thanks to the initiative HamleDT[4] (HArmonized Multi-LanguagE Dependency Treebank) it is now possible to compare the data from more than 30 languages in a uniform way [6].

The HamleDT family of treebanks is based on the dependency framework and technology developed for the Prague Dependency Treebank (PDT)[5] [7], i.e., large syntactically annotated corpus for the Czech Language. Here we focus on the so-called analytical layer, i.e., the layer describing surface sentence structure (relevant for studying word order properties). The framework and its language independence was verified within (the English

part of) the Prague Czech English Dependency Treebank (PCEDT)[6] [8] – within this project, syntactically annotated Penn Treebank[7] [13] was automatically transformed from the original phrase-structure trees into the dependency annotation.[8] Based on this experience, the HamleDT initiative goes further, syntactically annotated corpora for different languages are collected and transferred into the common format. Here we make use of the TIGER corpus[9] for the German language [9], the corpus with native phrase-structure annotation enriched with the information about the head for each phrase (and thus bearing also information on dependencies). Figures 2, 6 and 7 show sample trees for Czech, English and German, respectively, and Table 1 summarizes the size of these corpora.

corpus	# preds	lang	type	genre
PDT	79,283	Czech	manual	news
PCEDT	51,048	English	automatic	economy
TIGER	36,326	German	automatic	news

Table 1: Overview of all three treebanks (# preds represents the number of predicates in the given corpora)

2.2 HamleDT and PMLTQ Tree Query

For searching the data, we exploit a PML-TQ search tool,[10] which has been primarily designed for processing the PDT data. PML-TQ is a query language and search engine designed for querying annotated linguistic data [10] – it allows users to formulate complex queries on richly annotated linguistic data.

Having the treebanks in the common data format, the PML-TQ framework makes it possible to analyse the data in a uniform way – the following sample query gives us trees with an intransitive predicate verb (in a main clause), i.e. Pred node with Sb node and no Obj nodes among its dependent nodes, where Sb follows the Pred; the filter on the last line (>> for $n0.lemma give $1, count()) outputs a table listing verb lemmas with this marked word order position and number of their occurrences in the corpus, see also Figure 1.

```
a-node $n0 :=
[ afun = "Pred",
    child a-node $n1 :=
    [ afun = "Sb", $n1.ord > $n0.ord ],
    0x child a-node
    [ afun = "Obj"]]
>> for $n0.lemma give $1, count()
```

---

[4]http://ufal.mff.cuni.cz/hamledt
[5]http://ufal.mff.cuni.cz/pdt3.0

[6]http://ufal.mff.cuni.cz/pcedt2.0/cs/index.html
[7]https://www.cis.upenn.edu/ treebank
[8]This dependency-based surface annotation then served as a basis for deep syntactic dependency-based annotation of English; however, as for Czech, only surface structure is interesting for the studied phenomenon of word order.
[9]http://www.ims.uni-stuttgart.de/forschung/ressourcen/korpora/tiger.html
[10]https://lindat.mff.cuni.cz/services/pmltq/

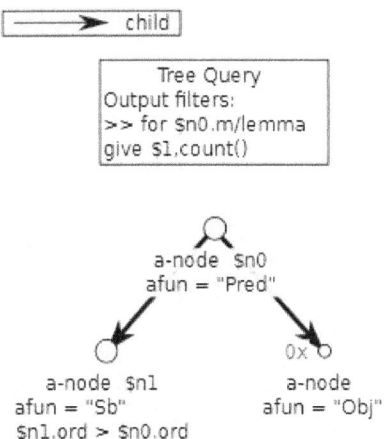

Figure 1: Visualization of the PML-TQ query

3 Analysis of Data

Let us now look at the syntactic typology of natural languages under investigation. We are going to take into account especially the mutual position of subject, predicate and direct object. After a thorough investigation of the ways how indirect objects are annotated in all three corpora, we have decided to limit ourselves – at least in this stage of our research – to basic structures and to extract and analyse only sentences without too complicated or mutually interlocked phenomena. Namely we focus on sentences with the following properties:

- A predicate under scrutiny belongs to the main clause (as e.g. in the sentence $Jsou_{Pred}$ *vám nejasná některá ustanovení daňových zákonů?* 'Are_{Pred} certain provisions of the tax laws unclear to you?', see the dependency tree in Fig. 2); i.e., we do not analyse word order of dependent clauses;

- We analyse only non-prepositional subjects and objects (compare e.g. with the sentence *V 2180 městech a obcích žije na 2.6 milionu obyvatel$_{Sb}$;* 'There are (about 2.6 milion of inhabitants)$_{Sb}$ living in 2 180 towns and villages;', see Fig. 3);

- Sentences may contain coordinated predicates (as, e.g., predicates *následoval* and *opakovalo* in the corpus sentence *Vzápětí následoval$_{Pred}$ další regulační stupeň a vše se opakovalo$_{Pred}$.* 'The next level of regulation immediately followed$_{Pred}$ and everything repeated$_{Pred}$ again.', see Fig. 4);

  However, sentences with common subjects (or objects) are not taken into account (thus sentences as, e.g., *Koupelna$_{Sb}$ nebo teplá voda$_{Sb}$ nejsou trvale k dispozici.* 'A bathroom$_{Sb}$ or hot water supply$_{Sb}$ are not at the permanent disposal.', see Fig. 5 are not counted in the tables).[11]

---

[11] Including coordination phenomena in all their complexity would require much robust queries in any dependency framework; thus we have decided to disregard this type of sentences at all.

Figure 2: Sample Czech dependency tree from PDT

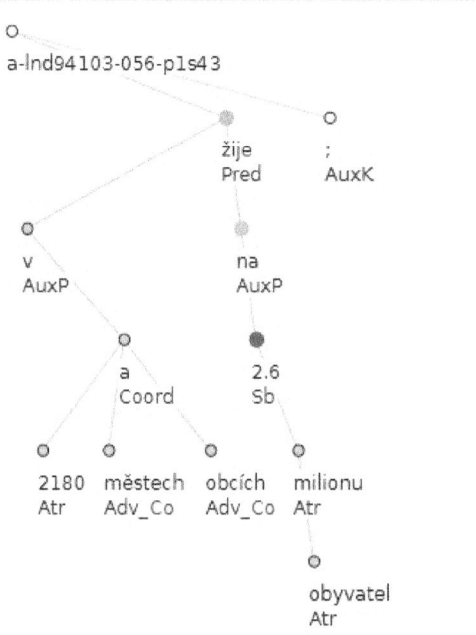

Figure 3: Sample Czech dependency tree from PDT with prepositional subject (excluded from the resulting tables)

3.1 Czech

The highest quality syntactically annotated Czech data can be found in the Prague Dependency Treebank; in fact, it is the only corpus we work with that has been manually annotated and thoroughly tested for the annotation consistency. The texts of PDT belong mostly to the journalism genre, it consists of newspaper texts and (in a limited scale) of texts from a popularizing scientific journal.

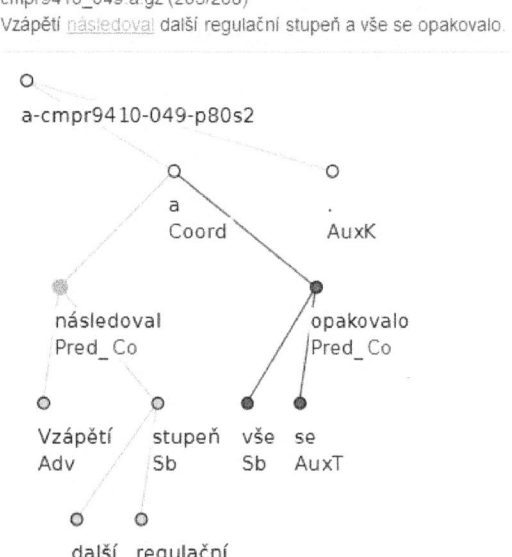

Figure 4: Sample Czech dependency tree from PDT with coordinated predicates (included in the resulting tables)

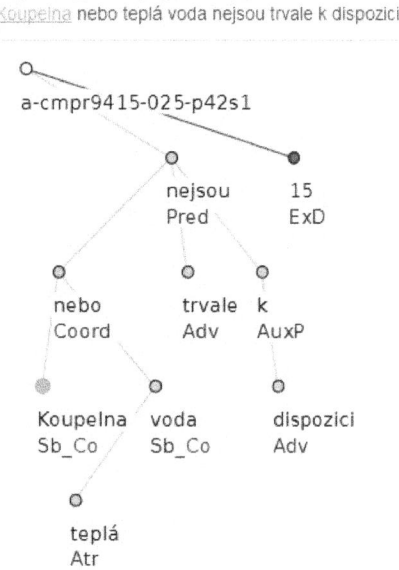

Figure 5: Sample Czech dependency tree from PDT with coordinated subject (excluded from the resulting tables)

The following Table 2 summarizes the number of sentences with intransitive verbs in main clauses in PDT with respect to the word order positions of Sb and Pred – we can see that the marked word order (verb preceding its subject) is quite common in Czech.[12]

The second table displays the distribution of individual combinations of a subject, predicate and a single object.

---

[12]In our settings, we do not checked the part of speech of the predicate; however, out of the 79,283 sentences conforming to the properties mentioned above, only 329 have other than verbal predicate.

Word order type	Number	%
SV	16,909	56.66
VS	12,932	44.34
Total	29,841	100.00

Table 2: Sentences with intransitive verbs

It is not surprising that the unmarked – intuitively "most natural" – word order type, SVO, accounts for only slightly more than half of cases. The relatively high degree of word order freedom is thus supported also quantitatively.

Word order type	Number	%
SVO	11,158	52.42
SOV	1,533	7.20
VSO	1,936	9.10
VOS	2,136	10.04
OVS	4,001	18.80
OSV	521	2.45
Total	21,285	100.00

Table 3: Sentences with a single object

Even more interesting (and also supporting the claim that the word order freedom of Czech is relatively high) are the results for sentences with at least two objects. They are summarized in Table 4. The distribution is even flatter than in Table 3 with all types being represented (even those starting with two objects, see the following example) and none of them exceeding 30%.

Plán mu v úterý ***předložil*** *velvyslanec USA v Chorvatsku Peter* ***Galbraith***.

Word order type	Number	%
SVOO	293	26.95
SOVO	223	20.52
SOOV	33	3.04
VSOO	45	4.14
VOSO	16	1.47
VOOS	27	2.48
OSVO	70	6.44
OSOV	10	0.92
OOSV	15	1.38
OOVS	124	11.41
OVSO	78	7.18
OVOS	153	14.08
Total	1,087	100.00

Table 4: Sentences with two objects

3.2 English

The statistics concerning the distribution of word-order types for English have been calculated on the English part of the Prague Czech English Dependency Treebank

(PCEDT). This corpus actually contains the same set of sentences as the Wall Street Journal section of Penn Treebank,[13] (see above for references) but unlike its predecessor, its syntactic structure has been annotated using dependency trees. As was mentioned above, the transformation on the surface syntactic layer was fully automatic, which has of course affected the quality of annotation.

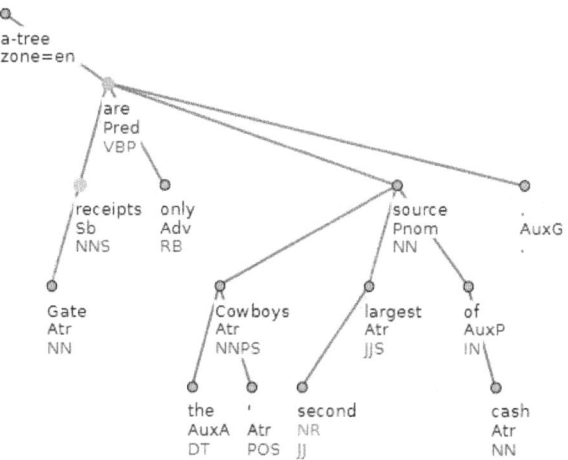

Figure 6: Sample English dependency tree from PCEDT

The statistics of different types of word order have been collected in the same manner as in the previous subsection. We have also applied identical filters as for Czech sentences from PDT. Table 5 contains data for sentences with intransitive verbs. Only as few as 40 sentences have other than verbal predicate.

Word order type	Number	%
SV	28,236	96.91
VS	900	3.09
Total	29,136	100.00

Table 5: English sentences with intransitive verbs

As we can see, the strict word order of English sentences manifests itself in a vast majority of sentences having the prototypical word order of the subject being followed by a predicate. The examples of the opposite word order include sentences containing direct speech with the following pattern:

"It's just a matter of time before the tide turns," says one Midwestern lobbyist.

Out of the 900 sentences with the reversed word order, as many as 630 contained the predicate *to say*, 121 *to be*. Each of all other verbs involved in these constructions

were represented less than 10 times. In total, 23 verbs appear in these sentences at least twice, out of them 16 can be classified as verbs of communication (verba dicendi) (in total, it means 678 occurrences out of 822, i.e., 82,5 % of all occurrences with at least two hits in the corpus).

The results for sentences containing one object also strongly confirm the fact that the order Subject - Predicate - Object (SVO) is practically the only acceptable order in standard sentences. The remaining types of word order (representing only 1.06% sentences in the corpus) mentioned in Table 6 actually represented annotation errors in a vast majority of cases (esp. auxiliary verbs which have been quite often incorrectly annotated as Objects).

Word order type	Number	%
SVO	12,481	98.94
SOV	77	0.61
VSO	9	0.07
VOS	1	0.01
OVS	2	0.02
OSV	45	0.36
Total	12,615	100.00

Table 6: English sentences with a single object

It turns out that for English, it does not make sense to construct a similar table as Table 4 sentences with more than one object. The automatic annotation of PCEDT is, unfortunately, biased in what should be considered an Object (in the original Penn Treeank annotation, the verbal complements are labeled just as noun (or prepositional) phrases (NPs and PPs), no distinction between Objects and Adverbials.) As a consequence, adverbial constructions are very often incorrectly annotated as Objects and thus it is impossible to rely on this distinction (and the analysis shows that the numbers would be highly misleading).

3.3 German

German has more constraints on word order than Czech and less than English, therefore it constitutes a very natural candidate for our experiment. On top of that, there are also numerous high quality resources which can be exploited. We have used the German treebank conforming to the HamleDT initiative, which is located in the Lindat repository.[14]

The statistics for German were collected in the same way and with the same constraints as Czech and English ones. The statistics for German sentences with intransitive predicates are presented in Table 7.

The almost equal number of sentences with SV and VS word order types is quite surprising. The fact that SV represents the typical word order in declarative sentences, while VS in interrogative ones provides an obvious explanation. Unfortunately, this explanation does not

---

[13]The Czech part had been created as translation of original English sentences.

[14]https://lindat.mff.cuni.cz/services/pmltq/hamledt_dt_de/

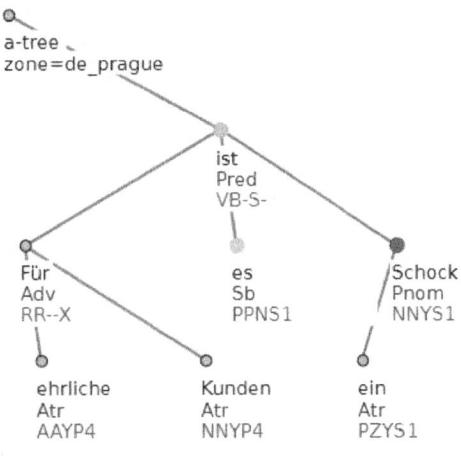

Figure 7: Sample German dependency tree from HamleDT

Word order type	Number	%
SV	6,165	56.67
VS	4,713	43.33
Total	10,878	100.00

Table 7: German sentences with intransitive verbs

cover all occurrences because the analyzed corpus (consisting mostly of newspaper texts) contains only a very small proportion of interrogative sentences. We have not investigated the reason for the surprisingly high number of VS sentences, but it definitely constitutes a very interesting topic for further research. The same is valid also for the results contained in Table 8, where we have found relatively high number of sentences having the word order of an interrogative sentence, too.

Word order type	Number	%
SVO	10,662	50.31
SOV	193	0.91
VSO	7,425	35.04
VOS	690	3.26
OVS	2,206	10.41
OSV	15	0.07
Total	21,191	100.00

Table 8: German sentences with a single object

Neither for German we have investigated the sentences with two or more objects due to annotation inconsistencies.

4 Proposed Measure of Word Order Freedom

The statistics presented in the previous section actually confirm the well known fact that Czech has the highest degree of word order freedom from all three languages investigated in our experiment. This fact is also reflected in the chart 8 comparing the results for sentences with one object for all three languages.

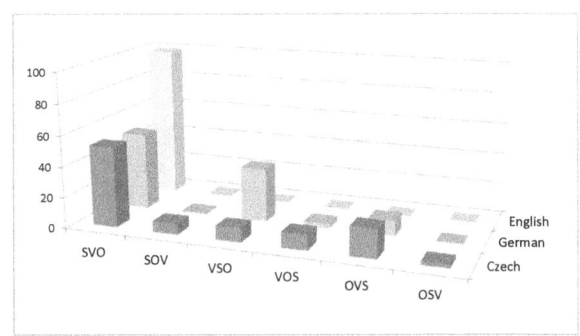

Figure 8: Comparison of results

Let us now try to suggest a formula which might allow to express the degree of word order freedom in a more precise way. Intuitively, the more free is the word order, the more equally distributed should be the results of all six word order types. The more strict the word order, the more distant are the values from the ideal (equal distribution). This leads directly to the application of a least squares method:

$$M = \frac{1}{6}\sqrt{\sum_{i=1}^{6}(V_i - Av)^2}, \qquad (1)$$

where M is the proposed measure, V_i the percentual value of the i-th word order type and Av is the average percentage for each word type (i.e., 100/6). For the three languages in our experiment we then get the following values:

- Czech: 6.82

- German: 19.20

- English: 36.79

These values seem to correspond to the intuitive feeling that the word order order of English is really strongly fixed, while German and Czech have more free word order with Czech having the highest degree of word order freedom. If we express the results in the form of percentages of the absolutely fixed word order (i.e., one of the word order types accounts for 100% and all others do not appear at all), we'll get the following results:

- Czech: 18.31%

- German: 51.52%

- English: 98.73%

5 Conclusions

The experiment described in this paper brought several interesting results which may be taken as a basis for further experiments. First of all, it shows that the endeavor to unify the annotation schemes used for various treebanks in the HamleDT project provides new opportunities for linguistic research. The treebank data can now be studied in a relation to other treebanks using the common search tool and obtaining results which are not dependent on peculiarities of individual annotation schemes.

These new opportunities have been demonstrated on a small-scale experiment involving three languages (Czech, German and English). We have managed to extract quantitative clues confirming the linguistic hypothesis about the degree of word order freedom of all three languages under consideration. The main advantage of our approach is the fact that our research is based on a large number of sentences of each language and thus it provides a representative sample of the actual language usage in a given genre. Contrary to theoretical linguistic research, our approach does not concentrate upon marginal (but definitely linguistically interesting) phenomena, but it is based upon the real language captured in the treebanks.

In the future we would like to continue the research in two directions. One will be the obvious endeavor to collect the statistics for more languages, the second one will be a more subtle treatment of linguistic phenomena appearing in treebanks, as, e.g. the investigation including also subordinated clauses or interrogative sentences.

Grant support

This paper exploits language data developed and/or distributed in the frame of the project MŠMT ČR LINDAT/CLARIN (project LM2010013).

References

[1] Saussure, F.: Course in general linguistics. Open Court, La Salle, Illinois (1983) (prepared by C. Bally and A. Sechehaye, translated by R. Harris)

[2] Saussure, F.: Kurs obecné lingvistiky. Academia, Praha (1989) (translated by F. Čermák)

[3] Sapir, E.: Language. An introduction to the study of speech. Harcourt, Brace and Company, New York (1921) (http://www.gutenberg.org/files/12629/12629-h/12629-h.htm).

[4] Skalička, V.: Vývoj jazyka. Soubor statí. Státní pedagogické nakladatelství, Praha (1960)

[5] Čermák, F.: Jazyk a jazykověda. Pražská imaginace, Praha (1994)

[6] Zeman, D., Dušek, O., Mareček, D., Popel, M., Ramasamy, L., Štěpánek, J., Žabokrtský, Z., Hajič, J.: HamleDT: Harmonized multi-language dependency treebank. Language Resources and Evaluation **48** (2014), 601–637

[7] Hajič, J., Panevová, J., Hajičová, E., Sgall, P., Pajas, P., Štěpánek, J., Havelka, J., Mikulová, M., Žabokrtský, Z., Ševčíková-Razímová, M.: Prague Dependency Treebank 2.0. LDC, Philadelphia, PA, USA (2006)

[8] Hajič, J., Hajičová, E., Panevová, J., Sgall, P., Bojar, O., Cinková, S., Fučíková, E., Mikulová, M., Pajas, P., Popelka, J., Semecký, J., Šindlerová, J., Štěpánek, J., Toman, J., Urešová, Z., Žabokrtský, Z.: Announcing Prague Czech-English Dependency Treebank 2.0. In: Proceedings of the 8th International Conference on Language Resources and Evaluation (LREC 2012), Istanbul, Turkey, ELRA, European Language Resources Association (2012), 3153–3160

[9] Brants, S., Dipper, S., Eisenberg, P., Hansen, S., König, E., Lezius, W., Rohrer, C., Smith, G., Uszkoreit, H.: TIGER: Linguistic Interpretation of a German Corpus. Journal of Language and Computation (2004), 597–620

[10] Pajas, P., Štěpánek, J.: System for querying syntactically annotated corpora. In: Proceedings of the ACL-IJCNLP 2009 Software Demonstrations, Suntec, Singapore, Association for Computational Linguistics (2009), 33–36

[11] Holan, T., Kuboň, V., Oliva, K., Plátek, M.: On complexity of word order. Les grammaires de dépendance – Traitement automatique des langues (TAL) **41** (2000) 273–300

[12] Kuboň, V., Lopatková, M., Plátek, M.: On formalization of word order properties. In: Gelbukh, A., (ed.), Theoretical Computer Science and General Issues, Computational Linguistics and Intelligent Text Processing, CICLing 2012, volume 7181 of LNCS., Berlin / Heidelberg, Springer-Verlag (2012) 130–141

[13] Mitchell P. Marcus, Mary Ann Marcinkiewicz, B.S.: Building a large annotated corpus of English: the Penn Treebank. Computational Linguistics **19** (1993)

Efficient Computational Algorithm for Spline Surfaces

Lukáš Miňo

Institute of Computer Science, Faculty of Science, P. J. Šafárik University in Košice
Jesenná 5, 040 01 Košice, Slovakia
lukas.mino@upjs.sk

Abstract: Many data mining tasks can be reformulated as optimization problems, in the solution of which approximation by surfaces plays a key role. The paper proposes a new efficient computational algorithm for spline surfaces over uniform grids. The algorithm is based on a recent result on approximation of a biquartic polynomial by bicubic ones, that ensures C^2 continuity of the corresponding four bicubic spline components. As a consequence of this biquartic polynomial based approach to constructing spline surfaces, the classical de Boor's computational task breaks down to a reduced task and a simple remainder one. The comparison of the proposed and classical computational algorithm shows that the former needs less multiplication operations resulting in non negligible speed up.

1 Introduction

Recent years a considerable effort has been seen to develop reliable and efficient data mining tools to discover hidden knowledge in very large data bases. The fundamental problem is proposing algorithms to extract some useful information from very large databases. Fortunately, many data mining tasks can be reformulated as optimization problems, where approximation by surfaces plays a key role [1], [7].

The goal of the paper is to show that even such standard result as de Boor's sequential algorithm for construction of interpolating spline surfaces can be improved. It suggests a computational algorithm based on new model equations, in derivation of which biquartic polynomials played an essential role.

The idea of using quartic and biquartic polynomials in cubic and bicubic spline construction comes from recent results of Török and Szabó [16], [19], [13]. They have proven a key interrelation of these polynomials using the IZA representation [18], [14], which can incorporate both interpolation and approximation. The IZA representation was obtained using an r-point transformation that was a generalization of its three point ancestor [4]. A three point transformation was successfully applied to various approximation problems. Works [15], [8] showed how it can be used to assess the unknown degree in regression polynomials. In [5] a three point method was developed to detect piecewise cubic approximation segments for data with moderate errors. The technique, based on which the IZA representation has been derived, was first used in [10]. The paper [17] showed how to properly use the IZA representation's reference points for segment connection and their relation to derivatives. Papers [6], [14] contain results on approximation of 3D data based on the reference point approach. The first remarkable asymptotic properties of the IZA representation based two-part approximation model were gained in [11] and [18]. These properties confirmed the validity of the two-part approximation model, which led first to [16], where the interrelation of quartic and cubic polynomials was shown, and then to papers [19] and [13], [9] that introduced the reduced system approach to spline curve construction and proved the interrelation of bicubic and biquartic polynomials.

The algorithm presented in this paper has a decreased number of equations and is based on the generalized results of [19] and [9].

The structure of the article is as follows. Section two is devoted to problem statement. To be self-contained, section three briefly describes de Boor's algorithm. The next section first provides the definition of bicubic and biquartic polynomials. Then it shortly discusses the interrelation of these bivariate polynomials and their role in the computational schema. Section five contains the proposed sequential computational algorithm based on reduced systems. Section six briefly compares the new and the classical algorithm. The efficiency of the proposed algorithm is shown in the last but one section by computing the theoretical speedup that is approximately 1.33.

2 Problem Statement

The section defines the inputs for the spline surface, and the requirements that it should fulfil and based on which it can be constructed.

Consider a uniform grid

$$[u_0, u_1, \ldots, u_{2m}] \times [v_0, v_1, \ldots, v_{2n}], \qquad (1)$$

where

$$u_i = u_0 + ih_x, \quad i = 1, 2, \ldots, 2m, m \in \mathbb{N},$$

$$v_j = v_0 + jh_y, \quad j = 1, 2, \ldots, 2n, n \in \mathbb{N}.$$

According to [3], the spline surface is defined by given values

$$z_{i,j}, \quad i = 0, 1, \ldots, 2m, \quad j = 0, 1, \ldots, 2n \qquad (2)$$

at the equispaced grid-points, and given first directional derivatives

$$d_{i,j}^x, \quad i = 0, 2m, \quad j = 0, 1, \ldots, 2n \qquad (3)$$

at boundary verticals,

$$d_{i,j}^y, \quad i = 0, 1, \ldots, 2m, \quad j = 0, 2n \qquad (4)$$

at boundary horizontals and cross derivatives

$$d_{i,j}^{x,y}, \quad i = 0, 2m, \quad j = 0, 2n \qquad (5)$$

at the four corners of the grid.

The task is to define a quadruple $[z_{i,j}, d_{i,j}^x, d_{i,j}^y, d_{i,j}^{x,y}]$ at every grid-point $[u_i, v_j]$, based on which a uniform bicubic clamped spline surface S of class C^2 can be constructed with properties

$$S(u_i, v_j) = z_{i,j}, \qquad \frac{\partial S(u_i, v_j)}{\partial y} = d_{i,j}^y,$$
$$\frac{\partial S(u_i, v_j)}{\partial x} = d_{i,j}^x, \qquad \frac{\partial^2 S(u_i, v_j)}{\partial x \partial y} = d_{i,j}^{x,y},$$

where the adjacent spline segments are twice continuously differentiable. Our aim is to solve this task with less equations and less multiplications than the standard spline construction algorithm [3]. We will achieve this by means of Hermite splines and using a recently derived relationship property between biquartic and bicubic polynomials.

3   Carl de Boor's Algorithm

Paper [3] is devoted to bicubic spline surface interpolation. It formulates the problem and gives the solution to it. We briefly reformulate the paper's main result for uniform splines to show the four main equations, based on which the numerous tridiagonal systems of de Boor's algorithm for solution of the unknown derivatives are constructed. Thanks to it the paper is self contained and the reader can count up the number of operational multiplications and so quantitatively compare de Boor's and the proposed new algorithm.

Lemma 1 (de Boor). *If the above z values and d derivatives are given, then the values*

$$\begin{array}{ll} d_{i,j}^x, & i = 1, \ldots, 2m-1, \quad j = 0, \ldots, 2n, \\ d_{i,j}^y, & i = 0, \ldots, 2m, \quad j = 1, \ldots, 2n-1, \\ d_{i,j}^{x,y}, & i = 1, \ldots, 2m-1, \quad j = 0, \ldots, 2n, \\ & and \ i = 0, \ldots, 2m, \quad j = 1, \ldots, 2n-1 \end{array}$$

are uniquely determined by the following $2(2m) + (2n) + 5$ linear systems of altogether $3(2m)(2n) + (2m) + (2n) - 5$ equations:
for $j = 0, \ldots, 2n$,

$$d_{i+1,j}^x + 4d_{i,j}^x + d_{i-1,j}^x = \frac{3}{h_x}(z_{i+1,j} - z_{i-1,j}), \qquad (6)$$

where $i = 1, \ldots, 2m-1$;
for $j = 0, 2n$,

$$d_{i+1,j}^{x,y} + 4d_{i,j}^{x,y} + d_{i-1,j}^{x,y} = \frac{3}{h_x}(d_{i+1,j}^y - d_{i-1,j}^y), \qquad (7)$$

where $i = 1, \ldots, 2m-1$;
for $i = 0, \ldots, 2m$,

$$d_{i,j+1}^y + 4d_{i,j}^y + d_{i,j-1}^y = \frac{3}{h_y}(z_{i,j+1} - z_{i,j-1}), \qquad (8)$$

where $j = 1, \ldots, 2n-1$;
for $i = 0, \ldots, 2m$,

$$d_{i,j+1}^{x,y} + 4d_{i,j}^{x,y} + d_{i,j-1}^{x,y} = \frac{3}{h_y}(d_{i,j+1}^x - d_{i,j-1}^x), \qquad (9)$$

where $j = 1, \ldots, 2n-1$.

4   Biquartic Polynomials and Bicubic Splines

The section begins with definition of bicubic and biquartic polynomials. Then it shortly discusses the interrelation of these bivariate polynomials and its role in the computational schema.

The tensor product formulas of bicubic Hermite spline components, see [12], and biquaric polynomials are given by the following two definitions.

Definition 1. *On grid (1) the bicubic Hermite spline components $S_{i,j}(x, y)$ for*

$$i = 0, 1, 2, \ldots, 2m-1, \qquad j = 0, 1, 2, \ldots, 2n-1,$$
$$x \in [u_i, u_{i+1}], \qquad y \in [v_j, v_{j+1}],$$

are defined as follows

$$S_{i,j}(x, y) = \boldsymbol{\lambda}^T(x, u_i, h_x) \cdot \boldsymbol{\varphi}_{i,j} \cdot \boldsymbol{\lambda}(y, v_j, h_y), \qquad (10)$$

where $\boldsymbol{\lambda}$ is a vector of basis functions

$$\boldsymbol{\lambda}(t, t_0, h) = \begin{bmatrix} \frac{(1 + 2\frac{t - t_0}{h})(t - t_1)^2}{h^2} \\ \frac{(t - t_0)^2(1 - 2\frac{t - t_1}{h})}{h^2} \\ \frac{(t - t_0)(t - t_1)^2}{h^2} \\ \frac{(t - t_0)^2(t - t_1)}{h^2} \end{bmatrix}^T,$$

$t_1 = t_0 + h$ and $\boldsymbol{\varphi}$ is a matrix of function values and first derivatives

$$\boldsymbol{\varphi}_{i,j} = \begin{pmatrix} z_{i,j} & z_{i,j+1} & d_{i,j}^y & d_{i,j+1}^y \\ z_{i+1,j} & z_{i+1,j+1} & d_{i+1,j}^y & d_{i+1,j+1}^y \\ d_{i,j}^x & d_{i,j+1}^x & d_{i,j}^{x,y} & d_{i,j+1}^{x,y} \\ d_{i+1,j}^x & d_{i+1,j+1}^x & d_{i+1,j}^{x,y} & d_{i+1,j+1}^{x,y} \end{pmatrix}.$$

For the spline components the following conditions hold

$$
\begin{aligned}
S_{i,j}(u_k, v_l) &= z_{k,l}, & k = i, i+1, & \quad l = j, j+1, \\
\frac{\partial S_{i,j}(u_k, v_l)}{\partial x} &= d_{k,l}^x, & k = i, i+1, & \quad l = j, j+1, \\
\frac{\partial S_{i,j}(u_k, v_l)}{\partial y} &= d_{k,l}^y, & k = i, i+1, & \quad l = j, j+1, \\
\frac{\partial^2 S_{i,j}(u_k, v_l)}{\partial x \partial y} &= d_{k,l}^{x,y}, & k = i, i+1, & \quad l = j, j+1.
\end{aligned}
$$

Based on (10) the second derivatives of $S_{i,j}(x,y)$ can be expressed effectively, e.g.

$$
\frac{\partial^2 S_{i,j}(x,y)}{\partial x^2} = \frac{\partial^2 \boldsymbol{\lambda}^T(x, u_i, h_x)}{\partial x^2} \cdot \boldsymbol{\varphi}_{i,j} \cdot \boldsymbol{\lambda}(y, v_j, h_y), \quad (11)
$$

where

$$
\frac{\partial^2 \boldsymbol{\lambda}(t, t_0, h)}{\partial t^2} = \begin{bmatrix} \frac{6(2t - 2t_0 - h)}{h^3} \\ \frac{6(-2t + 2t_0 + h)}{h^3} \\ \frac{2(3t - 3t_0 - 2h)}{h^2} \\ \frac{2(3t - 3t_0 - h)}{h^2} \end{bmatrix}^T .
$$

The biquartic polynomials are also defined by tensor product.

Definition 2. *On grid (1) the biquartic polynomials $F_{i,j}(x,y)$ for*

$$
\begin{aligned}
i &= 0, 2, 4, \ldots, 2(m-1), & j &= 0, 2, 4, \ldots, 2(n-1), \\
x &\in [u_i, u_{i+2}], & y &\in [v_j, v_{j+2}],
\end{aligned}
$$

are defined as follows

$$
F_{i,j}(x,y) = \mathbf{L}^T(x, u_i, h_x) \cdot \boldsymbol{\Phi}_{i,j} \cdot \mathbf{L}(y, v_j, h_y), \quad (12)
$$

where $\mathbf{L}$ is a vector of basis functions

$$
\mathbf{L}(t, t_0, h) = \begin{bmatrix} \frac{-(1 + 2\frac{t - t_0}{h})(t - t_1)(t - t_2)^2}{4h^3} \\ \frac{(t - t_0)^2(t - t_2)^2}{h^4} \\ \frac{(t - t_0)^2(t - t_1)(1 - 2\frac{t - t_2}{h})}{4h^3} \\ \frac{-(t - t_0)(t - t_1)(t - t_2)^2}{4h^3} \\ \frac{(t - t_0)^2(t - t_1)(t - t_2)}{4h^3} \end{bmatrix} ,
$$

$t_1 = t_0 + h$, $t_2 = t_0 + 2h$ and $\boldsymbol{\Phi}$ is a matrix of function values and first derivatives

$$
\boldsymbol{\Phi}_{i,j} = \begin{pmatrix}
z_{i,j} & z_{i,j+1} & z_{i,j+2} & d_{i,j}^y & d_{i,j+2}^y \\
z_{i+1,j} & z_{i+1,j+1} & z_{i+1,j+2} & d_{i+1,j}^y & d_{i+1,j+2}^y \\
z_{i+2,j} & z_{i+2,j+1} & z_{i+2,j+2} & d_{i+2,j}^y & d_{i+2,j+2}^y \\
d_{i,j}^x & d_{i,j+1}^x & d_{i,j+2}^x & d_{i,j}^{x,y} & d_{i,j+2}^{x,y} \\
d_{i+2,j}^x & d_{i+2,j+1}^x & d_{i+2,j+2}^x & d_{i+2,j}^{x,y} & d_{i+2,j+2}^{x,y}
\end{pmatrix} .
$$

For u_k, v_l defined in (1) the following conditions hold

$$
\begin{aligned}
F_{i,j}(u_k, v_l) &= z_{k,l}, & k = j, j+1, j+2, & \quad l = j, j+1, j+2, \\
\frac{\partial F_{i,j}(u_k, v_l)}{\partial x} &= d_{k,l}^x, & k = i, i+2, & \quad l = j, j+1, j+2, \\
\frac{\partial F_{i,j}(u_k, v_l)}{\partial y} &= d_{k,l}^y, & k = i, i+1, i+2, & \quad l = j, j+2, \\
\frac{\partial^2 F_{i,j}(u_k, v_l)}{\partial x \partial y} &= d_{k,l}^{x,y}, & k = i, i+2, & \quad l = j, j+2.
\end{aligned}
$$

The tensor product definition of $F_{i,j}(x,y)$ by (12) provides a compact way to express first derivatives, e.g.

$$
\frac{\partial F_{i,j}(x,y)}{\partial y} = \mathbf{L}^T(x, u_i, h_x) \cdot \boldsymbol{\Phi}_{i,j} \cdot \frac{\partial \mathbf{L}(y, v_j, h_y)}{\partial y}. \quad (13)
$$

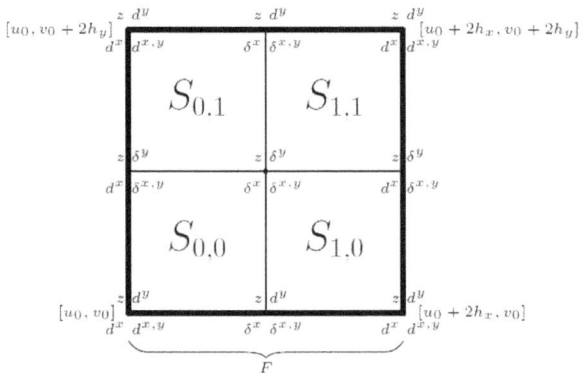

Figure 1: Schema of objects of a 2×2 - component bicubic Hermite spline surface.

Works [13], [9] prove how is a biquartic polynomial approximated by four bicubic polynomials. We want to apply this idea in our new approach to computing uniform bicubic splines of class C^2. The point of the approach is to solve only one half of derivatives from equations, and the second half of derivatives to compute from simple formulas that are derived from corresponding biquartic polynomials.

Unlike de Boor's lemma, we provide only the interpretation of the main result of [9]: a 2×2-component bicubic Hermite spline of class C^1 will be of class C^2, if the grid-points are equispaced and the unknown derivatives of the bicubic spline components at them are computed from a corresponding biquartic polynomial that is uniquely determined by the spline problem of Section 2 for the $[u_0, u_1, u_2] \times [v_0, v_1, v_2]$ grid.

This interrelation between a biquartic and four bicubic polynomials is illustrated by the schema in Fig. 1. The biquartic polynomial F over $[u_0, u_2] \times [v_0, v_2]$ is defined by given nine function values z and sixteen derivatives d that set up four quadruples $[z, d^x, d^y, d^{x,y}]$, two pairs $[z, d^x]$, two pairs $[z, d^y]$ and a single z. Every bicubic spline component is defined by four quadruples $[z, d^x, d^y, d^{x,y}]$. The nine quadruples in the figure are depicted around nine grid-points. Those eleven directional and cross first derivatives that are computed from the biquartic polynomial F and

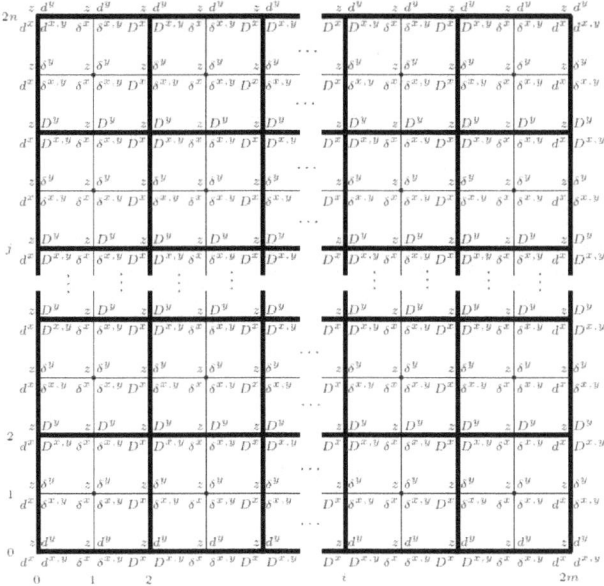

Figure 2: Schema of objects of a $(2m+1) \times (2n+1)$ - component bicubic Hermite spline surface.

that are needed to construct the four bicubic spline components $S = \{S_{0,0}, S_{0,1}, S_{1,0}, S_{1,1}\}$, are denoted by δ, see Fig. 1.

The below proposed algorithm was developed by generalizing the above described interrelation between biquartic and bicubic polynomials. First biquartic polynomials were handled and then based on them new model equations and formulas for unknown derivatives of the bicubic spline surface were derived.

Figure 2 illustrates the schema of the proposed computational algorithm for $(2m+1) \times (2n+1)$ bicubic spline surface of class C^2. There are $2m+1$ verticals and $2n+1$ horizontals. Rectangles and thick rectangles indicate the boundary of bicubic spline components and biquartic polynomials, respectively. There are two types of objects at every grid-point: known and unknown ones. The given values and derivatives are denoted by z, d and the unknown first derivatives by D, δ. Notice that z is provided at every grid-point and d only along the total grid's boundary. The most important is where are the unknown D and δ parameters. The D parameters are located only along the thick rectangles, but never in their center. As we shall see later the D parameters will be computed from equations and the δ parameters from explicit formulas. The equations were derived from the equality of second derivatives of spline components and the formulas from the biquartic polynomials.

The derived new model equations for the unknown D derivatives of the spline surface segments and parts of the explicit formulas for δ are generalization of model equations and formulas of the unknown derivatives of spline curve segments, see [16].

5   Reduced System Algorithm

This section presents a new sequential algorithm for computing a C^2-class uniform spline surface's unknown first derivatives. Its efficiency will be shown in the next section. The central part of the algorithm are three new model equations and five new explicit formulas. We do not derive these model equations and explicit formulas, only mention that for their derivation we had to (see Fig. 2) thoroughly analyse the structure of the bicubic and biquartic polynomials, specify which derivatives should be the D and which the δ parameters, understand which polynomials are critical for obtaining the equations and formulas, and for all this the following steps were needed:

1. construction of some biquartic polynomials $F_{i,j}$, see (12),
2. construction of δ parameters as functions, see (13),
3. construction of some appropriate Hermite spline components S, see (10), for comparing of their second derivatives, see (11).

The below proposed algorithm can be characterized from two aspects

- what it computes,
- the quality of its outcome.

The algorithm computes D and δ coefficients for bicubic spline surface components from inputs given at equispaced grid-points described in Section 2. The D coefficients are computed from linear systems based on equations (14), (16), (18) – (21). The δ coefficients are gained from explicit formulas (15), (17), (22) – (24). Since the equations for the D parameters were derived from the equality of second derivatives of spline components and the formulas for the δ parameters were gained from biquartic polynomials that as we know grant C^2 continuity of their components, the algorithm provides such coefficients that the uniform bicubic spline surface will be of class C^2.

Algorithm for computing the unknown first derivatives of the spline surface in three main steps with reduced systems.
Inputs: z and d values, see (2) – (5).
Step 1a. Computation of D^x parameters along the horizontals from equation systems.

For each horizontal we construct a system of linear equations to compute the D^x values, located on the inside odd grid-points. Each horizontal represents an independent tridiagonal system of linear equations.

For each horizontal, see Fig. 2, $j = 0, 1, \ldots, 2n$, a tridiagonal system is constructed based on equations

$$D^x_{2(i+1),j} - 14D^x_{2i,j} + D^x_{2(i-1),j} =$$
$$= \frac{3}{h_x}(z_{2(i+1),j} - z_{2(i-1),j}) - \frac{12}{h_x}(z_{2i+1,j} - z_{2i-1,j}), \quad (14)$$

where $i = 1, 2, \ldots, m-1$.

Step 1b. Computation of δ^x parameters from explicit formulas.

To finish the computation of all first partial derivatives with respect to x we have to calculate

$$\delta^x_{i,j} = \frac{3}{4h_x}(z_{i+1,j} - z_{i-1,j}) - \frac{1}{4}(d^x_{i+1,j} + d^x_{i-1,j}), \quad (15)$$

where $i = 1, 3, \ldots, 2m-1$, $j = 1, 3, \ldots, 2n-1$.

Step 2a. Computation of D^y parameters along the horizontals from equation systems.

For each vertical we construct a system of linear equations to compute the D^y values, located on the inside odd grid-points. Each vertical represents an independent system of linear equations.

For each vertical, $i = 0, 1, \ldots, 2m$, a tridiagonal system is constructed based on equations

$$D^y_{i,2(j+1)} - 14D^y_{i,2j} + D^y_{i,2(j-1)} =$$
$$= \frac{3}{h_y}(z_{i,2(j+1)} - z_{i,2(j-1)}) - \frac{12}{h_y}(z_{i,2j+1} - z_{i,2j-1}), \quad (16)$$

where $j = 1, 2, \ldots, n-1$.

Step 2b. Computation of δ^y parameters from explicit formulas

To finish the computation of all first partial derivatives with respect to y we have to calculate

$$\delta^y_{i,j} = \frac{3}{4h_y}(z_{i,j+1} - z_{i,j-1}) - \frac{1}{4}(d^y_{i,j+1} + d^y_{i,j-1}), \quad (17)$$

where $i = 1, 3, \ldots, 2m-1$, $j = 1, 3, \ldots, 2n-1$.

At this moment all directional derivatives are known: some were provided and the unknown D and δ directional ones were computed in Steps 1 and 2. In the further steps all directional derivatives will be denoted by d and contained in the right hand side of equations and formulas.

Step 3a. Computation of $D^{x,y}$ parameters along the bottom and top horizontals, and left vertical from equation systems.

We construct systems of linear equations for bottom and top horizontals and left verticals by [3]. The systems for the bottom boundary horizontal is

$$D^{x,y}_{i+1,0} + 4D^{x,y}_{i,0} + D^{x,y}_{i-1,0} = \frac{3}{h_x}(d^y_{i+1,0} - d^y_{i-1,0}), \quad (18)$$

where $i = 1, \ldots, 2m-1$;
for the top boundary horizontal is

$$D^{x,y}_{i+1,2n} + 4D^{x,y}_{i,2n} + D^{x,y}_{i-1,2n} = \frac{3}{h_x}(d^y_{i+1,2n} - d^y_{i-1,2n}), \quad (19)$$

where $i = 1, \ldots, 2m-1$;
and for the left boundary vertical is

$$D^{x,y}_{0,j+1} + 4D^{x,y}_{0,j} + D^{x,y}_{0,j-1} = \frac{3}{h_y}(d^x_{0,j+1} - d^x_{0,j-1}), \quad (20)$$

where $j = 1, 2, \ldots, 2n-1$.

Step 3b. Computation of $D^{x,y}$ parameters from the inside grid-points using systems of equations

For the odd verticals, $i = 2, 4, 6, \ldots, 2m$, a tridiagonal system is constructed based on equations

$$D^{x,y}_{i,j+2} - 14D^{x,y}_{i,j} + D^{x,y}_{i,j-2} =$$
$$= \frac{1}{7}(d^{x,y}_{i-2,j+2} + d^{x,y}_{i-2,j-2}) - 2d^{x,y}_{i-2,j} +$$
$$+ \frac{3}{7h_x}(d^y_{i-2,j+2} + d^y_{i-2,j-2}) + \frac{3}{7h_y}(-d^x_{i-2,j+2} + d^x_{i-2,j-2}) +$$
$$+ \frac{9}{7h_x}(d^y_{i,j+2} + d^y_{i,j-2}) + \frac{9}{7h_xh_y}(-z_{i-2,j+2} + z_{i-2,j-2}) +$$
$$+ \frac{12}{7h_x}(-d^y_{i-1,j+2} - d^y_{i-1,j-2}) + \frac{12}{7h_y}(d^x_{i-2,j+1} - d^x_{i-2,j-1}) +$$
$$+ \frac{3}{h_y}(d^x_{i,j+2} - d^x_{i,j-2}) + \frac{27}{7h_xh_y}(-z_{i,j+2} + z_{i,j-2}) +$$
$$+ \frac{36}{7h_xh_y}(z_{i-1,j+2} - z_{i-1,j-2} + z_{i-2,j+1} - z_{i-2,j-1}) -$$
$$- \frac{6}{h_x}d^y_{i-2,j} + \frac{12}{h_y}(d^x_{i,j+1} + d^x_{i,j-1}) + \frac{108}{7h_xh_y}(z_{i,j+1} - z_{i,j-1}) -$$
$$- \frac{18}{h_x}d^y_{i,j} + \frac{144}{7h_xh_y}(-z_{i-1,j+1} + z_{i-1,j-1}) + \frac{24}{h_x}d^y_{i-1,j}, \tag{21}$$

where $j = 4, 6, \ldots, 2n-4$.

Mention must be made, that this step was the most critical. At first after computation of $D^{x,y}$ unknowns along the bottom and top horizontals in Step 3a we got equations with six $D^{x,y}$ unknowns on the left side. Török suggested to compute the Dxy parameters along the left vertical using de Boors equation in Step 3a and thanks to this three of six $D^{x,y}$ parameters could be moved to the right side as computed.

Step 3c. Computation of $\delta^{x,y}$ parameters from explicit formulas

To finish the computation of all first cross derivatives we have to calculate for the even verticals and the even horizontals

$$\delta^{x,y}_{i,j} = \frac{1}{16}(d^{x,y}_{i+1,j+1} + d^{x,y}_{i+1,j-1} + d^{x,y}_{i-1,j+1} + d^{x,y}_{i-1,j-1}) -$$
$$- \frac{3}{16h_y}(d^x_{i+1,j+1} - d^x_{i+1,j-1} + d^x_{i-1,j+1} - d^x_{i-1,j-1}) -$$
$$- \frac{3}{16h_x}(d^y_{i+1,j+1} + d^y_{i+1,j-1} - d^y_{i-1,j+1} - d^y_{i-1,j-1}) +$$
$$+ \frac{9}{16h_xh_y}(z_{i+1,j+1} - z_{i+1,j-1} - z_{i-1,j+1}z_{i-1,j-1}), \tag{22}$$

where $i = 1, 3, \ldots, 2m-1$, $j = 1, 3, \ldots, 2n-1$;
for the even verticals and the odd horizontals

$$\delta^{x,y}_{i,j} = \frac{3}{4h_y}(d^x_{i,j+1} - d^x_{i,j-1}) - \frac{1}{4}(d^{x,y}_{i,j+1} + d^{x,y}_{i,j-1}), \quad (23)$$

where $i = 1, 3, \ldots, 2m-1$, $j = 2, 4, \ldots, 2(n-1)$;
and for the odd verticals and the even horizontals

$$\delta^{x,y}_{i,j} = \frac{3}{4h_y}(d^x_{i,j+1} - d^x_{i,j-1}) - \frac{1}{4}(d^{x,y}_{i,j+1} + d^{x,y}_{i,j-1}), \quad (24)$$

where $i = 2, 4, \ldots, 2m-1$, $j = 1, 3, \ldots, 2n-1$.

6 The Comparision of the New and de Boor's Algorithm

We sum up the complete spline task and how is it completed by the considered two algorithms. Then we give some details about the role of the biquartic polynomials that are absent in de Boor's algorithm but played a crucial task in the design of the proposed one.

In case of $(2m+1)(2n+1)$ grid-points, to fulfill the complete spline task means to construct $(2m)(2n)$ spline components using $(2m+1)(2n+1)$ various quadruples, where one quadruple looks the following way: $[z, d^x, d^y, d^{x,y}]$. The input comprises $(2m+1)(2n+1)$ z values, $2(2n+1)$ d^x, $2(2m+1)$ d^y and 4 $d^{x,y}$ derivatives. The unknown derivatives require computation.

In de Boor's algorithm every unknown derivative is computed from equations that are of four types, see (6)–(9) and [3]:

- from $2n+1$ systems with $2m-1$ variables $(2n+1)(2m-1)$ derivatives d^x are computed,

- from 2 systems with $2m-1$ variables $2(2m-1)$ derivatives $d^{x,y}$ are computed,

- from $2m+1$ systems with $2n-1$ variables $(2m+1)(2n-1)$ derivatives d^y are computed,

- from $2m+1$ systems with $2n-1$ variables $(2m+1)(2n-1)$ derivatives $d^{x,y}$ are computed.

The proposed algorithm's benefit is that lesser number of unknown derivatives (D parameters) are computed from systems of equations, see Tab. 3, compared to de Boor's algorithm [3]. The remaining derivatives (δ parameters) are computed from explicit formulas. This was achieved thanks to using mn biquartic polynomials $F_{i,j}(x,y)$ behind the scene, whose definitions use only $(m+1)(n+1)$ quadruples.

The algorithm's drawback is that it uses more types of relation: six types of equations and five types of explicit formulas. Nevertheless it has to compute approximately $12/5$ times less equations within systems – compare Tab. 2 and Tab. 3 from Section 7.

Let us have a closer look at biquartic polynomials and their role in the algorithm's design. One biquartic polynomial $F_{i,j}(x,y)$ needs 25 parameters, see Definition 2. We distinguish between four types of F polynomials, see Fig. 2,

1. at the corners,

2. at boundary horizontals,

3. at boundary verticals,

4. over the inside grid-points.

While for example for the biquartic polynomial F over inside grid-points all the sixteen derivatives are unknown,

they are D parameters, for F from the corners only seven are unknown and nine are given, these are the d parameters. After obtaining the D parameters the remaining derivatives are computed based on F. From every F eleven delta parameters can be obtained: two pairs of type $[\delta^x, \delta^{x,y}]$, two pairs of type $[\delta^y, \delta^{x,y}]$ and one triple $[\delta^x, \delta^y, \delta^{x,y}]$. Naturally, δ parameters are functions of D parameters, see [9].

After introducing the new algorithm in the previous section and giving a short insight into its design in this one, the next section is devoted to its quantitative characterization.

7 Number of Multiplications

The standard way of solving a tridiagonal linear system

$$\underbrace{\begin{bmatrix} b & 1 & 0 \\ 1 & b & 1 \\ 0 & 1 & b \\ & \ddots & \ddots & \ddots & \ddots \\ & & & & b \end{bmatrix}}_{A} \underbrace{\begin{bmatrix} d_1 \\ d_2 \\ d_3 \\ \vdots \\ d_K \end{bmatrix}}_{d} = \underbrace{\begin{bmatrix} r_1 - d_0 \\ r_2 \\ r_3 \\ \vdots \\ r_K - d_{K+1} \end{bmatrix}}_{r}$$

uses the LU factorization $A d = L \underbrace{U d}_{y} = r$, where

$$L = \begin{bmatrix} 1 & 0 \\ l_2 & 1 \\ 0 & l_3 & 1 \\ & & \ddots & \ddots & \ddots & \ddots \\ & & & & l_K & 1 \end{bmatrix},$$

$$U = \begin{bmatrix} u_1 & 1 & 0 \\ 0 & u_1 & 1 \\ & & u_2 \\ & & & \ddots & \ddots & \ddots \\ & & & & & u_K \end{bmatrix},$$

the u_i and l_i elements are computed as, see [2],

$$LU: \quad u_1 = b, \{l_i = \frac{1}{u_{i-1}}, u_i = b - l_i\}, i = 2, ..., K, \quad (25)$$

and the forward (Fw) and backward (Bw) steps of the solution are

$$\text{Forward:} \quad Ly = r, \quad (26)$$

where $y_1 = r_1$, $\{y_i = r_i - l_i y_{i-1}\}$, $i = 2, ..., K$;

$$\text{Backward:} \quad Ud = y, \quad (27)$$

where $d_K = \frac{y_K}{u_K}$, $\{d_i = \frac{1}{u_i}(y_i - x_{i+1})\}$, $i = K-1, ..., 1$.

The tridiagonal systems of equations for de Boor's and our algorithm are solved by LU decomposition. All the systems of these algorithms are diagonally dominant with

	(25)	(26)	(27)		
Dim.	**LU**	**Fw**	**Bw**	**RHS**	**Total mult.**
$N \times N$	γN	N	γN	βN	$2\gamma N + \beta N + N$

Table 1: Count of multiplications in one system of equations

de Boor	**System**	**Equation**	β
Step 1 (6)	$2n+1$	$2m-1$	1
Step 2 (7)	2	$2m-1$	1
Step 3 (8)	$2m+1$	$2n-1$	1
Step 4 (9)	$2m+1$	$2n-1$	1
Total equations	\multicolumn{3}{c}{$12mn+2m+2n-5$}		
Total mult.	\multicolumn{3}{c}{$24\gamma mn+24mn+4\gamma m+4\gamma n$ $+4m+4n-10\gamma-1$}		

Table 2: Count based characteristics – de Boor's algorithm

Proposed	**System**	**Equation**	β
Step 1a (14)	$2n+1$	$m-1$	2
Step 2a (16)	$2m+1$	$n-1$	2
Step 3a (18)	1	$2m-1$	1
Step 3a (19)	1	$2m-1$	1
Step 3a (20)	1	$2n-1$	1
Step 3b (21)	m	$n-1$	17
Total equations	$5mn+2m+n-5$		
Total mult.	$10\gamma mn+30mn+4\gamma m+2\gamma n$ $-13m+n-10\gamma-12$		

Table 3: Count based characteristics – proposed algorithm

Proposed	**Formula**	β
Step 1b (15)	$m(2n+1)$	2
Step 2b (17)	$(2m+1)n$	2
Step 3c (22)	mn	4
Step 3c (23)	$m(n-1)$	2
Step 3c (24)	mn	2
Total formulas	$7mn+n$	
Total mult.	$16mn+2n$	

Table 4: Count based characteristics – explicit formulas in proposed algorithm

elements 1, 4, 1 and 1, −14, 1. The LU and backward steps contain a division that is indicated by γ, the ratio between division and multiplication: the performance of a division operation is equivalent to γ multiplications.

The proposed and de Boor's algorithm differ

- in number of systems of equations,
- in number of equations within systems,
- in number of multiplication operations on right hand side (RHS) of equations.

Tab. 1 presents the number of multiplications for solving one general tridiagonal $N \times N$ matrix, where β denotes the number of multiplications on the right hand side of an equation.

The second and third columns of Tab. 2 and Tab. 3 provide the count of equations within the given steps (equations) and the count of equations within a system, respectively, for a grid of size $(2m+1) \times (2n+1)$. The last but one rows contain the total number of equations. The total count of multiplications to solve the tridiagonal systems within de Boor's and the proposed algorithm are in the last row.

In the proposed algorithm we evaluate in addition to the solution of the tridiagonal systems of equations, see Tab. 3, as well as δ parameters using explicit formulas. Therefore the total number of multiplications in the proposed algorithm based on tables 3 and 4 is

$$10\gamma mn + 46mn + 4\gamma m + 2\gamma n - 13m + 3n - 10\gamma - 12.$$

We can see that the number of multiplication in the proposed algorithm is less. The theoretical speed up for some various grid sizes were computed under assumption that $\gamma = 3.5$. Based on Tab. 5 we can conclude that the proposed model is asymptotically 1.33 times faster.

Grid	de Boor	Proposed	Speed up
11×11	2 835	2 033	1.394
101×101	271 755	203 003	1.339
1001×1001	27 017 955	20 255 453	1.333
$(10^6+1) \times (10^6+1)$	$27 \cdot 10^{12}$	$20.25 \cdot 10^{12}$	1,333
$(10^{12}+1) \times (10^{12}+1)$	$27 \cdot 10^{24}$	$20.25 \cdot 10^{24}$	1,333

Table 5: Speed up

8 Conclusion

We suggested a new efficient sequential algorithm for computation of a spline surface over an equispaced grid. Its theoretically evaluated asymptotic speed up over de Boors algorithm is approximately 1.33. The algorithm has also a very nice property from the view point of parallel computation: the computation of the second half of unknowns based on explicit equations can be parallelized automatically. Naturally, parallel methods, suggested for solving tridiagonal systems of de Boor's algorithm can be used for solution of the new algorithm's tridiagonal systems as well. Therefore, the proposed reduced system based algorithm is preferable over de Boor's algorithm not only for sequential, but also for parallel computation.

Acknowledgement

The author thanks Cs. Török for the problem statement, his helpful suggestions and support. This work was partially supported by the research grants VEGA 1/0073/15 and VVGS-PF-2015-477.

References

[1] Berry M. J., Linoff G. S.: Data mining techniques (Third Edition). For Marketing, Sales, and Customer Relationship Management, John Wiley and Sons, 2011

[2] Björck A.: Numerical methods in matrix computations. Springer, 2015

[3] de Boor C.: Bicubic spline interpolation. Journal of Mathematics and Physics, **41 (3)** (1962), 212–218

[4] Dikoussar N. D.: Function parametrization by using 4-point transforms. Comput. Phys. Commun. **99** (1997), 235–254

[5] Dikoussar N. D., Török C.: Automatic Knot Finding for Piecewise Cubic Approximation. Mat. Model., 2006, T-17, N.3

[6] Dikoussar N. D., Török C.: Kybernetika **43 (4)** (2007), 533–546

[7] Hegland M., Roberts S., Altas I.: Finite element thin plate splines for data mining applications. In: M. Daehlen, T. Lyche, and L. Schumaker, (eds.), Mathematical Methods for Curves and Surfaces II, pages 245–252, Nashville, TN, 1998. Vanderbilt University Press. Available as Mathematical Research Report MRR 057-97, School of Mathematical Sciences, Australian National University

[8] Matejčiková A., Török C.: Noise suppression in RDPT. Forum Statisticum Slovacum, 3/2005, Bratislava, ISSN 1336-7420, 199–203

[9] Miňo L., Szabó I., Török C.: Bicubic splines and biquartic polynomials, 2015, to appear

[10] Révayová M., Török C.: Piecewise approximation and neural networks. Kybernetika **43 (4)** (2007), 547–559

[11] Révayová M., Török C.: Reference points based recursive approximation. Kybernetika **49 (1)** (2013), 60–72

[12] Salomon D.: Curves and surfaces for computer graphics. Springer, 2006

[13] Szabó I.: Approximation algorithms for 3D data analysis. PhD Thesis, P. J. Šafárik University in Košice, Slovakia, 2015, to appear

[14] Szabó I., Török C.: Smoothing in 3D with reference points and Polynomials. 29th Spring Conference on Computer Graphics SCCG 2013, Smolenice–Bratislava, Comenius University, 39–43

[15] Török C.: 4-point transforms and approximation, Comput. Phys. Commun. **125** (2000), 154–166

[16] Török C.: On reduction of equations' number for cubic splines. Matematicheskoe modelirovanie **26 (11)** (2014), 33–36

[17] Török C.: Piecewise smoothing using shared parameters. Forum Statisticum Slovacum, 7/2009, 188–193

[18] Török C.: Reference points based transformation and approximation. Kybernetika **49 (4)** (2013), http://www.kybernetika.cz/content/2013/4/644/paper.pdf

[19] Török C.: Speedup of interpolating spline construction, 2015, to appear

J. Yaghob (Ed.): ITAT 2015 pp. 38–42
Charles University in Prague, Prague, 2015

About Security of the RAK DEK

Richard Ostertág

Department of Computer Science, Comenius University,
Mlynská dolina, 842 48 Bratislava, Slovakia
ostertag@dcs.fmph.uniba.sk

Abstract: The RAK DEK operating unit is a standalone access control system. This unit, and its more advanced versions, are widely used in Slovakia to protect entrance doors to block of flats. In this paper we have studied security of RAK DEK with respect to timing attack. We have tried two attack vectors. This system shows to be invulnerable to our first attack, but we have succeeded with the other attack vector. Now we are in state of finishing functional exploit using identified vulnerability and investigation of its applicability to the more advanced version of this family of access control systems.

1 Introduction and Basic Description of the RAK DEK

The RYS is a Slovak company that develops and sales access control and door communication systems. This company develops its own line of access control systems based on iButton (a.k.a. touch or digital electronic key – DEK) and the RAK DEK operating-memory units.

These systems were designed for the apartment buildings and became very popular. They are also used to provide access control in commercial or industrial settings (e.g. hotels, offices, stores, schools, server housing) [1].

We choose to discuss this system because of its popularity in Slovakia. We have already described cloning of DEK and generally applicable brute-force attack in [2]. In this paper we have exploited specific properties of RAK DEK, so our conclusions apply only to this specific system. However, described timing attack may be applicable even to the other systems using 1-Wire protocol and serial number iButtons, but actual applicability has to be individually investigated.

1.1 Operating-Memory Unit

The operating-memory unit, e.g. RAK-DEK (see figure 1) is the brain of RYS access control system.

This unit is connected through its RELE output with door's electromagnet and through 4-pin connector on back-side with an iButton touch probe. This unit is capable to store serial numbers for hundreds of iButtons. If a user touches the touch probe with a DEK, the iButton serial number is transferred from the DEK to the operating-memory unit. If the transferred number is stored in the unit, the unit temporarily deactivates the electromagnets (using the RELE output) and the user is allowed to enter.

Figure 1: The RAK-DEK operating-memory unit

We are interested in the communication between the DEK and the operating-memory unit. As the DEK is just a standard DS1990R serial number iButton® from Maxim Integrated Products, Inc., this communication uses standardized 1-Wire protocol.

1.2 Serial Number iButton

The DS1990R is a rugged button-shaped data carrier, which serves as an electronic registration number. It is produced in two basic sizes (F3 and F5) as is schematically depicted on figure 2.

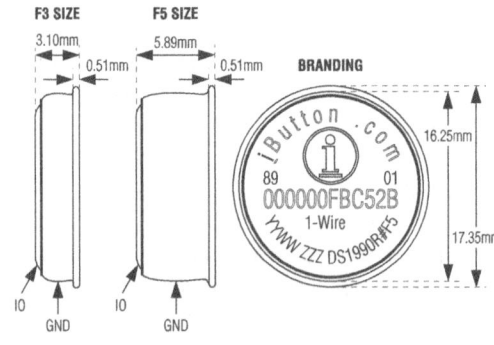

Figure 2: Schema of DS1990R serial number iButton

For the DEK an iButton of F5 size is used, together with a plastic holder for it (see figure 3). This holder can be put on a key chain and can be in different colors (but black is usually used).

Figure 3: Picture of DS1990R-F5 serial number iButton

Every DS1990R is factory lasered with a guaranteed unique 64-bit registration number that allows for the absolute traceability. This 64 bit registration (or serial) number has internal structure as depicted in figure 4.

high address	MSB		LSB	low address
CRC byte		6–byte serial number		family code 01

Figure 4: Data structure of a DS1990R serial number

It contains: six-byte device-unique serial number, one-byte family code and one-byte CRC verification. Every DS1990R have family code fixed to $(01)_{16}$. There are also another iButton devices with different family codes. E.g. $(10)_{16}$ is a temperature iButton, but they are not usually used in this kind of systems. Therefore every DEK can be considered as a 48 bits long factory set unique number (analogous to unique MAC addresses of network cards).

1.3   Communication Protocol between RAK-DEK and iButton

All iButton devices utilizes the 1-Wire protocol, which transfers data serially, half-duplex, through a single data lead (1-wire) and a ground return (GND).

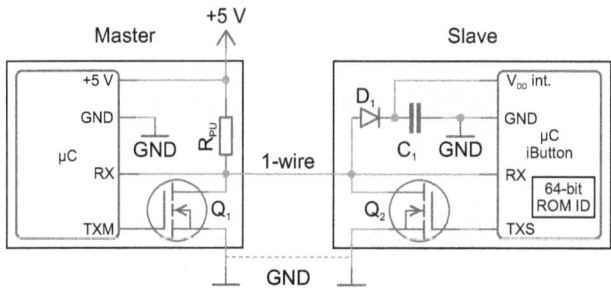

Figure 5: Simplified schema of an iButton and a master

Figure 5 depicts simplified implementation of the 1-wire communication using two micro-controllers with two unidirectional ports. The slave (in this case iButton) has no power source and is powered from an operating-memory unit using the parasite power system on data lead. This system consists of diode D_1 and capacitor C_1 and provides power to iButton during low voltage states of 1-wire bus.

The master uses input port RX to sense value on 1-wire bus. The slave uses its RX input port the same way. In the idle state 1-wire bus is pulled up to 5 V by resistor R_{PU}. In this state all RX ports read logical one. Standard defines that voltage should be at least 2.2 V to be interpreted as logical one.

If any device wants to set 1-wire bus to logical zero, it uses its output port (TXM or TXS) to activate its internal MOSFET switch (Q_1 or Q_2) to connect the data lead to the ground. As a result of this action, 1-wire voltage falls down to near 0 V. Standard defines that voltage should be at most 0.8 V to be interpreted as logical zero.

If device wants to set 1-wire bus to logical one, it just deactivates its internal MOSFET switch. If more devices set 1-wire bus state at the same time, then resulting state is logical AND of all states. In other words: if at least one device is setting 1-wire bus to logical zero, then resulting state is logical zero.

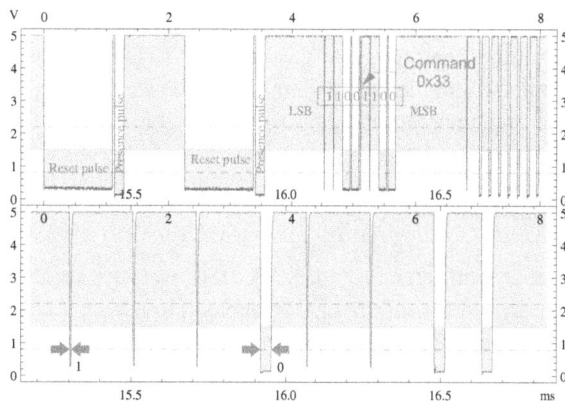

Figure 6: Example of real 1-wire communication

Communication always starts by the reset pulse issued by the master. The reset pulse is just long enough (in this case 1.1 ms) logical zero state of 1-wire bus (see figure 6). After this reset pulse all slave devices are reseted to well-known initial state. All slave devices respond to the reset pulse by the presence pulse, in this case with length of 0.149 ms. If no presence pulse is detected by the master, then no iButton is connected to the master. In this situation RAK-DEK waits for 100 ms and then tries again with another reset pulse. After successful detection of iButton, RAK-DEK makes a new, unnecessary, reset pulse for unknown reasons (again followed by the presence pulse).

After presence pulse, the master will send a command. RAK-DEK always sends the command 0x33, i.e. the "read

ROM" command. This command is transferred from the master to the slave by serial transfer within defined time slots. Any time slot is initiated by the master (in this case RAK-DEK) and starts by falling edge on the data lead. After 0.025 ms (after this falling edge), the iButton read state of the 1-wire bus. If it is at least 2.2 V, the master sends bit 1, otherwise bit 0. Bits are always sent from the least significant bit to more significant bits.

After receiving the "read ROM" command, the iButton is ready to send its 64-bit serial number stored in its ROM. Again, transfer is done in time slots initiated by the master from LSB to MSB. So, the slave is waiting for the falling edge. After 0.004 ms (after this falling edge) RAK-DEK turns off the switch Q_1 and the pull up resistor will raise the data lead to 5 V. So if iButton wants to send bit 1, it has just to wait. If iButton wants to send bit 0, then in this 0.004 ms interval iButton activates its switch Q_2 for 0.032 ms. In either case RAK-DEK reads state of 1-wire bus about 0.02 ms from the beginning of time slot. And again, if it is at least 2.2 V, then master receives bit 1, otherwise bit 0. In figure 6 we can see first 8 bits of serial number after command 0x33. In the case of DEK it is always 0x01 (family code). Lower half of figure 6 zooms to the last but one byte of serial number (in case of this specific key it is $(00110111)_2 = (37)_{16}$.

Communication ends when RAK-DEK receives whole 64-bit serial number. If received number is on internal list of authorized DEKs, then RAK-DEK releases electromagnet holding the doors. At this point RAK-DEK sends the reset pulse and the whole communication starts again. For more implementation details of the protocol see [3].

2   Hardware

To be able to interact with RAK-DEK we need to implement an iButton emulator. We decided to use an Arduino compatible hardware platform developed at Slovak University of Technology – Acrob [4], depicted on figure 7.

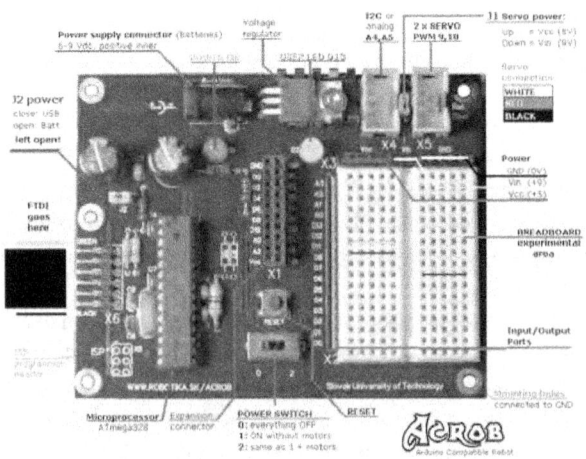

Figure 7: Acrob – an educational robotic platform

This hardware platform uses the Atmel ATmega328P microcontroller running on 16 MHz, which we programmed in C++ like language, using standard Arduino IDE [5].

In contrast to our previous paper [2], where we have simulated operating-memory unit by Acrob, now we have to buy a real RAK-DEK operating-memory unit, because timing attacks are very sensitive to implementation details. We still use one Acrob device for emulation of iButtons.

The 1-Wire protocol uses only one data line. We implement this line by connecting together digital pin 12 of Acrob, with the center pad of touch probe (this is equivalent to connecting directly with pin 2 of the RAK-DEK). This probe is connected to the RAK-DEK using 4-pin connector on the back-side of PCB. To establish a ground return we connect Acrob GND pin with outside ring of touch probe (this is equivalent to connecting directly with pin 1 on the RAK-DEK).

The touch probe gives us one more information channel – the LED. RAK-DEK is blinking with this LED to make it easier to locate the touch probe at night. Also the LED lights up for some time when iButton touches the probe.

To be able to analyze even this source of information we decided to use a photoresistor facing to the LED in the touch probe. We used a photoresistor module with an opamp used as a comparator and a potentiometer for setting a threshold. When light intensity is over the threshold, then DO pin of the module is on logical 0 level (near 0 V), otherwise it is on logical 1 level (near 3.3 V because we have used 3.3 V as V_{cc} for the module). We have connected DO pin on the photoresistor module to pin 8 on Acrob.

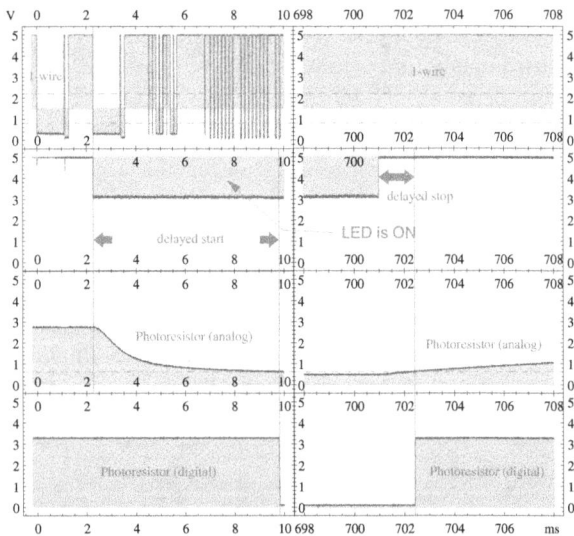

Figure 8: Calibration of the photoresistor module

Photoresistors are slow and that is why we can see a delayed start and a delayed stop in figure 8. We have rotated the potentiometer to set the threshold around 620 mV. By this calibration we obtained a small stop delay at cost of longer start delay and hight sensitivity to ambient light. In this case it was not problem. We know, that LED starts to lit at the start of second reset pulse and the ambient light was shielded. In fact, the length of the stop delay is not important, we only need it to be constant. If smaller delays are needed then phototransistor can be used.

3 The Brute Force Attack

If we omit the predictable parts of serial numbers (i.e. family code and CRC), we have to find six bytes. Our empirical observations suggest that serial numbers are allocated in sequence. All keys we have seen so far had zeros in two most significant bytes of these six bytes. Therefore for a brute force attack it would be sufficient to try all 2^{32} serial numbers of the form mentioned above.

In our experiments we have observed that RAK-DEK is issuing the reset pulse every 100 ms when waiting for DEK. But if DEK is found, then next rest pulse does not come immediately, but always after 700 ms from the first. This does not leave any space for timing attack and substantially increases time for the bruteforce attack that we have estimated in [2]. If we assume 700 ms as an upper bound to try one serial number, we will need $700\,\text{ms} \times 2^{4\times8}/60/60/24/365.5 \approx 95$ years for a successful brute force attack in the worst case.

4 The Timing Attack

As a last resort we have tried to analyze time that elapses from the moment we send 64-bit serial number to the moment LED goes off. Ours idea was to store one key, e.g. 0x0000000000000000 into RAK-DEK unit and then emulate two keys, e.g. 0xFF00000000000000 and 0x00000000000000FF, and measure time needed for the LED to go off in both cases. Through this experiment we have realized that RAK-DAK is firstly validating CRC and family code. It is not possible to do tests with an unrealistic DEK. Therefore we choose one valid DEK and make modifications only to its 6 inner bytes in such way to not modify resulting CRC. Then we tried to send four different keys to RAK-DEK with different positions of the first discrepancy from stored key. Resulting times are depicted in figure 9.

From this figure we can see, that RAK-DAK is clearly comparing DEK bytes form LSB to MSB, because time is increasing as position of first discrepancy goes to more significant bytes. Also we can see a nice linear relationship between the position and the time. Using a linear regression we estimated it to be:

$$f(p) = (1.33\,\text{ms})p + 307.96\,\text{ms}$$

Based on this liner regression we can say that test of one byte from electronic key takes approximately 1.3 ms. To verify correctness of this hypothesis we loaded some random DEKs into RAK-DEK. Then we tried to identify

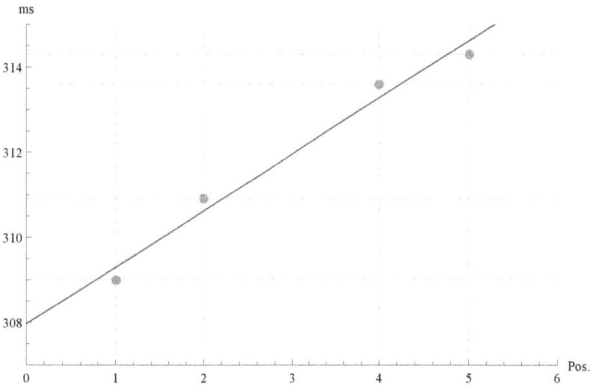

Figure 9: Position of first discrepancy vs. LED lit time. Positions are numbered from right (LSB) to left (MSB).

value on position 1 (position 0 always has value of 0x01). But our implementation did not work. Finally, we found that RAK-DAK is comparing key bytes from LSB to MSB, but firstly it checks if CRCs are equal. This is probably an optimization to speed up comparison of long byte sequences in case we have their CRCs already precomputed.

Using this information, we can do much better then brute force attack. We still need to search through the key space, but we can do it byte by byte now. Starting from CRC (at position 8) and then going from position 1 to 5, calculating value at position 6 in such way not to change resulting CRC. If we see that system response delayed by 1.3 ms we know, that we hit correct value for actual position and we can advance to next position, until correct DEK is found. Using this technique and our experience of position 5 and 6 to be zero on all known DEKs we can estimate time of successful attack, in worst case, as:

$$700\,\text{ms} \times 4 \times 2^8/60 \approx 12 \text{ minutes}.$$

5 Conclusion

We have investigated possibilities of timing attacks on RAK-DEK. We identified timing attack vulnerability exploiting LED on the touch probe. We are now in state of finishing a functional exploit using identified vulnerability and investigation of its applicability to more advanced version of this family of access control systems. This attack requires only access to an Arduino compatible device and a photoresistor (cost around 30.00 €). The time needed for this attack is less than 12 minutes.

On the other hand, this attack can easily be mitigated by disconnecting LED in the touch probe from RAK-DEK. Better solution would be to modify firmware of RAK-DEK to turn off LED with next reset pulse (which is already fixed to 700 ms after beginning of communication).

This work was supported by VEGA grant 1/0259/13.

References

[1] RYS: Access control and door entry systems. (`http://www.rys.sk/html_eng/english.htm`) [Online; accessed 8-July-2015].

[2] Ostertág, R.: About security of digital electronic keys. In: ITAT 2013: Information Technologies – Applications and Theory, North Charleston: CreateSpace Independent Publishing Platform (2013) 122–124 ISBN: 978-1490952000.

[3] Maxim Integrated Products, Inc.: Book of iButton standards (application note 937). `http://www.maximintegrated.com/en/app-notes/index.mvp/id/937` (2002) [Online; accessed 8-July-2015].

[4] Balogh, R.: Acrob - an educational robotic platform. AT&P Journal Plus **10** (2010) 6–9 ISSN 1336-5010. `http://ap.urpi.fei.stuba.sk/balogh/pdf/10ATPplusAcrob.pdf` [Online; accessed 8-July-2015].

[5] Arduino: Arduino software. (`http://www.arduino.cc/en/main/software`) [Online; accessed 8-July-2015].

Redukční analýza a Pražský závislostní korpus*

Martin Plátek[1], Dana Pardubská[2], and Karel Oliva[3]

[1] MFF UK Praha, Malostranské nám. 25, 118 00 Praha, Česká Republika
martin.platek@ufal.mff.cuni.cz
[2] FMFI UK Bratislava, Mlynská dolina, 84248 Bratislava
pardubska@dcs.fmph.uniba.sk
[3] UJČ ČAV Praha, Letenská, 118 00 Praha, Česká Republika
oliva@ujc.cas.cz

Abstrakt: Cílem tohoto příspěvku je uvést, formálně zavést a exaktně pozororovat větnou redukční analýzu svázanou s redukční analýzu D-stromů. Tímto způsobem upřesníme strukturální vlastnosti D-stromů se závislostmi a koordinacemi z Pražského závislostního korpusu (PDT). Zvýrazňujeme vlastnosti, kterými se závislosti a koordinace liší. Snažíme se pracovat metodou, která je blízká metodám matematické lingvistiky, a to především těm, které formulují omezující podmínky pro syntaxi přirozených jazyků. Ukazujeme nové možnosti takových formulací.

1 Úvod

Postupně se věnujeme *větné redukční analýze* (RA) a její vazbě na *redukční analýzu D-stromů* (RADS), abychom získali nové formální prostředky vhodné pro studium strukturálních vlastností D-stromů. Na základě těchto prostředků formulujeme pozorování o D-stromech v Pražském závislostním korpusu (PDT viz [1]). Tento článek vznikl ve spolupráci s Markétou Lopatkovou, která nám pomocí vybíraných příkladů zprostředkovala přístup do PDT a často s námi diskutovala, zvláště o problematice redukcí stromů z PDT s koordinacemi.

1.1 Neformální úvod do (manuální) redukční analýzy českých vět a redukční analýzy jejich D-stromů

V této sekci se pokusíme čtenáře neformálně uvést do problematiky manuální redukční analýzy vět a poukázat na souvislosti s redukční analýzou D-stromů, které těmto větám odpovídají. Redukční analýzou českých vět a jejímu modelování se zabýváme již delší dobu (viz např. [3, 5]), naopak explicitní zmínky o redukční analýze D-stromů se objevují ponejprv na loňském ITATU (viz [4, 2]). Při formalizaci obou typů redukčních analýz zvýrazňujeme jejich minimalistický charakter a využíváme ho při strukturální charakterizaci D-stromů.

RA je založena na postupném zjednodušování analyzované věty po malých krocích, viz [3, 5]. RA definuje možné posloupnosti větných redukcí – každá redukce RA spočívá ve *vypuštění* několika slov, nejméně však jednoho

slova analyzované věty. V některých redukcích může být kromě vypouštění použita operace *shift*, která přesune nějaké slovo na novou pozici ve větě.

Metoda (manuální) redukční analýzy, studovaná v tomto příspěvku, dodržuje následující zásady:

(i) tvary jednotlivých slov (i interpunkčních znamének), jejich morfologické charakteristiky i jejich syntaktické kategorie se nemění během RA;

(ii) gramaticky správná věta (přesněji její čtení) musí zůstat správná i po redukci;

(iii) vynecháme-li z libovolné redukce jednu či více operací vypuštění nebo shift, nastane porušení principu zachování správnosti (ii);

(iv) předložkové vazby (např. 'o otce'), se vynechávají celé (jinak je možný posun významu, často i změny v pádech);

(v) věta, která obsahuje správnou větu (nebo její permutaci) jako svoji (případně nesouvislou) podposloupnost, musí být dále redukována;

(vi) redukce používají operaci shift jenom v případech vynucených principem zachování korektnosti, tedy v případech, kdy vynechání shiftu by vedlo k nekorektnímu větnému slovosledu;

(vii) syntaktická struktura věty po redukci zachovává strukturu věty před redukcí.

Novým prvkem mezi zásadami pro větnou redukční analýzu oproti [5] je položka (vii). Syntaktická struktura zde znamená větný rozbor odpovídající stromům z Pražského závislostního korpusu (D-strom). Tato zásada fakticky formuluje základní vztah mezi větnou redukční analýzou a redukční analýzou D-stromů. Výše uvedené zásady postupně upřesníme ve formální části příspěvku.

V následujících odstavcích uvedeme serii příkladů ilustrujících prvky redukční analýzy, které se týkají redukcí zjednodušující jak závislosti, tak především koordinace. Všimněme si, že redukce koordinací budou ve dvou aspektech složitější než redukce závislostí. Pozorování koordinačních jevů a formalizace těchto pozorování je hlavní novinkou a přínosem tohoto příspěvku.

*Příspěvek prezentuje výsledky dosažené v rámci projektu agentury GAČR číslo GA15-04960S.

D-stromy na našich obrázcích se liší od D-stromů z PDT jen ve dvou aspektech. Za prvé: neobsahují identifikační uzel, který nenese žádnou syntaktickou informaci a neodpovídá žádnému slovu věty. Za druhé: značka 'Coord' je nahrazena značkou 'Cr'.

Příklad 1.

(1) Petr.Sb se.AuxT bojí.Pred o.AuxP otce.Obj ..AuxK

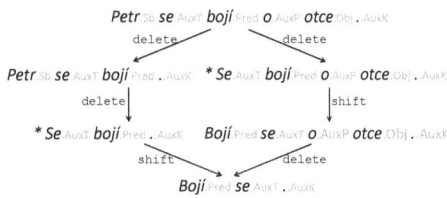

Obrázek 1: Schema RA pro větu (1).

Z obrázku 1 vidíme, že věta (1) může být v prvním kroku redukována dvěma způsoby:
(i) buď vypuštěním předložkové vazby 'o otce'; této větné redukci odpovídá redukce D-stromu T_1 z obrázku 2 na D-strom T_2 z obrázku 3,
(ii) nebo vypuštěním podmětu (subjektu) 'Petr', to však vede k větě se špatným slovosledem. Gramatické české věty nemohou začínat klitikou. To vede k použití přesunu klitiky 'se' na druhou pozici ve větě. Získáme tak korektní větu 'Bojí se o otce.' Této větné redukci odpovídá redukce D-stromu T_1 na D-strom T_4 z obrázku 5.

Potom pokračují redukce podobným způsobem v obou větvích, až dospějeme k neredukovatelné správné větě 'Bojí se.'. Této fázi odpovídají redukce D-stromů T_2 a T_4 na D-strom T_3 z obrázku 4.

Předchozí příklad ilustruje přirozenou souvislost mezi větnou redukční analýzou věty (1) a redukční analýzou D-stromu se závislostní strukturou téže věty z obrázku 2.

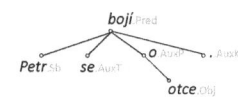

Obrázek 2: Závislostní strom T_1.

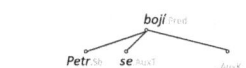

Obrázek 3: T_2, vzniklé redukcí z T_1.

Příklad 2.
Na obrázku 6 vidíme schema redukční analýzy věty (2). Věta (2) obsahuje trojnásobnou koordinaci předmětů. Povšimněme si, že dalšímu zjemnění schematu zabraňují kategorie (značky), použité podle vzoru PDT. Značka 'Cr' znamená koordinující symbol (slovo), 'Co' značí koordinované slovo, či symbol. Schematu na obrázku odovídají redukce D-stromů, které reprezentují obrázky 7

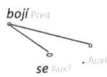

Obrázek 4: T_3 vzniklé redukcí se shiftem z T_2 nebo redukcí bez shiftu z T_4.

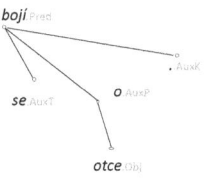

Obrázek 5: T_4, vzniklé redukcí z T_1.

až 11. Všechny tři redukce D-stromu Tc_1 odstraňují (při zjednodušování trojnásobné koordinace na dvojnásobnou) dva nesouvisející uzly (podstromy). Třetí redukce navíc používá shift. Tyto redukce se liší od předchozího příkladu, kde všechny redukce odtrhly jediný úplný souvislý podstrom. Zbylé redukce dvojnásobných koordinací se realizují odtržením souvislého úplného podstromu, určeného jejich vrcholem, podobně jako u redukcí v předchozím příkladě, týkající se závislostí.

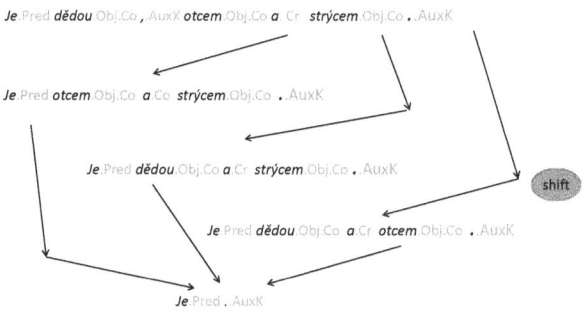

Obrázek 6: RA věty (2) s vícenásobnou koordinací.

Příklad 3.
Na obrázku 12 vidíme schema redukční analýzy věty (3). Toto schema znázorňuje jedinou redukci, která odstraňuje koordinovaná příslovečná určení, která jsou závislá na koordinovaných predikátech. Odpovídající redukci D-stromu ilustrují obrázky 13 a 14.

Příklad 4.
Na obrázku 15 vidíme schema redukční analýzy věty (4). Věta (4) je věta s vloženou koordinací. D-stromy zachycující odpovídající redukční analýzu D-stromů jsou na obrázcích 16 až 18. Vložená koordinace se v D-stromě Tcz_3 zjednodušuje tak, že se vyjme jedna hrana s řídícím uzlem se značkou 'Cr.Co' (ve složitějších případech i to co na ní visí). To odpovídá dvěma redukcím ve větné redukční analýze z obrázku 15. Tento typ redukce je nový oproti předchozím případům a je vynucen principy zachování korektnosti a minimality ve větné redukční analýze.

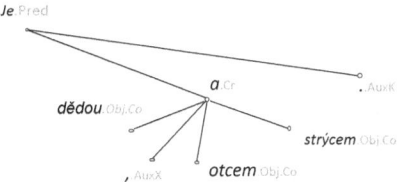

Obrázek 7: D-strom Tc_1.

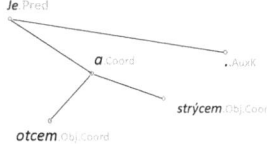

Obrázek 8: Tca_2, vzniklé redukcí z Tc_1.

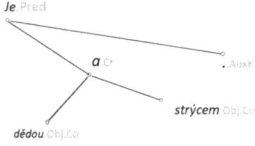

Obrázek 9: Tcb_2, vzniklé redukcí z Tc_1.

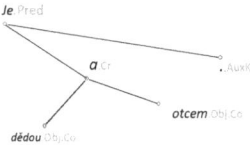

Obrázek 10: Tcc_2, vzniklé redukcí z Tc_1.

Obrázek 11: Tc_3, vzniklé redukcí z Tca_2, Tcb_2 a Tcc_2.

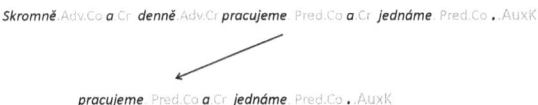

Obrázek 12: RA závislé koordinace na řídící koordinaci.

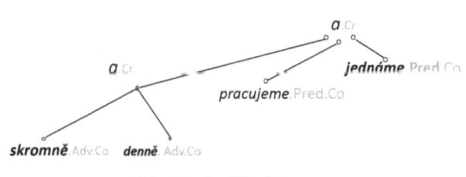

Obrázek 13: Tcz_2

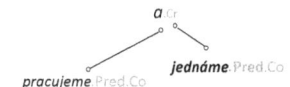

Obrázek 14: Tcz_{22}, vzniklé redukcí z Tcz_2.

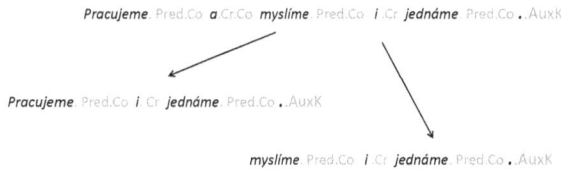

Obrázek 15: AR věty s vloženou koordinací.

2 Formalizace

Formalizace RA přirozených jazyků začíná formalizováním lexikální analýzy těchto jazyků. Lexikální analýza kromě jiného umožňuje rozlišovat možnosti uplatnení jednotlivých typů redukcí.

2.1 Lexikální analýza

Při formalizaci lexikální analýzy pracujeme se třemi abecedami (slovníky)- konečnými množinami slov. Σ_p, tzv. slovník [1], se využívá na modelování jednotlivých slovních forem. Σ_c označuje abecedu kategorií, například syntaktických značek v PDT. Kombinací dostáváme hlavní slovník $\Gamma \subseteq \Sigma_p \times \Sigma_c$, který umožňuje odstraňovat lexiko-morfologické nejednoznačnosti jednotlivých slovních forem. Lexiko-morfologicky zjednoznačněná věta tedy vstupuje do RA jako retězec nad slovníkem Γ.

Projekce z Γ^* do Σ_p^* resp. do Σ_c^* přirozeně definujeme pomocí homomorfismů: *slovníkovým homomorfismem* h_p : $\Gamma \to \Sigma_p$ a *kategoriálním homomorfismem* h_c : $\Gamma \to \Sigma_c$: $h_p([a,b]) = a$ a $h_c([a,b]) = b$ pro všechny $[a,b] \in \Gamma$.

Příklad 5. *Definované pojmy ilustrujeme na příklade, který vychází z příkladu 1*
Slovník: $\Sigma_p^1 = \{$ Petr, se, bojí , o, otce, . $\}$
Abeceda kategorií: $\Sigma_c^1 = \{$ Sb, AuxT, Pred, AuxP, Obj, AuxK$\}$
Hlavní slovník: $\Gamma^1 = \{b_1 = [Petr, Sb], b_2 = [se, AuxT], b_3 = [bojí, Pred], b_4 = [o, AuxP], b_5 = [otce, Obj], b_6 = [., AuxK]\}$

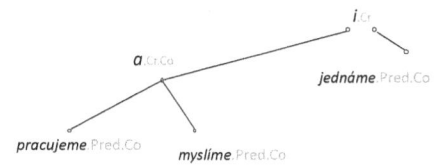

Obrázek 16: Tcz_3

---

[1]Index p při označení abecedy se vztahuje na anglickou verzi, kde se používá slovo proper

Obrázek 17: Tcz_{31}, vzniklé redukcí z Tcz_3.

Obrázek 18: Tcz_{32}, vzniklé redukcí z Tcz_3.

V abecedě kategorií v tomto příkladě jsou jen závislostní kategorie (ne všechny). Koordinační kategorie vznikají kombinacemi se značkami 'Cr', 'Co'.

2.2   Formální RA

V této sekci zavádíme postupně formální redukční analýzu vět (řetězů) RA a formální redukční analýzu pro D-stromy.

Nejprve zavedeme na jazyce L tzv. DS-redukci $\succ_L$. Nechť u, v jsou řetězce. Říkáme, že u je větší než v vzhledem k jazyku L a označujeme $u >_L v$ pokud:

- $u, v \in L$ a $|u| > |v|$;
- v je permutace nějaké podposloupnosti u.

Říkáme, že v je DS-redukce u vzhledem k jazyku L a označujeme $u \succ_L v$ pokud:

- $u >_L v$ a neexistuje žádné $z \in L$ takové, že $u >_L z >_L v$, t.j., platí princip minimality redukcí.

Reflexívní a tranzitívní uzávěr relace $\succ_L$ označujeme $\succ_L^*$. Částečné uspořádání $\succ_L$ přirozeně definuje

- $L_\succ^0 = \{v \in L \mid \neg \exists u \in L : v \succ_L u\}$ - množinu ireducibilních vět jazyka L
- $L_\succ^{n+1} = \{v \in L \mid \exists u \in L_\succ^n : u \succ_L v\} \cup L_\succ^n, n \in N$ - množina těch vět z jazyka, které je možné zredukovat na ireducibilní větu z jazyka posloupností DS-redukcí délky nanejvýš $n + 1$.

Množinu $\succ_L = \{u \succ_L v \mid u, v \in L\}$ nazveme množinou DS-redukcí jazyka L. Analogicky pro větu w jazyka L nazveme $\succ_L(w) = \{u \succ_L v \mid w \succ_L^* u\}$ DS-redukční množinou věty w.

Fakt: $\succ_L$ aj $\succ_L(w)$ jsou jednoznačne určené L, resp. w a L.

Přistupme k formalizaci (minimalistické) redukční analýzy. Říkáme, že relace $\rhd_L \subseteq \succ_L$ je *DS-(redukční) analýza jazyka L* pokud $L = L_\succ^0 \cup \{v \mid \exists u, z : v \rhd_L u \rhd_L^* z \in L_\succ^0\}$. Analogicky definujeme DS-analýzu $\rhd_L(w)$ pro $w \in L$; $\rhd_L(w) = \{u \rhd_L v \mid w \rhd_L^* u\}$.

Uvědomme si, že zatím co jazyk L je jednoznačne určený pomocí $\rhd_L$ a $L_\succ^0$, věta $w \in L$ může mít více DS-analýz. Různé DS-analýzy věty w v lingvistice odpovídají různému čtení (porozumění) této věty.

Relace $\rhd_L$ určuje velikost zkrácení, které je možné dosáhnout jedním krokem redukce. Říkáme, že $\rhd_L$ a L jsou

k-*omezené* pokud délka slov z $L_\succ^0$ je nejvýše k a $|u| - |v| \leq k$ pro všechny $u \rhd_L v \in \rhd_L$.

Bylo by zvláštní, kdyby v DS-redukci přirozeného jazyka byly ireducibilní věty dlouhé, přičemž všechny redukce z $\rhd_L$ by zkracovaly věty jen málo. Zajímáme se proto hlavně o takové DS-analýzy, v kterých $\forall w \in L_\succ^0$ existují $u, v, u \rhd_L v$ takové, že $|u| - |v| \geq |w|$. Takovým DS-analýzám říkáme *proporcionální*.

Všimněme si, že redukční analýza české věty z příkladu 1 vyhovuje podmínkám kladeným na proporcionální 2-omezenou DS-analýzu, zatímco redukční analýza české věty z příkladu 2 je proporcionální 3-omezenou DS-analýzou.

DS-analýzu budeme považovať za relevantní model skladby přirozených i umělých jazyků, pokud to bude DS-analýza konečných, anebo nekonečných semi-lineárnych jazyků, které jsou proporcionální a k-ohraničené pro nejaké nevelike k.

2.3   D-struktury a D-stromy

V následující části zavedeme tzv. D-struktury a D-stromy, ktoré jsou grafovou reprezentací struktury vět a jejich odvození.[2] D-struktura reprezenuje syntaktické jednotky (slova a jejich kategorie použité v príslušné větě) jako vrcholy grafu a vzájemné syntaktické vztahy mezi nimi hranami; pořadí slov je určené totálním uspořádáním vrcholů.

D-struktura na Γ je trojice $D = (V, E, ord(V))$, kde (V, E) je orientovaný acyklický graf, V konečná množina jeho vrcholů a $E \subset V \times V$ konečná množina jeho hran. *Vrchol* $u \in V$ je dvojice $u = [i, a]$, kde $a \in \Gamma$ je symbol (slovo) spolu s přiřazenými kategoriemi, i (*index/ identifikační číslo*) je přirozené číslo sloužící pro jednoznačnou identifikaci vrcholu u a $ord(V)$ je totální uspořádání na V, obvykle popsané uspořádaným seznamem prvků z V.

Hrany D-struktury interpretujeme jako syntaktické vztahy mezi odpovídajícími lexikálními jednotkami, uspořádání $ord(V)$ reprezentuje pořadí slov v modelované větě. Je-li $ord(V) = \{[i_1, a_1], \cdots, [i_n, a_n]\}$, tak $w = a_1 \cdots a_n$ je řetězec (resp. věta), který označujeme $St(D) = w$, a říkáme, že je projekcí D-struktury D.

Říkáme, že D-struktura $D = (V, E, ord(V))$ je *normalizovaná*, pokud $ord(V) = ([1, a_1], [2, a_2], \cdots, [n, a_n])$ pro nejaké $a_1, \cdots, a_n$. *Normalizace* D-struktury $D = (V, E, ord(V))$ je taková normalizovaná D-struktura $D_1 = (V_1, E_1, ord(V_1))$, pro kterou (V, E) a (V_1, E_1) jsou izomorfní a $St(D) = St(D_1)$. Všimněme si, že normalizace D-struktury je jednoznačne daná.

Dve D-struktury jsou ekvivalentní pokud mají stejnou normalizaci. Ekvivalentní D-struktury obvykle nebudeme rozlišovat. Uvidíme, že nenormalizované D-struktury (stromy) získáme z normalizovaných pomocí operací, které zavedeme.

---

[2]prefix Dje převzatý z anglických pojmů Delete a Dependency.

Vzhledem k charakteru zkoumané problematiky budeme většinou pracovat se stromovými D-strukturami. Říkáme, že D-struktura $D = (V, E, ord(V))$ nad Γ je D-strom nad Γ pokud (V, E) je kořenový strom (t.j., všechny maximální cesty (V, E) začínají v listech a končí v jediném kořeni).

Budeme pracovat s redukcemi D-stromů - relace $\sqsupset$ a $\vdash$ definované na D-stromech souvisí s realizací různých typů redukcí. Nechť $D = (V, E, ord(V)), D_1 = (V_1, E_1, ord(V_1))$ jsou D-stromy.

$D \sqsupset D_1$ pokud

(1) (V_1, E_1) je podstrom (V, E)

(2) V_1 obsahuje kořen D

(3) $ord(V_1)$ je permutace podposloupnosti $ord(V)$.

$D \vdash D_1$, pokud podmínku (1) nahradíme dvěma podmínkami

(1a) $V \subset V_1$

(1b) $\forall v_1, v_2 \in V_1$ platí, že pokud existuje cesta z v_1 do v_2 ve stromě (V, E) tak existuje také cesta z v_1 do v_2 i ve stromě (V_1, E_1).

Příklad 6. *Následuje popis D-stromů T_1 a T_2, které reprezentují obr. 2 a obr. 3:*

$T_1 = (V_1, E_1, ord(V_1))$, *pričemž*

$V_1 = \{[1, b_1], [2, b_2], [3, b_3], [4, b_4], [5, b_5], [6, b_6]\}$

$E_1 = \{([1, b_1], [3, b_3]), ([2, b_2], [3, b_3]), ([4, b_4], [3, b_3]),$
$([5, b_5], [4, b_4]), ([6, b_6], [3, b_3])\},$

$ord(V_1) = ([1, b_1], [2, b_2], [3, b_3], [4, b_4], [5, b_5], [6, b_6])$

$T_2 = (V_2, E_2, ord(V_2))$, *pričemž*

$V_2 = \{[1, b_1], [2, b_2], [3, b_3], [6, b_6]\}$

$E_2 = \{([1, b_1], [3, b_3]), ([2, b_2], [3, b_3]), ([6, b_6], [3, b_3])\}$

$ord(V_2) = ([1, b_1], [2, b_2], [3, b_3], [6, b_6])$

Je snadno vidět, že $T_1 \sqsupset T_2$.

Takřka všechny neformální redukce z kapitoly jedna vedou k realizaci relace $\sqsupset$. Neplatí to jen pro redukce na obr. 17 a 18. Tyto redukce splňují obecnější relaci $\vdash$. Tyto dvě relace reprezentují dvě varianty zachování zbylé D-struktury, vzniklé zmenšením při uplatnění redukcí redukční analýzy na D-stromech.

Nechť T je nějaká množina D-stromů na Γ. Říkáme, že T tvoří T-jazyk na Γ a píšeme $T \subseteq T(\Gamma)$. Analogicky, množinu $St(T) = \{St(t) \mid t \in T\}$ nazýváme *projekcí* T, množina $h_p(St(T)) = \{h_p(St(t)) \mid t \in T\}$ je vlastní jazyk pro T, a $h_c(St(T)) = \{h_c(St(t)) \mid t \in T\}$ je kategoriální jazyk pro T.

Zavedeme tři operace pro práci s D-stromy. Umožní nám realizovat typ redukcí čistě závislostních i redukce různých typů koordinací.

Najjednodušší operací je tzv. *shift*, což je takový posun některého vrcholu D-stromu $D = (V, E, ord(V))$ na nové

místo v $ord(V)$, který zachová stromovou strukturu D, tedy zachová všechny uzly z V a všechny hrany z E.

Druhou operaci nazveme *UNC*, z anglického *upper-node-cut*. Je typická pro redukce závislostí a při jejím zavádění si pomůžeme jednodušší operací *LNC*, z anglického *lower-node-cut*. Operace UNC i LNC jsou určené uzlem u D-stromu různým od kořene. Tento uzel jednoznačně určuje rozklad D-stromu D na dva podstromy:

1) $T_L(u, D)$ označuje výsledek LNC aplikovaného na D v uzlu u ; je to podstrom stromu D, který tvoří uzly ležící na nějaké cestě z listu do u (včetně u). Pořadí uzlů v $T_L(u, D)$ je určené pořadím v D.

2) $T_U(u, D)$ označuje výsledek UNC aplikovaného na D v uzlu u; je to maximální podstrom D obsahující kořen D a všechny uzly mimo $T_L(u, D)$. Pořadí uzlů je určené pořadím v D. UNC tedy transformuje D na D-strom $T_U(u, D)$.

Poslední operací je *UEC*, z anglického *upper-edge-cut*. Použití této operace jsme viděli při redukci (odstraňování) vložených koordinací z obr. 17 a 18. Nechť (u, v) a (v, v_1) jsou takové hrany D-stromu D, že existuje právě jeden uzel $u_1 \neq u$ a hrana (u_1, v) vedoucí do v. Operace UEC aplikovaná na D podle hrany (u, v) vytvoří D-strom $T_E((u, v), D)$. $T_E((u, v), D)$ získáme následujícím způsobem: nejprve aplikací UNC-operace vytvoříme $T_U(u, D)$ a následně z něj odstraníme uzel v spolu s hranami (u, v) a (v, v_1). Potom spojíme vrcholy u_1, v_1 novou hranou (u_1, v_1) a získáme tak D-strom, který označujeme $T_E((u, v), D)$.

Nyní zavádíme formální redukce a redukční analýzu na D-stromech tak, abychom pokryli jak závislostní, tak koordinační jevy z PDT.
Nechť $T \subseteq T(\Gamma)$, $t_1, t_2 \in T$. Symbolem $\vdash_T$ budeme označovat zúžení operace $\vdash$ na T^3

Říkáme, že t_1 je *NES-redukované* na $t_2 \in T$ a označujeme $t_1 \hookrightarrow_{NES} t_2$, pokud redukci $t_1 \vdash_T t_2$ umíme popsat pomocí množiny O_N UNC-operací a/nebo množiny O_E UEC-operací, případně následovanými množinou shiftů O_S. Navíc, $O_N \cup O_E$ je neprázdná, každý uzel je operací z O_s přesouvaný nejvýše jednou, O_s může být prázdná.

Pokud v predchozí definici nepovolíme UEC-operace, budeme říkat, že t_1 je *NS-redukované* na $t_2 \in T$ a označovat $t_1 \hookrightarrow_{NS} t_2$.

Pokud při redukci nepovolíme ani shifty, budeme hovořit o *N-redukci* a označovat $t_1 \hookrightarrow_N t_2$.

Redukce typu NES, NE a N mohou být, v principu, aplikované na libovolné D-stromy. Nás však zajímají redukce D-stromů daného T-jazyka, proto vyžadujeme, aby i po aplikování zmíněných redukcí byl vzniklý strom platným D-stromem zkoumaného jazyka. Při definování pojmu redukce proto přidáváme parametr T.

Nechť $X \in \{NES, NS, N\}$, $T \subseteq T(\Gamma)$. Říkáme, že t_1 je (X, T)-redukované na t_2 a píšeme $t_1 \gg_{(T,X)} t_2$ pokud:

- $t_1, t_2 \in T$
- $t_1 \hookrightarrow_X t_2$ a neexistuje $z \in T$ tak, aby $t_1 \hookrightarrow_X z \hookrightarrow_X t_2$, t.j., platí princíp minimality redukcí.

---

[3] Při $\vdash_T$ tedy vyžadujeme, aby t_1 i t_2 byli z T.

Tranzitivní, reflexívní uzávěr $\gg_{(T,X)}$ označujeme $\gg^*_{(T,X)}$. Tranzitivní, anti-reflexívní uzávěr $\gg_{(T,X)}$ označujeme $\gg^+_{(T,X)}$. V situaci, kdy je T zřejmé z kontextu, hovoříme jen o NES-, NE, resp. N-redukci.

Uvědomme si, že T a X jednoznačně určují množinu $\gg_{(T,X)} = \{u \gg_{(T,X)} v \mid u,v \in T\}$, ktorou považujeme za redukční analýzu T-jazyka T. Říkáme, že $\gg_{(T,X)}$ je X-redukcí T. Všimněme si rozdílu oproti DS-analýze retězcových jazyků, která nebývá jednoznačně určená svým jazykem.

Nechť $X \in \{NES,NS,N\}$.
$(T,X)^0_{\gg} = \{t \in T \mid \neg \exists s \in L : t \gg_{(T,X)} s\}$,
$(T,X)^{n+1}_{\gg} = \{v \in T \mid \exists u \in (T,X)^n_{\gg} : v \gg_{(T,X)} u\} \cup T^n_{\gg}$.
Nechť $t \in T$. Píšeme $\gg_{(T,X)} (t) = \{u \gg_{(T,X)} v \mid t \gg^*_{(T,X)} u\}$. Říkáme, že $\gg_{(T,X)} (t)$ je X-analýza (redukční) D-stromu t.

V následující sekci budeme navíc ještě vázat použití jednotlivých typů operací na (ne)přítomnost koordinačních značek v určujících hranách a uzlech těchto operací. O takových typech omezení uplatnění operací jsme zatím nemluvili.

2.4 Principy, vlastnosti a pozorování

Zde zavedeme principy, které nám umožní formulovat požadavky na redukční analýzu na D-stromech a formulovat pozorování o jejich plnění na stromech z PDT. Při těchto pozorováních uplatníme možnost porovnávat NES-analýzy, NS-analýzy a N-analýzy D-stromů a využijeme tato porovnání pro charakterizaci (klasifikaci) těchto D-stromů.

Princip S-kompatibility. Nech $X \in \{NES,NS,N\}$. Pokud platí, že $t_1 \gg_{(T,X)} t_2$ a zároveň platí, že $Str(t_1) \succ_{Str(T)} Str(t_2)$, tak říkáme, že redukce $t_1 \gg_{(T,X)} t_2$ je S-kompatibilní. Neformálně řečeno, pokud redukci D-stromů odpovídá řetězová redukce na řetězech získaných projekcí ze stromů, která je vztažena k jazyku řetězů $Str(T)$, daných množinou stromů T.

Podobně říkáme, že $\gg_{(T,X)} (t)$ je S-kompatibilní, pokud všechny jeho X-redukce jsou S-kompatibilní a pokud za předpokladu $u \in (T,X)^0_{\gg}$ a $t \gg^*_{(T,X)} u$ platí, že $Str(u) \in Str(T)^0_{\succ}$.

Říkáme, že X-analýza $\gg_{(T,X)}$ je S-kompatibilní pokud všechny její D-stromy mají S-kompatibilní X-analýzu.

Fakt. Vidíme, že $\gg_{(T,X)} (t)$ je S-kompatibilní pokud $Str(\gg_{(T,X)} (t)) = \{Str(u) > Str(v) \mid u \gg_{(T,X)} v \in \gg_{(T,X)} (t)\}$ tvoří DS-analýzu věty $Str(t)$ vzhledem k jazyku $Str(T)$.

Princip S-kompatibility je tak požadavkem, který zaručuje přirozený vztah mezi větnou DS-analýzou a X-analýzami na D-stromech.

Fakt. Uvažujeme NS-analýzu A D-stromu t. Platí, že uzel u, který je ve stromě t na cestě ke kořeni blíže než uzel v, nemůže být v žádne větvi NS-analýzy A vypuštěn dříve než v.

Tento fakt přímo vyplývá z definice UNC-operace.

Předchozí fakt zpřesňuje intuitivně vnímané vlastnosti (ne)závislostí v (čistě) závislostních stromech.

Následující dva principy jsou blízké algebraickému principu konfluence.

Princip Tl-kompatibility. Požadujeme, aby všechny větve v NES-analýze A stromu t byly stejně dlouhé a v každé větvi byl použit stejný počet UNC-operací a UEC-operací.

Následující princip je přísnější. Odlišuje čistě závislostní D-stromy od D-stromů s koordinacemi.

Princip Ta-kompatibility (Formulace závislostniho principu). Tento princip uvažuje pouze D-stromy t, které nemají koordinační znaky, a jejichž NES-analýzy jsou i NS-analýzami a zároveň splňují princip Tl-compatibility. Dále zde požadujeme, aby množina UNC-operací užitých v dané NS-analýze A byla určena libovolnou větví z A (t.j. v každé větvi byla ta množina stejná) a aby všechny větve z A končily stejnou neredukovatelnou větou (algebraickou terminologií A tvoří svaz).

Další dva principy formulují volnější předpoklady, jak by měla redukční analýza reprezentovat tvar analyzovaného D-stromu, ve kterém jsou i koordinační značky.

Princip Tb-kompatibility. Pokud máme NES-analýzu A D-stromu t a dva různé uzly u, v D-stromu t, které jde redukovat jako určující uzly dvěma UNC-operacemi a přitom nevede cesta mezi u a v, tak požadujeme, aby během A mohla být dříve provedena kterákoliv z těchto UNC-operací (tj. aby existovaly dvě větve z A, kde v první větvi je provedena dříve redukce s u a v té druhé větvi je dříve provedena redukce s v.)

Princip Tc-kompatibility. Nechť máme NES-analýzu A D-stromu t, dvě hrany e_1, e_2 stromu t, které neleží (oběma uzly) na jedné cestě v t a e_1, e_2 jde obě redukovat jako určující hrany UEC-operací. Požadujeme, aby během A mohla být dříve provedena kterákoliv z těchto UNC-operací (tj. existují dvě větve z A, kde v první je provedena dříve redukce s e_1 a v té druhé je redukována dříve e_2. Poznamenejme, že v jedné větvi nemusí být nutně provedeny obě tyto redukce.

Říkáme, že X-analýza $\gg_{(T,X)}$ je k-omezená, pokud počet vypuštěných uzlů v jednotlivých X-redukcích z $\gg_{(T,X)}$ nepřesahuje k a $(T,X)^0_{\gg}$ neobsahuje D-strom s více uzly než k.

Analogicky lze zavést k-omezenou X-analýzu jednotlivého stromu.

Říkáme, že X-analýza $\gg_{(T,X)} (t)$ D-stromu t je proporcionální, pokud $Str(\gg_{(T,X)} (t))$ je proporcionální.

Máme také možnost měřit složitost X-redukcí pomocí počtu operací užitých v jednotlivých X-redukcích.

Příklad 7. D-*strom reprezentující obrázek 4:*
$T_3 = (\{[2,b_2],[3,b_3],[6,b_6]\},$
$\{([2,b_2],[3,b_3]), ([6,b_6],[3,b_3])\}, ([3,b_3],[2,b_2],[6,b_6]))$
D-*strom reprezentující obrázek 5:*

$$T_4 = (\{[2,b_2],[3,b_3],[4,b_4],[5,b_5],[6,b_6]\},$$
$$\{([2,b_2],[3,b_3]),\ ([4,b_4],[3,b_3]),\ ([5,b_5],[4,b_4]),$$
$$([6,b_6],[3,b_3])\},$$
$$([3,b_3],[2,b_2],[4,b_4],[5,b_5],[6,b_6]))$$

Příklad 8. *Vidíme, že D-strom T_1 má jen značky odpovídající závislostem (nemá značky Cr, Co pro koordinace). Let $R_2 = \{T_1,T_2,T_3,T_4\}$, kde D-stromy T_1,T_2,T_3,T_4 byly popsány v předchozích příkladech.*

Vidíme, že

$\gg_{(R_2,NES)} = \{T_1 \gg_{(R_2,NES)} T_2,\ T_2 \gg_{(R_2,NES)} T_3,\ T_1 \gg_{(R_2,NES)} T_4, T_4 \gg_{(R_2,NES)} T_3\}$,

a dále že $\gg_{(R_2,NES)}$ je rovno nejen $\gg_{(R_2,NES)} (T_1)$ ale, i $\gg_{(R_2,NS)} (T_1)$.

Platí, že $(R_2,NES)^0_\gg = \{T_3\}$.

$\gg_{(R_2,NES)} (T_1)$ *je tedy NS-analýzou věty T_1, ale není její N-analýzou, jelikož NS-redukce $T_2 \gg_{(R_2,NS)} T_3$ a $T_1 \gg_{(R_2,NS)} T_4$ používají shift.*

Vidíme také, že $\gg_{(R_2,NS)} (T_1)$ je S-kompatibilní, a že její redukce používají jedinou UNC-operaci a maximálně jeden shift.

$\gg_{(R_2,NS)} (T_1)$ *je také Ta-kompatibilní, Tb-kompatibilní (a triviálně Tc-kompatibilní a Tl-kompatibilní), 2-omezená, a proporcionální.*

Vymezení čistě závislostních D-stromů. Podobné vlastnosti jako má NS-analýza D-stromu T_1 požadujeme po všech čistě závislostních D-stromech (obsahují jen hrany (uzly) se závislostními kategoriemi (značkami)). Čistě závislostní D-stromy mají NS-analýzu, jejíž redukce obsahují jedinou operaci UNC a nejvýše tři shifty. Každá NS-analýza čistě závislostního D-stromu má být S-kompatibilní, Ta-kompatibilní, Tb-kompatibilní (triviálně i Tc-kompatibilní a Tl-kompatibilní) a proporciální vzhledem k množině všech korektních NS-redukcí korektních čistě závislostních stromů. Toto formální vymezení závislostních stromů odpovídá rozšířenému intuitivnímu vnímání závislostí a je logickým vzorem i pro vymezení D-stromů s koordinacemi.

Pozorování a poznámka. V PDT jsme nezpozorovali žádnou odchylku proti předchozímu vymezení u D-stromů s čistě závislostními značkami. Pokud však budeme uvažovat jen N-analýzu D-stromu T_1, tak ta není ani S-kompatibilní, ani Ta-kompatibilní. Pozorování příkladů tohoto typu nás vedla k rozšíření původně užívané N-analýzy na vhodnější NS-analýzu, kterou lze uplatňovat zřejmě na celou třídu čistě závislostních D-stromů při zachování výše požadovaných principů.

Příklad 9. *V tomto příkladě budeme pozorovat D-strom Tc_1 z obrázku 9, jeho NES-analýzu A_1 na obrázcích 10 až 13 a jeho DS-analýzu z obrázku 6. Tc_1 neobsahuje uzel s dvojicí značek Cr, Co, ani hranu, která má oba uzly se značkou Co.*

Vidíme, že A_1 D-stromu Tc_1 je NS-analýzou (nepoužívá UEC-operace).

A_1 je S-kompatibilní, Tl-kompatibilní a Tb-kompatibilní (triviálně i Tc-kompatibilní) a proporciální.

A_1 je NS-analýzou věty (D-stromu) s trojnásobnou (nezapuštěnou) koordinací.

A_1 není Ta-kompatibilní, protože množiny UNC-operací v jednotlivých větvích nejsou stejné.

A_1 obsahuje redukce, které používají dvě UNC-operace. Tím se liší od závislostních redukcí, které používají jen jednu UNC-operaci.

Všimněme si, že určující uzly dvou UNC-operací v jedné redukci visí na stejném uzlu (se značkou Cr) a odstraněné podstromy tvoří souvislý úsek v uspořádání uzlů.

Povšimněme si ještě, že budeme-li uvažovat N-analýzu A_2 D-stromu Tc_1, tak přijdeme o poslední větev se shiftem. A_2 je také S-kompatibilní, Tl-kompatibilní, Tb-kompatibilní a proporciální. A_2 má tedy také pěkné vlastnosti.

Vymezení závislostně-koordinačních D-stromů bez vložených koordinací. Podobné vlastnosti jako má NS-analýza D-stromu Tc_1 požadujeme po všech D-stromech bez vložených koordinací. Má to být NS-analýza, která je S-kompatibilní, Tl-kompatibilní a Tb-kompatibilní (triviálně i Tc-kompatibilní). Může používat dvě UNC-operace v jedné redukci, které odstraňují dva vedlejší podstromy visící na jednom uzlu.

Pozorování. V PDT jsme zatím nezpozorovali žádnou odchylku proti předchozímu vymezení. Pokud však budeme uvažovat jen NS-analýzu D-stromu Tc_1, která bude pracovat s jedinou UNC-operací v redukci, tak ta není S-kompatibilní.

Poznamenejme, že malou technickou změnou v metodě zobrazování vícenásobných koordinací v PDT bychom dosáhli toho, že by pro zachování S-kompability u redukcí tohoto jevu by nebylo třeba použít více než jednu UNC-operaci.

Příklad 10.

V tomto příkladě budeme pozorovat D-strom Tcz_3 z obrázku 16, jeho NES-analýzu A_3 na obrázcích 16 až 18 a jeho DS-analýzu z obrázku 8.

Tcz_3 obsahuje uzel s dvojicí značek Cr, Co i hranu, která má oba uzly se značkou Co.

A_3 je S-kompatibilní, Tl-kompatibilní a Tb-kompatibilní i Tc-kompatibilní.

A_3 je NES-analýzou věty (D-stromu) s vloženou koordinací, kde UEC-operace jsou uplatněny na hrany u kterých mají oba uzly značku Co, tedy hrany vložené koordinace. Řídící uzel těchto hran mívá ještě značku Cr.

Uvažujeme-li NS-analýzu A_4 D-stromu Tcz_3, tak vidíme, že A_4 není S-kompatibilní, jelikož nemá na rozdíl od odpovídající DS-analýzy z obrázku 8 žádné redukce.

Vymezení závislostně-koordinačních D-stromů. Podobné vlastnosti jako má NES-analýza D-stromu Tcz_3 požadujeme po všech D-stromech s koordinacemi a závislostmi. Má to být NES-analýza, která je S-kompatibilní, Tl-kompatibilní a Tb-kompatibilní (triviálně i Tc-kompatibilní). Může používat UEC-operace s určující hranou jejíž oba uzly nesou značku Co (jiné UEC-operace nejsou povoleny).

Pozorování. V PDT jsme zatím nezpozorovali žádnou odchylku proti předchozímu vymezení.

Pozorování obr. 19. Na obrázku 19 je jeden z autentických stromů z PDT. Podle vzoru tohoto stromu vznikly naše obrázky 12 až 18 pro tři různé typy redukcí koordinací.

Připomeňme, že symbol Coord z obrázku 19 je symbol Cr na našich obrázcích, symbol Coord_Co je v našich obrázcích nahrazen symbolem Cr.Co. Symbol Coord_Co je značkou, která má označovat řídící uzel (otce) vložené koordinace. V obrázku 19 je tento symbol jednou užit nesprávně, a to pro frázi 'skromě a Coord_Co každodenně'. Tato fráze zde není vloženou koordinací, ale koordinovanou závislostí podobně jako na obrázku 13.

NES-analýzou získáme z obrázku 19 několik dále neredukovalných vět s koordinacemi, které mají bez identifikačního uzlu a uzlu pro tečku jen tři uzly. NES-analýza bude S-kompatibilní, Tl-kompatibilní a Tb-kompatibilní i Tc-kompatibilní.

Použijeme-li na stejný D-strom jen NS-analýzu, nedostaneme se u redukovaných a dále neredukovatelných D-stromů pod sedm uzlů. Toto poslední pozorování připomíná pozorování z [2], kde se implicitně uvažují redukce, používající maximálně jednu UNC-operaci a žádnou UEC-operaci.

NS-analýza D-stromu z obrázku 19 nemůže být S-kompatibilní.

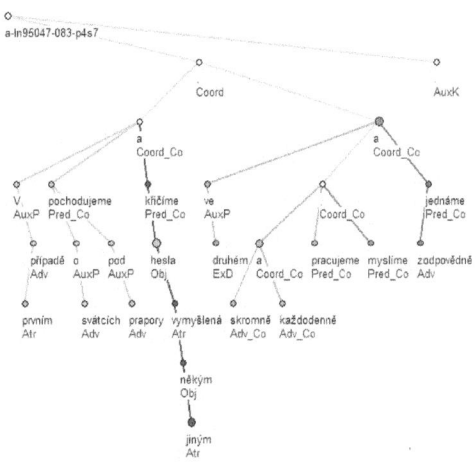

Obrázek 19: Autentický D-strom z PDT.

2.5 Shrnutí

V tomto příspěvku jsme exaktně zavedli pojmy větné redukční analýzy a tři typy redukční analýzy D-stromů. Formulovali jsme požadavky na kompabilitu větné redukční analýzy a redukční analýzy D-stromů. Našli jsme operace a typy redukcí, které dovolí provádět redukční analýzu D-stromů se závislostmi a koordinacemi stejně jemně a se stejnými k-omezeními jako větnou redukční analýzu. To je hlavní přínos tohoto příspěvku. Při formulaci typů redukčních analýz pro D-stromy jsme vycházeli z pozoro-

vání D-stromů z Pražského závislostního korpusu (PDT) a to D-stromů, které kromě modelování závislostí, modelují také složené koordinace. Tři (přesněji čtyři) typy redukčních analýz D-stromů nám dávají přirozenou taxonomii závislostních a koordinačních jevů zachycených D-stromy z PDT.

Domníváme se, že zavedený aparát dovolí hlouběji porozumět neprojektivitě a jejím mírám a volnosti slovosledu. To bude jedno z témat, kterým se budeme zabývat v blízké budoucnosti.

Dále se domníváme, že uvedená metoda by měla pomoci při odhalování nekonzistencí (či chyb) v PDT, podobně jako to bylo v případě D-stromu z obrázku 19.

V blízké budoucnosti bychom také rádi zahrnuli do metody redukční analýzy zbylé syntaktické jevy, které jsou v PDT rozlišeny. Máme na mysli hlavně koordinace s elipsami.

Na závěr děkujeme Markétě Lopatkové za poskytování informací o PDT i za komentáře k poskytnutému materiálu a ochotu o něm diskutovat.

Reference

[1] Hajič, J. Panevová, J., Hajičová, E., Sgall, P., Pajas, P., Štěpánek, J., Havelka, J., Mikulová, M., Žabokrtský, Z., Ševčíková-Razímová, M.: Prague Dependency Treebank 2.0. Linguistic Data Consortium, Philadelphia, 2006.

[2] Lopatková, M., Mírovský, J., Kubon, V.: Gramatické závislosti vs. koordinace z pohledu redukční analýzy. In: Proceedings of the Main Track of the 14th Conference on Information Technologies – Applications and Theory (ITAT 2014), with selected papers from Znalosti 2014 collocated with Znalosti 2014, Demanovska Dolina – Jasna, Slovakia, September 25–29, 2014., pages 61–67, 2014.

[3] Lopatková, M., Plátek, M., Kuboň, V.: Modeling syntax of free word-order languages: dependency analysis by reduction. In: Matoušek, V. et al., editor, Proceedings of TSD 2005, volume 3658 of *LNCS*, pages 140–147. Springer, 2005.

[4] Plátek, M.: Analysis by reduction of d-trees. In: Proceedings of the main track of the 14th Conference on Information Technologies – Applications and Theory (ITAT 2014), with selected papers from Znalosti 2014 collocated with Znalosti 2014, Demanovska Dolina – Jasna, Slovakia, September 25–29, 2014., pages 68–71, 2014.

[5] Plátek, M., Pardubská, D., Lopatková, M.: On minimalism of analysis by reduction by restarting automata. In: Formal Grammar – 19th International Conference, FG 2014, Tübingen, Germany, August 16–17, 2014. Proceedings, pages 155–170, 2014.

J. Yaghob (Ed.): ITAT 2015 pp. 51–58
Charles University in Prague, Prague, 2015

A Versatile Algorithm for Predictive Graph Rule Mining

Karel Vaculík

KD Lab, FI MU Brno, Czech Republic
xvaculi4@fi.muni.cz

Abstract: Pattern mining in dynamic graphs has received a lot of attention in recent years. However, proposed methods are typically limited to specific classes of patterns expressing only a specific types of changes. In this paper, we propose a new algorithm, DGRMiner, which is able to mine patterns in the form of graph rules capturing various types of changes, i.e. addition and deletion of vertices and edges, and relabeling of vertices and edges. This algorithm works both with directed and undirected dynamic graphs with multiedges. It is designed both for the single-dynamic-graph and the set-of-dynamic-graphs scenarios. The performance of the algorithm has been evaluated by using two real-world and two synthetic datasets.

1 Introduction

Data mining of complex structures has been extensively studied for quite a while. Recently, more attention has been drawn to the area of dynamic graphs, i.e. graphs evolving through time. A lot of research has been carried out both from the global and local perspectives of dynamic graphs. For instance, global characteristics such as density and diameter were studied in real dynamic graphs in [6]. On the other hand, graph mining on the local level is most frequently focused on pattern mining. Such patterns are represented by subgraphs and their evolution through a short period of time [1, 2, 8].

Nevertheless, most of the pattern mining methods for dynamic graphs impose various restrictions, such as the type of the dynamic graphs or the type of changes captured by such patterns. For example, GERM algorithm [1] assumes that vertices and edges are only added and never deleted in the input dynamic graph. Furthermore, it mines patterns representing only edge additions.

In this paper, we propose a new algorithm DGRMiner for mining frequent patterns that can capture various changes in dynamic graphs. Specifically, the patterns are in the form of predictive rules expressing how a subgraph can be changed into another subgraph by adding new vertices and edges, deleting specific vertices and edges, or relabeling vertices and edges. An example of a dynamic graph and two predictive rules are illustrated in Fig. 1.

The algorithm is able to mine patterns from a single dynamic graph and also from a set of dynamic graphs. Such graph rules are useful for prediction in dynamic graphs, they can be used as pattern features representing dynamic graphs or simply for gaining an insight into internal processes of the graphs.

The paper is organized as follows. Section 2 gives the necessary definitions and presents the representation of dynamic graphs. Section 3 then provides a short description of the gSpan algorithm [13], whose framework is employed in our new algorithm. In Section 4, we describe the new algorithm for graph rule mining. Experimental evaluation is presented in Section 5. Finally, related work and conclusion can be found in Section 6 and 7, respectively.

2 Predictive Graph Rules

In this section, we provide definitions of a dynamic graph and predictive rules. Definitions of significance measures support and confidence are also part of this section.

2.1 Dynamic Graphs

Before defining a dynamic graph, we need to consider the notion of a static graph. In this paper, by a static graph we will denote a directed labeled multigraph without loops and with a restriction that no two edges with the same source and target vertices can have the same label. The proposed mining algorithm, however, can work with undirected edges too. For the sake of simplicity, we will restrict ourselves only to directed graphs in this section.

Definition 1 (Static graph). *A static graph is a 5-tuple $G = (V_G, E_G, f_G, l_{G,V}, l_{G,E})$, where V_G is a set of vertices, E_G is a set of edges, $f_G : E_G \to V_G \times V_G$ is a map assigning a pair of vertices (u, v), $u \neq v$, to every edge, $l_{G,V}$ and $l_{G,E}$ are two maps describing labeling of the vertices and edges, respectively. Furthermore, $\forall e_1, e_2 \in E_G(f(e_1) = f(e_2) \Rightarrow l_E(e_1) \neq l_E(e_2))$.*

A dynamic graph is then given by a finite sequence of static graphs in which no two adjacent graphs are identical as we want to capture only the changes in the dynamic graph. Moreover, we extend each static graph G by *timestamp* functions $t_{G,V} : V_G \to T$ and $t_{G,E} : E_G \to T$, which map each vertex and edge to a point in time. We will work with discretized time and thus T will be the set of integers, i.e. $T = \mathbb{Z}$.

Definition 2 (Dynamic graph). *A dynamic graph is a finite sequence $DG = (G_1, G_2, ..., G_n)$, where G_i is a static graph extended by timestamp functions $t_{G_i,V}$, $t_{G_i,E}$ for all $1 \leq i \leq n$, and $G_j \neq G_{j+1}$ for all $1 \leq j \leq n$-1. Graph G_i is referred to as the snapshot of DG at time i. In addition, for each $1 \leq i \leq n$, the timestamp functions $t_{G_i,V}$, $t_{G_i,E}$ assign to*

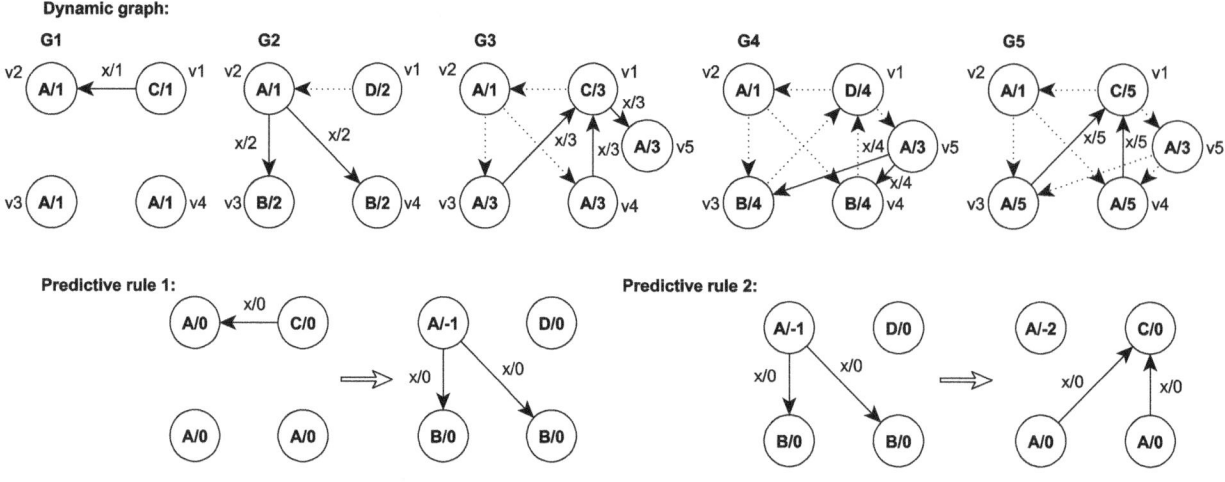

Figure 1: An example of a dynamic graph and two predictive graph rules. Numbers after slash symbols represent timestamps and dotted edges represent deleted edges.

each vertex and edge the time from which they have their current label, i.e. $t_{G_i,V}(v) = min(j|1 \le j \le i \wedge \forall k, j \le k \le i, v \in V_{G_k} \wedge l_{G_k,V}(v) = l_{G_i,V}(v))$ and similarly for $t_{G_i,E}(e)$.

As we want to capture patterns with rich information, we will also assume that the dynamic graph keeps track of the deleted vertices and edges, but only until they are added back into the graph. For example, consider the dynamic graph in Fig. 1 with five snapshots. Then $t_{G_1,V}(v_1) = 1$, $t_{G_4,V}(v_5) = 3$, etc. Also notice the edge between vertices $v3$ and $v1$ in snapshot G_3. It is deleted in snapshot G_4 but we keep information about this deleted edge in this snapshot. This information is discarded in snapshot G_5 because a new edge with exactly the same information is added there.

2.2 Predictive Rules

The aim of the mining algorithm is to find predictive graph rules, i.e. rules expressing how a subgraph of a snapshot will most likely change in future. As we want to incorporate the time information into the rules and at the same time we are interested in mining general patterns which are not tied to absolute time, we need to use relative timestamps for rules. A relative timestamp equal to 0 will denote a current change and timestamp equal to $-t$ will denote a change that happened t snapshots earlier. Now we can define predictive graph rules as follows.

Definition 3 (Predictive Graph Rule). *Let G_A, G_C be two static graphs with timestamp functions $t_{G_A,V}$, $t_{G_A,E}$, $t_{G_C,V}$, $t_{G_C,E}$ restricted to range $(-\infty, 0]$ such that union graph[1] of G_A and G_C is a connected graph and exactly one of the following conditions holds:*

i. $V_{G_A} = \emptyset \wedge V_{G_C} \ne \emptyset \wedge \forall v \in V_{G_C}(t_{G_C,V}(v) = 0) \wedge \forall e \in E_{G_C}(t_{G_C,E}(e) = 0)$

ii. $V_{G_A} \ne \emptyset \wedge V_{G_C} = \emptyset$

iii. $(V_{G_A} \cap V_{G_C} \ne \emptyset) \wedge$
$(\exists v \in V_{G_C}(t_{G_C,V}(v) = 0) \vee \exists e \in E_{G_C}(t_{G_C,E}(e) = 0)) \wedge$
$(\forall v \in V_{G_C} \smallsetminus V_{G_A}(t_{G_C,V}(v) = 0)) \wedge$
$(\forall e \in E_{G_C} \smallsetminus E_{G_A}(t_{G_C,E}(e) = 0)) \wedge$
$(\forall v \in V_{G_A} \cap V_{G_C}((t_{G_C,V}(v) = t_{G_A,V}(v) - 1 \wedge l_{G_C,V}(v) = l_{G_A,V}(v)) \vee (0 = t_{G_C,V}(v) \ge t_{G_A,V}(v) \wedge l_{G_C,V}(v) \ne l_{G_A,V}(v)))) \wedge$
$(\forall e \in E_{G_A} \cap E_{G_C}(f_{G_C}(e) = f_{G_A}(e) \wedge ((t_{G_C,E}(e) = t_{G_A,E}(e) - 1 \wedge l_{G_C,E}(e) = l_{G_A,E}(e)) \vee (0 = t_{G_C,E}(e) \ge t_{G_A,E}(e) \wedge l_{G_C,E}(e) \ne l_{G_A,E}(e)))))$

Then we say that $G_A \Rightarrow G_C$ is a predictive graph rule, where G_A is called antecedent *and G_C* consequent.

The first two conditions in the above definition cover situations in which the rules express either addition of an isolated graph into a dynamic graph or a deletion of a subgraph from a dynamic graph. The third condition covers situations in which one subgraph is transformed into another subgraph. Here, we require $V_{G_A} \cap V_{G_C} \ne \emptyset$ because we are not interested in rules consisting of unrelated graphs. In addition, we require the rule to contain at least one change related to a vertex or an edge. Vertices and edges occurring only in consequent must have timestamp equal to 0 as they represent an addition. For vertices and edges common for both graphs we require that they either were not changed a thus their relative timestamps differ by one, or they were changed and thus their timestamp cannot be lower in the consequent. Moreover, we cannot change edges by reorienting them, i.e. we would have to delete the original edge and add a new one with the opposite orientation. Lastly, as we keep track of the deleted edges and

---
[1] A graph created from union of vertices and edges.

vertices in the dynamic graph, the predictive rules can also contain these deleted vertices and edges. It does not pose any restriction to the patterns, contrariwise it can only help us capture more information in the patterns in case such information is present in the dynamic graph.

In Fig. 1 we can see two examples of graph prediction rules. Both rules depict changes in connection and also label changes. There is also a vertex with label A in both rules which is not changed and thus its timestamp is decreased by one in the consequent.

In order to select only interesting rules, various measures of significance are typically used. Here, we use support and confidence. As we use relative timestamps for graphs in rules, we also need to provide the notion of an occurrence of such a graph in the given dynamic graph.

Definition 4 (Occurrence of a rule graph). *Let G be a graph used in a rule, say in antecedent without loss of generality, with timestamp functions $t_{G,V}$, $t_{G,E}$, and let $DG = (G_1, ..., G_i, ..., G_n)$ be a dynamic graph. We say that G occurs in snapshot G_i, written as $G \sqsubseteq G_i$, if there exists a function $\varphi : V_G \rightarrow V_{G_i}$ such that:*

i. $\forall u \in V_G(\varphi(u) \in V_{G_i} \wedge l_{G,V}(u) = l_{G_i,V}(\varphi(u)) \wedge t_{G,V}(u) = t_{G_i,V}(\varphi(u)) - i)$,

ii. $\forall e \in E_G(f_G(e) = (u,v) \Rightarrow \exists! e' \in E_{G_i}(f_{G_i}(e') = (\varphi(u), \varphi(v)) \wedge l_{G,E}(e) = l_{G_i,E}(e') \wedge t_{G,E}(e) = t_{G_i,E}(e') - i)$

Definition 5 (Support and Confidence). *Let $G_A \Rightarrow G_C$ be a rule and $DG = (G_1, ..., G_i, ..., G_n)$ a dynamic graph. We define* support *of $G_A \Rightarrow G_C$,* support *of G_A, and* confidence *of $G_A \Rightarrow G_C$ as follows:*

$$\sigma_{DG}(G_A \Rightarrow G_C) = |\{i | G_A \sqsubseteq G_i, G_C \sqsubseteq G_{i+1},$$
$$1 \leq i \leq n-1\}|/(n-1)$$
$$\sigma_{DG}(G_A) = |\{i | G_A \sqsubseteq G_i, 1 \leq i \leq n-1\}|/(n-1)$$
$$conf_{DG}(G_A \Rightarrow G_C) = \sigma_{DG}(G_A \Rightarrow G_C)/\sigma_{DG}(G_A)$$

For a set of dynamic graphs $DGS = \{DG_1, DG_2, ..., DG_m\}$ we extend these definitions as follows:

$$\sigma_{DGS}(G_A \Rightarrow G_C) = |\{i | \sigma_{DG_i}(G_A \Rightarrow G_C) > 0,$$
$$1 \leq i \leq m\}|/m$$
$$\sigma_{DGS}(G_A) = |\{i | \sigma_{DG_i}(G_A) > 0, 1 \leq i \leq m\}|/m$$
$$conf_{DGS}(G_A \Rightarrow G_C) = \sigma_{DGS}(G_A \Rightarrow G_C)/\sigma_{DGS}(G_A)$$

Thus, support of a rule for a single dynamic graph expresses the fraction of snapshots that were changed by the rule. For a set of dynamic graphs, we count the fraction of dynamic graphs that had at least one snapshot changed by the rule. Confidence expresses the frequency of such a change if we observe an occurrence of the antecedent. For example, both rules in Fig. 1 have support equal to 0.25 and confidence equal to 1.

Given a minimum support value σ_{min} and a minimum confidence value $conf_{min}$, the task is to find all predictive graph rules for which $\sigma \geq \sigma_{min}$ and $conf \geq conf_{min}$.

3 gSpan Revisited

The novel algorithm for mining predictive graph rules employs the framework of the gSpan algorithm [13]. We modified and further extended this framework for the purpose of mining graph rules from dynamic graphs. First, we revise the main ideas of gSpan and then we provide the details of the new algorithm.

gSpan [13] is an algorithm for mining frequent patterns (subgraphs) from a set of simple undirected static graphs. Simple means that it does not handle multiedges. The output frequent subgraphs are connected.

gSpan starts from single-edge patterns and extends these patterns edge by edge to create larger patterns. Each such pattern can be encoded by a *DFS (Depth-First Search) code*. A DFS code of a pattern represents a specific DFS traversal of the pattern and it is represented by a list of 5-tuples $(i, j, l_i, l_{(i,j)}, l_j)$. Such a 5-tuple represents an edge between the i-th and j-th discovered vertices by the DFS traversal, l_i and l_j are labels of those vertices, and $l_{(i,j)}$ is the label of the edge. Thus, the first 5-tuple has always $i = 0$ and $j = 1$, and it holds for other 5-tuples that $i < j$ if it is a forward edge in the DFS traversal and $i > j$ if it is a backward edge.

As there are more ways the DFS traversal can be performed on a single pattern, there are also more DFS codes for each pattern. A lexicographic order is defined on DFS codes and the minimum one is maintained for each pattern. This lexicographic order is also applied on codes of different patterns to represent the search space as a tree, called *DFS Code Tree*. In this DFS Code Tree, each vertex represents one DFS code and children of a vertex can be obtained by all possible single-edge extensions of the vertex. Therefore all codes on the same level of the tree have the same number of edges. Moreover, children of a vertex are ordered according to the lexicographic order. gSpan generates larger patterns in such a way that it corresponds to a depth-first search traversal of this DFS Code Tree, i.e. it generates patterns according to the lexicographic order. gSpan does not have to extend each pattern in all possible ways, it is enough to grow edges only from vertices on the rightmost path[2]. Specifically, it grows either a backward edge from the rightmost vertex[3] to another vertex on the rightmost path or a forward edge from a vertex on the rightmost path to a newly introduced vertex. When traversing the search space, gSpan checks whether the pattern of the considered DFS code is frequent. If not, it prunes the search space tree on this vertex and backtracks. This is possible because of the anti-monotonicity of the support measure. It also checks whether the considered code is the minimum one for the corresponding pattern. If it is not minimum, the search space tree is pruned on this vertex because all patterns in this pruned subtree were already found earlier.

---

[2]The *rightmost path* is given by the DFS code and it is the path from the root to the lastly discovered vertex by the DFS traversal.

[3]The last vertex on the rightmost path from root.

Algorithm 1 gSpan($\mathbb{D},\mathbb{S}$)

1: sort the labels in $\mathbb{D}$ by their frequency;
2: remove infrequent vertices and edges;
3: relabel the remaining vertices and edges;
4: $\mathbb{S}^1 \leftarrow$ all frequent 1-edge graphs in $\mathbb{D}$;
5: Sort $\mathbb{S}^1$ in DFS lexicographic order;
6: $\mathbb{S} \leftarrow \mathbb{S}^1$;
7: **for each** edge $e \in \mathbb{S}^1$ **do**
8:      initialize s with e, set $s.D$ by graphs containing e;
9:      Subgraph_Mining($\mathbb{D},\mathbb{S},s$);
10:     $\mathbb{D} \leftarrow \mathbb{D} - e$;
11:     **if** $|\mathbb{D}| < \sigma_{min}$ **then**
12:         **break**;

Algorithm 2 Subgraph_Mining($\mathbb{D},\mathbb{S},s$)

1: **if** $s \neq min(s)$ **then**
2:      **return**;
3: $\mathbb{S} \leftarrow \mathbb{S} \cup \{s\}$;
4: enumerate s in each graph in $\mathbb{D}$ and count its children;
5: **for each** c, c is s' child **do**
6:      **if** $\sigma(c) \geq \sigma_{min}$ **then**
7:          $s \leftarrow c$;
8:          Subgraph_Mining($\mathbb{D}_s,\mathbb{S},s$);

The pseudocode of gSpan is given in Algorithm 1. By removing the infrequent vertices and edges, the input graphs can be significantly reduced and the overall efficiency increased. Frequent vertices are appended to results as the smallest frequent patterns. The main part of the algorithm starts from single-edge patterns. Specifically, Subgraph_Mining 2 procedure is recursively called on each such pattern. This procedure first tests whether the code s is minimum. If it is minimum, it enumerates its children by taking single-edge extensions. The procedure is then called on the frequent children.

4   DGRMiner Algorithm

In this section we describe the new algorithm called DGR-Miner. It is based on the framework of gSpan, however, the framework is modified and extended. First, we provide necessary details about main modifications used in DGR-Miner and then we present the pseudocode of the whole algorithm with description of remaining building blocks.

4.1   Static Representation of the Dynamic Graphs by Union Graphs

The first step is a transformation of an input dynamic graph[4] into a data structure that can be considered as a set of static graphs. The idea is that we are able to represent the graph rules by single graphs and the input dynamic

---

graph as a set of static graphs in such a way that a modified static subgraph mining algorithm can be employed.

Let us start with the transformation of rules. In order to create a single graph from a rule, we take the union of the vertices and edges from its antecedent and consequent. Edges and vertices that do not represent any change in the rule will keep their labels and timestamps from the consequent. Let us remind that rules have relative timestamps less than or equal to 0 and thus these edges and vertices will have timestamps less than 0. Edges and vertices representing addition will keep their consequent timestamp, which is 0, but their labels will contain flag representing addition, for example label A will be changed to $+A$. However, timestamps of vertices and edges that were deleted or relabeled will have timestamps that are opposites of the antecedent timestamps. We know that consequent timestamps of such changes are equal to 0 so we can easily get the original value of the antecedent. We take the opposite values because later it will help us recognize current changes simply by timestamps greater than or equal to 0. Vertices and edges that were deleted or relabeled will also have new labels that can be easily decoded, for example $-A$ for deletion of an object with label A and $A => B$ for relabeling from A to B. Transformed rules from Fig. 1 are shown in Fig. 2.

Transformation of the input dynamic graph is very similar. Suppose that we have n snapshots in the dynamic graph. As the first snapshot does not represent any changes by itself, we create $n - 1$ new graphs in the following way. When creating the k-th graph, consider union of vertices and edges from snapshots 1 to k as an antecedent, where vertices and edges have their last assigned labels and timestamps of last changes relative to k. We can assume that all vertices and edges from the first snapshot had timestamps equal to 1. Similarly, use snapshots 1 to $k + 1$ to create a consequent. Then we use the method for rule transformation to create the k-th graph. All $n - 1$ graphs can be computed in a single pass as we can update the i-th graph to get the $(i + 1)$-th one.

Union of all graphs from the beginning may contain vertices and edges with very old changes that are not useful for the predictive rules. We use a window parameter to remove such vertices and edges from the union graphs. As edges cannot exist without their adjacent vertices, we do not remove old vertices adjacent to edges that are not old. Union graphs of the dynamic graph from Fig. 1 are shown in Fig. 2. In this case, window size is not set.

4.2   Modified DFS Code

Now that we have the dynamic graph represented by union graphs, which can be viewed as a set of static graphs, we made a large step towards mining the graph rules. There are, however, still several issues to be addressed.

Let us start with a richer representation of edges. In Section 3, we showed that gSpan uses 5-tuples of the

---

[4]Here, we assume that there is only one dynamic graph on the input. Extension to a set of dynamic graphs is described in Subsection 4.4.

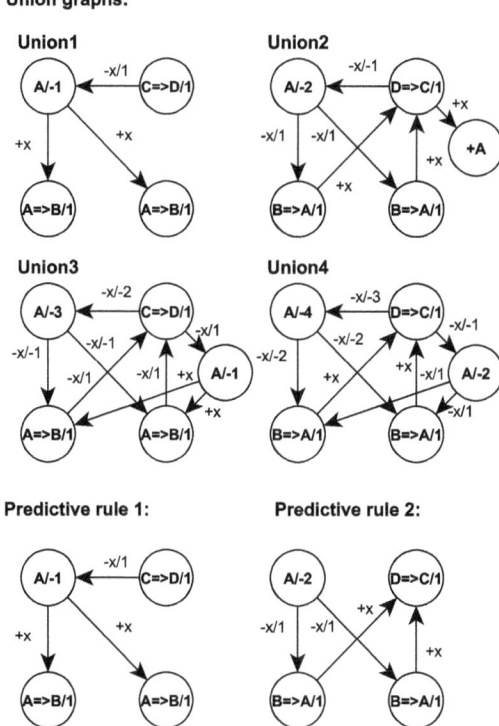

Figure 2: The union graph representation of the dynamic graph and the rules from Fig. 1.

The first method helps us to ignore timestamps of vertices. Specifically, we apply the signum function to all relative timestamps of vertices. Thus, negative timestamps become equal to -1 and positive timestamps become equal to 1. This method is useful for dynamic graphs in which all or almost all changes are caused by edges and vertices remain more or less intact.

The second method also uses the signum function but now it converts timestamps of both vertices and edges. It is useful in situations where patterns in dynamic graphs are very diverse and it is not possible to find many frequent patterns with exact timestamps.

4.4 The Complete DGRMiner

In this section we provide remaining details and a description of the whole DGRMiner algorithm for predictive graph rule mining. The pseudocode of DGRMiner is given in Algorithm 3.

First, DGRMiner converts the input dynamic graph into a set of union graphs as described in Section 4.1. In the case of a set of dynamic graphs, the algorithm simply computes union graphs for each one of them and then concatenates the results. It only needs to keep the mapping of those union graphs into the original dynamic graphs in order to be able to compute their support and confidence correctly. Optional application of an abstraction method, described in Section 4.3, follows next. Then the algorithm removes infrequent vertices and edges but only those that represent changes as the other ones may be used later for confidence computation. When computing frequencies, it takes labels, timestamps, and edge orientations into account. Before moving to single-edge patterns, DGRMiner outputs frequent single-vertex patterns with high enough confidence. To compute confidence, it needs to decode antecedents of the patterns and then compute their support. After that, the algorithm takes frequent initial edges and sort them according to the extended version of the DFS lexicographic order of gSpan. An *initial* edge is such an edge that represents a change or at least one of its vertices does.

Now for each initial edge we recursively call subprocedure DGR_Subgraph_Mining, which searches for patterns growing from a given initial edge. This subprocedure, described in Algorithm 4, uses several arguments. s denotes the current pattern, which is represented by its DFS Code. In $\mathbb{D}$ and $\mathbb{A}$ we keep union graphs in which current pattern and its antecedent can be found. Finally, when growing patterns from the i-th initial edge, we keep the first i initial edges in $\mathbb{E}_{start}$. This last argument is used in function *min*, which can be found in the first line of Algorithm 4. The purpose of this function is to check whether the DFS code of the given pattern is minimum, i.e. it was not found earlier when traversing the search space. Because all patterns grow only from the initial edges $\mathbb{S}^1$, it is enough to check whether we cannot represent the current pattern by a smaller DFS Code which starts by one of the edges in

form $(i, j, l_i, l_{(i,j)}, l_j)$ to represent edges of patterns. In order to incorporate relative timestamps of rules and orientation of the edges, we simply extend these 5-tuples to 9-tuples of the form $(i, j, l_i, t_i, d_{(i,j)}, l_{(i,j)}, t_{(i,j)}, l_j, t_j)$. It is the same as the original 5-tuple except for the new elements. Specifically, t_i, t_j, $t_{(i,j)}$ are used for the relative timestamps of vertex i, vertex j, and the edge between i and j. Element $d_{(i,j)}$ represents the orientation of the edge between i and j, and it is one of the following: $\leftarrow$, $\rightarrow$, $-$. The last one is used for undirected edges. Each pattern, i.e. graph rule in the condensed representation, can be represented as a list of such 9-tuples. Furthermore, it is easy to extend the gSpan's DFS Code for these 9-tuples and thus we can create ordering between patterns and find a minimum DFS Code for each pattern. For example, suppose that we obtained the following order of vertex labels: A, +x, C=>D, A=>B, -x. Then the minimum code of the Predictive Rule 1 from Fig. 2 is $(0, 1, A, -1, \rightarrow, +x, 0, A => B, 1), (0, 2, A, -1, \rightarrow, +x, 0, A => B, 1), (0, 3, A, -1, \leftarrow, -x, 1, C => D, 1)$.

4.3 Time Abstraction

In order to be able to deal with a broader class of dynamic graphs, we extended the mining algorithm to include two time abstraction methods. By time abstraction we mean usage of coarser timestamp values of union graphs in situations where exact values are not required or suitable.

Algorithm 3 DGRMiner($\mathbb{DG}$)

1: convert the input dynamic graph(s) $\mathbb{DG}$ into the union graph representation $\mathbb{D}$;
2: optional: apply a time abstraction method on union graphs;
3: remove infrequent vertices and edges;
4: output frequent change vertices with high enough confidence;
5: $\mathbb{S}^1 \leftarrow$ all frequent initial edges in $\mathbb{D}$ sorted in DFS lexicographic order;
6: **for** $i \leftarrow 1$ **to** $|\mathbb{S}^1|$ **do**
7:    $p \leftarrow i$-th edge from $\mathbb{S}^1$
8:    $p.D \leftarrow$ graphs which contain p;
9:    $p.A \leftarrow$ graphs which contain antecedent of p;
10:   $\mathbb{E}_{start} \leftarrow$ first i edges from $\mathbb{S}^1$
11:   DGR_Subgraph_Mining($p,p.D,p.A,\mathbb{E}_{start}$);

Algorithm 4 DGR_Subgraph_Mining($s,\mathbb{D},\mathbb{A},\mathbb{E}_{start}$)

1: **if** $s \neq min(s,\mathbb{E}_{start})$ **then**
2:    **return**;
3: enumerate s in each graph in $\mathbb{D}$ and count its children;
4: remove children of s which are infrequent;
5: enumerate antecedent of s in graphs given by $\mathbb{A}$;
6: set s.A by graphs which contain antecedent of s;
7: $conf \leftarrow$ confidence of s;
8: **if** $conf \geq conf_{min}$ **then**
9:    output s;
10: sort remaining children in DFS lexicographic order;
11: **for each** child c **do**
12:    DGR_Subgraph_Mining($c,c.D,s.A,\mathbb{E}_{start}$);

$\mathbb{E}_{start}$. If we can find such a smaller code, then the pattern must have been discovered earlier a thus we backtrack.

If the code is minimum, we continue by enumerating the pattern in relevant graphs given by $\mathbb{D}$ and searching for its children candidates. This step is similar to the one in gSpan. Also, all infrequent children are removed. Before saving the current pattern, we need to compute its confidence. As we described in Section 4.1, we are able to extract the antecedent from the current pattern and then count its occurrences. Set $\mathbb{A}$ represents a set of candidate graphs, in which we should search for the antecedent occurrences. The actual set of graphs containing the antecedent is then saved to $s.A$, where $s.A \subseteq \mathbb{A}$, and it used as the $\mathbb{A}$ set for the pattern's children.

Before recursive processing of the children of the current pattern, we need to sort the children according to the extended version of the DFS lexicographic order. Set $c.D$ was created when the pattern s was enumerated and its children were counted.

5 Experiments

In this section we present results of experiments on three datasets. All the experiments were conducted by a C++

Dataset	Dynamic graphs	Snapshots
ENRON	1	895
RESOLUTION	103	2911
SYNTH	1	101
SYNTH 20	20	2020

Table 1: Datasets used for experiments.

Rank	Vertex label
Employee	Emp
Vice President	VP
Director	Dir
President	Pres
Manager	Man
Trader	Trad
CEO	CEO
Managing Director Legal Department	MDLP
In House Lawyer	Law
Managing Director	MD

Table 2: Ranks of employees in the Enron dataset and the corresponding vertex labels.

implementation of DGRMiner on a PC equipped with CPU Intel i5-4570, 3.2GHz, 16GB of main memory, and running 64-bit version of Windows 8.1. For all experiments, we set $conf_{min} = 0$ and window size for union graphs equal to 10.

5.1 Enron

The first dataset used in experiments is based on the email correspondence in the Enron company [3]. For our experiments we used preprocessed version of this email traffic [10]. Specifically, we used the data containing information about time, sender, receiver, and LDC topic. From the set of senders and receivers we created vertices of our dynamic graph. These vertices do not change through time. Each email message sent between a sender and a receiver is represented by addition of a directed edge in the dynamic graph. If the graph already contains the same edge, we just update its time of addition. Snapshots of the dynamic graph corresponds to days. As there were messages with anomalous dates, we removed all messages sent before 1998 and got 894 days of activity. With one extra day for vertex initialization we got 895 snapshots, as can be seen in Table 1. We used LDC topics as edge labels. There are 32 regular LDC topics expressing the topics of the messages plus two special topics used to label outlier messages and messages with non-matching topic. We also used rank of employees [10] to label vertices. Vertices with unknown rank were removed from the graph and thus only 130 vertices remained. Ranks and corresponding labels are available in Table 2.

Results on ENRON with $\sigma_{min} = 0.1$ can be found in the first row in Table 3. We decided to apply the time abstrac-

Dataset	Union vertices	Union edges	σ_{min}	$conf_{min}$	Time abstraction		1-vertex rules	All rules	Running time (sec)
					vertices	all			
ENRON	46290	182720	0.10	0	✓	✗	0	187	24.6
ENRON DEL	47612	196653	0.10	0	✓	✗	0	233	29.3
RESOLUTION	15966	5275	0.05	0	✗	✗	26	36	0.3
RESOLUTION	15966	5275	0.05	0	✗	✓	17	321	3.4
SYNTH	1124	2404	0.10	0	✗	✓	6	82	0.3
SYNTH 20	31455	52112	0.10	0	✗	✗	121	1604	50.9

Table 3: Results of experiments. Number of union vertices and edges is taken over all union graphs of the given dataset. 1-vertex rules are rules whose union graph consists of only one vertex. Running time is averaged over five runs. For all experiments we set window of size 10 when building union graphs.

tion method for vertices because they are only added in the first snapshot and never changed. We found 187 frequent rules, none of which was a single-vertex rule.

We also modified the previous dataset by deleting edges which were not updated immediately the next day. This modified dataset is named ENRON DEL in Table 3. The change allows us to capture patterns which could not be captured only by edge additions. Examples of two rules from this dataset are shown in Fig. 3.

5.2 Resolution Proofs in Propositional Logic

We used the set of graphs representing resolutions in propositional logic from [11] as the second dataset. Vertices of these graphs contain lists of literals in propositional logic. All edges are directed and have the same label in these graphs. These dynamic graphs are evolving by vertex and edge addition or deletion, and by change of vertex labels. Time of these events was transformed into a discrete sequence. Because there were 19 different assignments in total, the dynamic graphs had quite distinct vertex labels. In order to find frequent patterns, we restricted the dataset to only one assignment. Specifically, we took the assignment with the greatest number of solutions. This set of graphs contained 103 dynamic graphs with 2911 snapshots in total, see RESOLUTION dataset in Table 1. The initial snapshot of each dynamic graph in an empty graph. For more details about this data refer to [11].

We conducted two experiments on this dataset, both with $\sigma_{min} = 0.05$ as there were not many frequent patterns. One with no time abstraction, and one with time abstraction of both vertices and edges. From Table 3 we can see that the time abstraction helped us to find ten times more frequent rules for the same value of the minimum support. Furthermore, most of the rules were 1-vertex rules when the abstraction was not applied. Such rules capture vertex additions, deletions and relabelings without any context and may not be very informative.

5.3 Synthetic Datasets

Besides real-world data, we also tested our method on a synthetic datasets. One of this dataset, SYNTH in Table 1, was generated in the following way. First, a graph

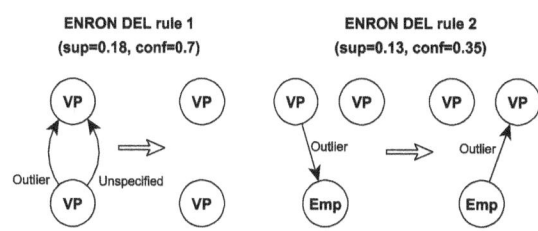

Figure 3: Examples of two rules from ENRON DEL.

with 10 vertices and 20 randomly assigned edges was created. Then we iteratively built 100 snapshots, each snapshot from the previous one by randomly chosen changes. The number of changes ranged uniformly 0–1 for vertex deletion, 0–1 for edge deletion, 0–1 for vertex addition, 0–3 for edge addition, 0–2 for vertex label change, and 0–2 for edge label change. All newly selected vertex (edge) labels were chosen from a uniform distribution over set $\{A, B\}$ ($\{y, z\}$). Each new snapshot had to be different from the previous one. In order to keep approximately the same number of vertices (edges) through time, additions or deletions of vertices (edges) were suspended if the number of vertices (edges) was not in the $[k/2, 2k]$ interval, where k = 10 (and k = 20 for edges). The second dataset, SYNTH 20, was created from 20 dynamic graphs, each one of them built by the process just described.

For SYNTH and $\sigma_{min} = 0.1$, the time abstraction of both vertices and edges was applied because there were almost no frequent patterns without the abstraction. On the other hand, experiments on SYNTH 20 with 20 dynamic graphs did not require any time abstraction and approximately 1600 frequent rules were found for the same value of the minimum support. This suggests that the support definition for a single dynamic graph is stricter than the one for a set of graphs.

6 Related Work

Several algorithms have been proposed for graph rule mining. However, a lot of these algorithms are expecting that the only changes in a dynamic graph are caused by edge additions, or by vertex additions if the vertices belong to the edges being added. These algorithms include GERM [1], LFR-Miner [7], and the algorithms presented in [5, 9]. These algorithms also pose various restrictions on the form of the rule graphs.

The work most related to ours is probably the algorithm for mining interesting patterns and rules proposed in [8]. This algorithm allows multiedges and also performs similar time abstraction, however, it also supposes that labels do not change and rules express only edge addition. Moreover, antecedents of the rules have to be connected. This is a stricter requirement than the one given by our algorithm, which requires only the union graphs of rules to be connected.

Predictive graph rules can be also seen as subgraph sequences of length two. Mining of subgraph subsequences from a set of graph sequences (dynamic graphs) is considered in [4], where algorithm FRISSMiner is proposed. This algorithm also works with union graphs, however, it builds the union graph from the whole sequence, i.e. the whole dynamic graph. FRISSMiner allows all types of changes in the input graph sequences but the subgraph sequences are not required to include the changes and thus user would still need to search for patterns representing changes in graphs. In addition, the algorithm is not designed for a single-dynamic-graph scenario.

Dynamic GREW [2] and the algorithm from [12] mine also patterns capturing information from several snapshots. They assume that the input dynamic graph has a fixed set of vertices, and edges are inserted and deleted over time, which makes them quite restricted.

Except for FRISSMiner, the algorithms presented in this section are designed only for the single-dynamic-graph scenario. On the contrary, FRISSMiner is designed only for the set-of-dynamic-graphs scenario.

Experimental comparison of the above algorithms with DGRMiner is difficult due to the fact that each algorithm mines different type of patterns. Also, different definitions of support and confidence affect which patterns are frequent. For example, GERM computes absolute support and counts multiple occurrences of a pattern in a snapshot. On the other hand, DGRMiner computes support of a pattern as a fraction of snapshots containing at least one occurrence of the pattern. Another difficulty arises from the fact that implementations of the algorithms, with the exception of GERM, are not freely available for download. Lastly, the datasets used for experimental evaluation of the above algorithms are often either not available or it is not possible to reproduce the same data.

7 Conclusion

We proposed a new algorithm DGRMiner for mining frequent predictive graph rules from both single dynamic graphs and sets of dynamic graphs. This algorithm is able to capture various changes in dynamic graphs. Edges in dynamic graphs are allowed to be directed or undirected multiedges. DGRMiner uses support and confidence as significance measures of the rules. Such graph rules are useful for prediction in dynamic graphs, they can be used as pattern features representing dynamic graphs or simply for gaining an insight into internal processes of the graphs. We evaluated the algorithm on two real-world and two synthetic datasets.

As future work, we plan to further investigate time abstraction methods and other ways of support computation. We also intend to modify the method for mining frequent closed patterns.

Acknowledgments. We would like to thank to Luboš Popelínský and other members of KDLab FI MU for their help and comments.

References

[1] Berlingerio, M., Bonchi, F., Bringmann, B., Gionis, A.: Mining graph evolution rules. In: ECML PKDD'09. Springer-Verlag, Berlin, Heidelberg, 2009, 115–130

[2] Borgwardt, K. M., Kriegel, H. -P., Wackersreuther, P.: Pattern mining in frequent dynamic subgraphs. In: IEEE ICDM'06. Washington, DC, USA, 2006, 818–822

[3] Cohen, W. W.: Enron email dataset. Web. Accessed 2 June 2015. urlhttps://www.cs.cmu.edu/~./enron/.

[4] Inokuchi, A., Washio, T.: Mining frequent graph sequence patterns induced by vertices. In: SIAM SDM, 2010.

[5] Kabutoya, Y., Nishida, K., Fujimura, K.: Dynamic network motifs: evolutionary patterns of substructures in complex networks. In: Proceedings of the APWeb'11. Springer-Verlag, Berlin, Heidelberg, 2011, 321–326

[6] Leskovec, J., Kleinberg, J., Faloutsos, C.: Graphs over time: densification laws, shrinking diameters and possible explanations. In: ACM KDD'05, New York, NY, USA, 2005.

[7] Leung, C. W. -K., Lim, E. -P., Lo, D., Weng, J.: Mining interesting link formation rules in social networks. In: ACM CIKM'10, New York, NY, USA, 2010, 209–218

[8] Miyoshi, Y., Ozaki, T., Ohkawa, T.: Mining interesting patterns and rules in a time-evolving graph. In: IMECS'11, Vol. I, March 16–18, 2011.

[9] Ozaki, T., Etoh, M.: Correlation and contrast link formation patterns in a time evolving graph. In: IEEE ICDMW'11, Washington, DC, USA, 2011, 1147–1154

[10] Priebe, C. E., Conroy, J. M. , Marchette, D. J., Park, Y.: Scan statistics on Enron graphs. Web. Accessed 8 June 2015. <http://www.cis.jhu.edu/ parky/Enron>

[11] Vaculík, K., Nezvalová, L., Popelínský, L.: Graph mining and outlier detection meet logic proof tutoring. In: G-EDM 2014, London, CEUR-WS.org, 2014, 43–50

[12] Wackersreuther, B., Wackersreuther, P., Oswald, A., Böhm, C., Borgwardt, K. M.: Frequent subgraph discovery in dynamic networks. In: MLG'10, ACM, New York, NY, USA, 2010, 155–162

[13] Yan, X., Han, J.: gSpan: Graph-based substructure pattern mining. In: IEEE ICDM'02, Washington, DC, USA, 2002.

Extension of Business Rule Sets Using Data Mining of GUHA Association Rules

Stanislav Vojíř

Department of Information and Knowledge Engineering
University of Economics, Prague
W. Churchill Sq. 4, Prague 3, 130 67, Czech Republic

Abstract. *The following paper is intended to introduce three suitable ways of using data mining of GUHA association rules in conjunction with existing set of business rules. The integration can be realized using full integration, as black box classification model and also using dynamic integration with data mining system. These ways are illustrated by demo use case based on data from a health insurance company.*

1 Introduction

Business rules are not only an effective way for modeling of business structure and descriptions of operations, definitions and constrains in an organization, but also an efficient way for separation of business logic from the application code of information systems. The separation of business logic, mainly "decision-making points" from the implementation of applications is very important, especially in today's rapidly changing world. For this reason, it can be observed an increasing number of applications of rule engines and business rules system.

In this paper, the presented approach of extension of a business rules base is illustrated using examples from a health insurance company. From this domain, examples of business rule could be: *"If the doctor has specialization 001, then the diagnosis AAA is OK."* or *"The child emergency cannot treat the adult patients."* Such rules are usually saved and managed by a business rule management system. The rule set in conjunction with the related terms dictionary can be called "knowledge base".

However, the applicability of the rule-based systems greatly depends on the complexity and completeness of their knowledge base. In addition to the manual input of business rules by domain experts, there have been discovered also some methods of obtaining business rules from the business data – for example from unstructured texts or from operational data store of the company. A suitable method for "learning" of business rules from the working or historical business data is application of data mining methods and reusage of the gained data mining models.

1.1 Related Work

From the relevant works and papers, the "semi-automatic learning of business rules" has been a subject of research activities for relatively long period. But there are still not too many real applications. The most relevant existing application of "data mining of business rules" is the component *RuleLearner*, which is a part of the business rules system *OpenRules*.[1] This system works with knowledge base in the form of decision tables in Excel worksheets. According to the information from the company OpenRules, Inc., the component RuleLearner is still non-public. It is based on data mining using open

source system Weka, but the conversion from data mining results to the form of classification tables for the OpenRules system can be realized only by experts from the authors´ company.

1.2 Business Rules

In this paper, the author describes three suitable ways of direct integration of data mining results into an existing business rule set. *Business rules* is not the name of one specification or system. The term "business rules" covers the relatively great area of rule-based systems and applications. It is mainly the name of modeling approach. In this approach, the modeling of the business behavior and decisions leads from the definition of basic entities and terms to the definition of standalone business rules. These rules are collected info rule sets in one complex knowledge base of the company.

The business rules approach has been applied in many specifications of languages for definition of business rules. The specifications can be divided by their main focus in two groups – specifications suitable for inference engines and specifications suitable for sharing of knowledge in human-friendly form. The work presented in this paper is more suitable for implementation in automatic inference (business rules) engines – JBoss Drools, Jess, Jena etc. The execution component takes the set of business rules and the base of facts, evaluates the conditions of business rules and activates the proprietary rules.

1.3 GUHA Association Rules

One of the possible and suitable methods for extension of knowledge base in the form of business rules is the application of data mining methods on the historical data of the company. It seems that the suitable data mining models are association and decision rules. The association rules can be discovered not only using the mostly known algorithm *APRIORI*, but also using the procedure *ASSOC* of the *GUHA method.*[1]

The GUHA method is original Czech data mining method for data mining of association rules with "rich semantic". The basic form of GUHA association rules is

$$\varphi \approx \psi$$

where φ *(antecedent),* ψ *(consequent*[2]*)* and possibly are logical combinations of attributes (with concrete values) and $\approx$ is the quantifier – function defined on the four feet table. Examples of the 4ft-quantifiers are *founded*

---

[1] In this paper, the rules founded using application of GUHA procedure ASSOC are called „GUHA association rules".

[2] In the GUHA method, *consequent* is called *succedent*

implication (combination of interest measures *confidence* and *support*) and *above average dependence* (this quantifier is convertible to the combination of interest measures *lift* and *support*). [5]

The GUHA association rules for the approaches presented in this paper are discovered using the data mining system LISp-Miner.[3] This software supports data mining of GUHA association rules also with the "dynamic binning of values in attributes". This feature extends the pattern of requested association rules (task definition). The attributes can contain the set of values – for example the rule attribute *age([0;1),[1;5))* is interpreted as age in interval from 0 to 5 years (without the request for redefinition of the data preprocessing). The dynamic binning can be defined as subsets of the given length, left or right cuts, intervals etc.

An example of the founded GUHA association rule:

age([20;40]) & city(Prague) & clinic(A, B) →
procedure(C) | confidence 0.6, support 0.01

The interpretation of this rule: If the age is in the interval from 20 to 30 years, city is Prague and the clinic is A or B, then the applied procedure is C. The confidence of this rule is 60% and support is 1%.

1.4 Structure of this Paper

This work is focused on the use of association rules obtained by application of *GUHA method* (below in text called "GUHA association rules"), but the principles are generalizable also for the usage of simpler association rules obtained using the algorithm APRIORI (for example in the system R). This paper follows the previous work of preparation classification business rule sets using GUHA association rules [2] and is also related to currently solved TAČR project TA04011691 "Automated extraction of business rules with feedback" [3].

The paper is organized as follows. Section 2 gives a walk through three suitable models of integration data mining model into business rule set. Section 3 contains example use cases motivated by real data. The conclusion summarizes the paper and outline for future work.

2 Integration of Data Mining Models into Existing Business Rule Set

Within this section, there are described three model ways of integration GUHA association rules into an existing business rule set. The suitability of their use differs according to the requested level of the integration and also to the analytical questing solved with the data mining task. All these ways are fully implementable (and have been practically verified) using business rule engine JBoss Drools [4] and data mining system LISp-Miner [5].

2.1 Direct Ttransformation of GUHA Association Rules into Business Rules

First variant of the involvement of founded association rules into an existing business rule set is the *direct transformation of them*. Within this transformation, every founded GUHA association rule is transformed into a separate business rule. From the GUHA association rule, antecedent and condition parts are transformed into condition of the business rule, consequent[4] of the association rule is "implemented" in the body of the business rule. The body of the business rule executes the requested action – returns the result of the classification task in suitable form (set of attributes with values, adds new data in the base of facts etc.) For this transformation, some constrains of the solved data mining tasks has to been considered.

Antecedent, condition and even consequent of a GUHA association rule can consists from multiple "partial cedents" (brackets in logical representation), containing conjunctions, disjunctions and negations. In case of mining using LISp-Miner system, every attribute in the rule can also contain multiple values, connected during the mining process using the "dynamic binning" feature. For the possibility of transformation from association rules to business rules, it is not necessary to apply any limits or constrains to antecedent and condition part of association rules. However, it is necessary to solve the *problem of the data dictionary*. The data dictionary has to be mapped to shared terms dictionary used in organization. If the data mining process has been initialized using data from operational data store of the organization, it is possible to use the default names of data attributes (columns) in the operational data store as the terms dictionary for definition of business rules.[5]

From the perspective of transformation to the form of business rules for the system JBoss Drools, condition of the rule can consist from logical expressions similar to native java code. The transformation consists from these steps:

1. Perform *reverse preprocessing* of used data. In data mining process it is common to prepare attributes from the data columns from the original data matrix. These attributes have different names and preprocessed values (during the preprocessing phase of data mining process, the original data values are grouped into named sets or intervals of original data values). The transformation itemizes the attributes included in association rules to the original names and values.

2. Remove unnecessary cedents from antecedent and condition part of GUHA association rule – because of the data mining task configuration and LISp-Miner export, the GUHA association rules saved in PMML[6] form often contain unnecessary partial cedents (multiple brackets without any added logical expression).

3. Transform antecedent and condition of every GUHA association rule into condition of a business rule. Dependently on the handling method of null values in the data set for data mining task, negation in association rule can be interpreted as *inequality* or

---

[3] http://lisp-miner.vse.cz

[4] In GUHA method is „consequent" called „succedent".

[5] Alternativelly in the organization maybe exists a mapping for data attributes from operational data store to an ontology or other "terms dictionary".

[6] Predictive Model Markup Language – XML-based format (technical standard) for saving of data mining models; developer by Data Mining Group

negation of the checking condition. For preparation of a classification business rule set, it is more suitable to use the interpretation as inequality (by testing results). Negation in association rule expression should be interpreted as *inequality*. In case of mining of GUHA association rules with condition, the condition can be appended to antecedent part (using conjunction), or could be interpreted as group condition for conditioned subset of business rules.

4. Prepare business rules´ bodies from the consequents association rules cedents. Semiautomatic acquisition of business rules from data mining results is suitable for solving of "classification" tasks. These tasks cannot return value of one "result" attribute. The limitations of consequent of the association rules for following automatic processing of results are as follows: Each consequent should contain one or more attributes with values, which were not preprocessed in data mining process. In case of more attributes in consequent part of association rule, these attributes should be connected within conjunction.

5. Use requested conflict resolution strategy.

Business rules in DRL form (format suitable for JBoss Drools) are based on Java classes, which represents the terminological dictionary. For support of solving classification problems using association rules, in most cases it is necessary to select the best result consequent (the resulting recommendation) in case of more business rules with matching antecedent/condition. Good conflict resolution strategy is to prefer classification rules with better values of confidence, support and shorter condition.[6] In DRL, the suitable strategy is implemented in one conflict resolution function written in DRL.

The result of recommendation/classification task can be processed with other part of information system of the organization, or can be processed with other business rules. Based on testing use cases, it can be said, that the following processing of the results using other business rules contributes to the clarity of the full knowledge base of the organization. From the perspective of knowledge management in organization in context of business rules, it is appropriate to build one shared knowledge base in form of business rules based on one shared terms dictionary. [7]

In implementation using JBoss Drools, it is suitable to (temporarily) insert results of classification subtask into the base of facts and continue in the business rules execution.

Great advantage of the transformation of each one association rules into a separate business rules is the possibility of their subsequent management and administration using tools from the business rules management system. It is easy to edit these rules, their priority and behavior.

In case of automatic transfer of the complete results of data mining of GUHA association rules into business rules, there can be also found some disadvantages. First big disadvantage of full integration is a large increase of the number of business rules. For solving of classification tasks using association rules without pruning algorithms, it is suitable to use data mining tasks with a really low requested minimal threshold value of support. Such tasks,

however, return a lot of founded rules (possibly thousands of rules). In case of their integration into the main knowledge base, it is appropriate to identify these rules with specific "tag".

In terms of practical evaluation, the options of this model of integration were verified in [2] and [8]. It is suitable to generate business rule set in DRL form from GUHA association rules. The classifier obtained by this method can achieve even better results than reference classifiers. [2] According to realized tests, dependently on the solved data set, the greater "expression language" of GUHA association rules can contribute to better results (but at the cost of more rules).

2.2 Black Box Classification Component

The second suitable variant for the inclusion of data mining results into an existing knowledge base in form of business rule set is the integration as "black box". In this way, the connected component is suitable for solving of classification tasks. The integration schema should be as follows:

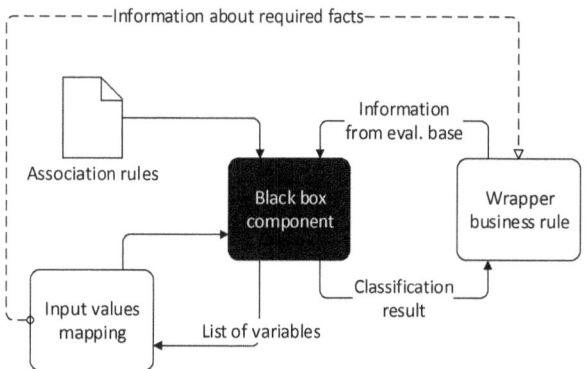

Fig. 1. Schema of black box component integration.

The user connects the black box component as one "part" of the knowledge base. It can be connected to body of a business rule, or as a partial condition. In the definition phase of the issued business rule, the user has to follow steps of a simple connection process:

1. Select results of a data mining task and export them into a standardized form (usually PMML).

2. Define *wrapper rule* – one rule, which initializes the evaluation of a classification black box component.

3. Import data mining results into the classification black box component. The component checks the structure of the uploaded model and detects all connecting points. The connecting points could be defined as input, output or shared. Input connecting points should include a definition of mapping between facts in the evaluation base and attributes used in conditions of the classification model.

4. Define mapping for the connecting points: In case of classification model based on GUHA association rules, the user defines 1:1 mapping between attributes used in antecedents and conditions of association rules and fields from the terms dictionary, the output connecting point is usually a variable for the result of

classification. The result variable could be immediately captured and processed in the wrapper business rule, or added into the evaluation base of facts used in the inference algorithm. For all the mappings, the black box component detects required data types for individual attributes and checks the mapping at least on the level of data type, at best on level of the definition range.

The involvement of a data mining model as the black box component brings many benefits. This way of integration has the lowest requirements for interaction with other rules in the knowledge base and it is applicable not only for data mining models consisting of rules but also for other suitable types of data mining models. For example, there can be considered decision trees or neural networks, too.

From the perspective of management or domain experts, this integration does not have too big impact on other business rules saved in the knowledge base. It is really easy interpretable: "In the condition of this rule matches the characteristic of client, the body of the rule returns the statistically most probable next offer for the client." The "most probable next offer" is determined with the black box component, so the management expert does not have to know the hidden algorithms used for this recommendation.

This integration has also disadvantages. The most of them is the problematic of "recycling" of specified data mining models for usage in more business rules. The data mining model is usually connected at only one point (in the black box component integrated in wrapper rule"). In case of usage models based on rules is a disadvantage also the exclusion of the evaluation of contained rules out of the main RETE network.[7]

In case of implementation of the black box component in the system JBoss Drools, it is possible to use external implementation in Java code, or implementation using separated, conditioned subset of business rules, which is evaluated only "on demand" (separated with special condition).

This model of integration has recently been implemented in TAČR project mentioned in Introduction of this paper.

2.3 Data Mining Initialized by Business Rules

Although the use of data mining models for solving of classification tasks integrated in business rule set is appropriately interpretable and user comprehendible, it is not suitable to limit the possible use cases for using only this way. The main reason for finding other, alternative approach is absence of the target attribute for classification in the operational data of the organization. Particularly in the case of usage data mining methods for finding of exceptions it is suitable to use dynamic data mining initialized by business rules. This process of definition the appropriate wrapper for initialization of data mining thought business rules engine in combination with LISp-Miner system could be defined as follows:

1. Define export from the operational data store of the organization. This export can be realized for example using SQL query and should be "repeatable" for later usages. The best way is definition of a view.

2. Define data mining task for selection of GUHA association rules in the data mining system LISp-Miner. Execute the task and check the results for the corresponding form. There are no limits for definition of the data mining task except of the "*final attribute*", which should be returned as result. This attribute should contain values from the original data matrix (without the use of values grouping in preprocessing phase or dynamic binning).

3. Export definition of the data mining task in PMML.

4. Define the wrapper business rule including the definition of data mining task, mapping of terms dictionary at least for "final attribute", database connection string and limits for counts of requested results.

   For some use cases, it is possible to map not only the final attribute, but also another attributes with fixed value for the definition of a condition.

5. Define period or condition for activation of the defined wrapper business rule. Within implementation using JBoss Drools, both these options are possible.

The wrapper business rule initializes the execution of data mining task. It is possible to run the LISp-Miner system not only from the graphical user interface, but also from the command line. After receiving the results from the data mining system, the wrapper rule compares the count of founded association rules. If the count is within the requested interval, the wrapper business rule extracts values of the final attribute in the founded rules and adds them as new facts in the evaluation base for processing using other business rules.

In case of inappropriate count of founded data mining results, the wrapper business rule can reinitialize the data mining task with modified thresholds of interest measures. To find association rules the user usually defines thresholds of two interest measures (usually confidence and support, for some cases also lift and support).[8] If the system founds too many rules, it is possible to increase the minimal requested thresholds of interest measures and execute the data mining task again.

This method of integration is suitable for interaction between business rules saved in the knowledge base and data mining systems for detection of exceptions in the operational data. Whether the exception can be negative or positive. For example detection of an increase in staff performance. The advantage of the application of data mining methods is the better performance than in case of evaluation the data matrix using set of specific business rules. However, also a disadvantage has to be considered. The separation of statistical evaluation of the operational data matrix from the knowledge base for execution using

---

[7] Most systems for execution of business rules are based on usage of RETE algorithm, which allows quickly inference evaluation.

[8] In the GUHA method, confidence and support are included in 4ft quaintifier „Founded implication", lift is compatible with „Above average dependance" quantifier (AAD).

external system can be founded either as advantage or disadvantage. It depends on the specialization of the domain expert. From the point of view of marketing or business specialists, it will be probably evaluated as an advantage – it is really simplification of the knowledge base.

2.4 Terms Dictionary for Definition of Business Rules

For a definition of business rules, it is required to use a *terms dictionary*. This terms dictionary should contain declaration of basic entities used in the organization. These terms composites to *facts*, and facts composites into *business rules*. For expanding of a business rule set using data mining results, the "good" terms dictionary can be the schema of the main operational database used in the organization.

The mapping techniques are not subject of this paper. For the integration of data mining results into existing business rule set, the best way is a definition of mapping in the mode 1:1 not only at level of data attributes, but also at level of their values.

In using of business rules, the mapping can be realized on basis of usage of specific *mapping rules*. In JBoss Drools, it is possible to define rules with conditional validity. So if the "mapping rule" detects in the evaluation base, it adds one or more other facts (instances of Java object) representing the mapped fact. The added fact is present and valid only while the mapping rule is active (it´s condition is evaluated as true).

3 Demo use case

For better illustration of the appropriateness of ways of integrating data mining results into a business rule set, it is suitable to explain them on a demo use case. In this paper, the author represents them on use cases defined on data from a health insurance company.

In every insurance company, it is necessary to collect the most possible data from the real life and reuse them for the risks analysis and for detection of fraud techniques. In the domain of health insurance, the medical facilities send lists of performed procedures and request a financial compensation for them. Every request composites from identification of the medical facility, the concrete medical worker, identification of the patient and details about the diagnosis and performed procedures. After composition in one "data row", respectively one data matrix, there are tens of data attributes.

The health insurance company has contracts with individual medical facilities, but it does not mean, that every facility requests only really performed procedures. The reason may be a mistake, of course, but also attempt to fraudulently acquire some finances. The insurance company should have a list of rules (optionally a knowledge base in form of business rules) for detection of patently false requests. For example if a family doctor requests finance for a surgical operation. But is it necessary to detect not only obvious errors in requests. The insurance company want to detect also unusual growths of performed procedures, which could be potentially evaluated as untruthful.

3.1 Direct acquisition of business rules

Most business rules for evaluation the correctness of requests from medical facilities is inputted manually by domain experts. For founding of unobvious relations in data, it is suitable to use data mining methods. The user can select some founded association rules, convert them into business rules and use them for following manual editing of the knowledge base.

To use data mining techniques, it is necessary to have access to the archive with operational data received in the past. In terms of medical procedures it is also necessary to respect the specificities of different seasons and impact of weather. For example, there are differences in frequency and types of illnesses and injuries between the summer and the winter.

On the basis of this data mining analysis, it is also possible to detect potentially interesting areas for application of models for automatic learning of business rules.

3.2 Classification model learning

Suitable analytical question for processing of the incoming data is the detection of facilities, which require probably too much procedures or unusual combinations of them. A concrete example could be redundant performing of laboratory analysis of blood or automatically request for RTG for all patients of a surgery. These unnecessary procedures are no benefit not only for the insurance company, but also for the patients.

To solve this task, it is possible to use historical data about the checks previously made in medical facilities in combination with results from these tasks. Based on these data, it is possible to prepare a classification model for recommending suitable facilities for the future check.

The classification model can be included into the knowledge base as native business rules, or better in form of black box component. The advantage of separated black box component is the simpler replacement of the full classification model with a newer version.

3.3 Periodically solved data mining task

Another interesting task suitably solvable using data mining methods is detection of unusual increase or decrease of performed medical procedures in a concrete medical facility compared to other facilities of the same type. This task cannot be resolved in the "flow check" system, but it is possible to solve it using archive of the incoming data.

It is suitable use case for application of periodical solving of a predefined data mining task. The domain experts defines a data mining task for founding GUHA association rules for example in form:

diagnosis(A) & facility(*) → *procedure(B) / clinicType(A)*

where *clinicType(A)* is condition of founded rules, the task is defined using $\land\land D$ quantifier (interest measures are *lift* and *support*) and the expert want to process as results the values of the attribute *facility*. The expert defines interval of minimal threshold of interest measures and maximal count of requested rules.

The data mining is then executed periodically once per month and the business rules system initializes the request for the check in the indicated medical facilities.

4 Conclusion and future work

In this paper, the author presented three suitable ways of integration data mining results (mainly GUHA association rules) into a knowledge base in form of business rules, which are suitable for automatically execution. These models are applicable not only in conjunction with JBoss Drools system, they are generally applicable with all "execution oriented" business rules systems. For example, there can be mentioned systems Jess, Jena or ERIAN.

Within the further work, it is necessary to propagate methods of automatic integration of data mining results into business rule sets. Another task is finalization of a model of knowledge base for combination data mining tasks with definitions of business rules. The demo implementation of the knowledge base, which concept was presented in [9], should be extended to a public methodology.

Acknowledgment

This paper was processed with contribution of long term institutional support of research activities and by IGA project 20/2013 by Faculty of Informatics and Statistics, University of Economics, Prague.

References

[1] OpenRules, Inc., "Rule Learner," *Open Rules* [online] http://openrules.com/rulelearner.htm [cit. 2015-01-28]

[2] Kliegr, T., Kuchař, J., Sottara, D., Vojíř, S.: Learning business rules with association rule classifiers. Rules on the Web. From Theory to Applications, Springer, 2014, 236-250

[3] Vysoká škola ekonomická v Praze and KOMIX s.r.o., TA04011691 - Automatizovaná extrakce byznys pravidel se zpětnou vazbou (2014-2016, TA0/TA), 2013

[4] Red Hat, Inc, "Drools," Drools - Business Rules Management System (Java™, Open Source) [online] http://www.drools.org/, [cit. 2015-04-21]

[5] Rauch, J., Šimůnek, M.: Dobývání znalostí z databází LISp-Miner a GUHA. Oeconomica Praha, 2015

[6] Thabtah, F. A.: A review of associative classification mining. Knowledge Engineering Review **22(1)** (2007),. 37-65

[7] Ross, R.G.: Principles of the Business Rule Approach. Addison-Wesley Professional, 2003

[8] Vojíř, S., Kliegr, T., Hazucha, A., Škrabal, R., Šimůnek, M.: Transforming association rules to business rules: EasyMiner meets Drools. RuleML Challenge 2013, CEUR-WS.org, vol. 1004, 2013

[9] Vojíř, S.: Concept of semantic knowledge base for data mining of business rules. Znalosti 2014 Exhibice, Edukace a nacházení Expertů - Exhibition, Education and Expert finding. Praha: KIZI FIS, 2014, 132-136

Slovenskočeský NLP workshop (SloNLP 2015)

Workshop Program Committee

Petra Barančíková, Charles University in Prague
Rudolf Rosa, Charles University in Prague
Vladimír Benko (JÚLŠ SAV)
Radek Cech (FF OU)
Ján Genci (KPI TUKE)
Aleš Horák (FI MUNI)
Pavel Ircing (KKY ZCU)
Markéta Lopatková, Charles University in Prague
Petr Štrossa (UI VŠE)

J. Yaghob (Ed.): ITAT 2015 pp. 66–72
Charles University in Prague, Prague, 2015

Resource-Light Acquisition of Inflectional Paradigms

Radoslav Klíč[1] and Jirka Hana[2]

[1] Geneea Analytics, Velkopřevorské nám. 1, 118 00 Praha 1
radoslav.klic@gmail.com
[2] MFF UK, Malostranské nám. 25, 118 00 Praha 1
jirka.hana@gmail.com

Abstract: This paper presents a resource-light acquisition of morphological paradigms and lexicon for fusional languages. It builds upon Paramor [10], an unsupervised system, by extending it: (1) to accept a small seed of manually provided word inflections with marked morpheme boundary; (2) to handle basic allomorphic changes acquiring the rules from the seed and/or from previously acquired paradigms. The algorithm has been tested on Czech and Slovene tagged corpora and has shown increased F-measure in comparison with the Paramor baseline.

1 Introduction

Morphological analysis is used in many computer applications ranging from web search to machine translation. As Hajič [6] shows, for languages with high inflection, a morphological analyzer is an essential part of a successful tagger.

Modern morphological analysers based on supervised machine learning and/or hand-written rules achieve very high accuracy. However, the standard way to create them for a particular language requires substantial amount of time, money and linguistic expertise. For example, the Czech analyzer by [7] uses a manually created lexicon with 300,000+ entries. As a result, most of the world languages and dialects have no realistic prospect for morphological analyzers created in this way.

Various techniques have been suggested to overcome this problem, including unsupervised methods acquiring morphological information from an unannotated corpus. While completely unsupervised systems are scientifically interesting, shedding light on areas such as child language acquisition or general learneability, for many practical applications their precision is still too low. They also completely ignore linguistic knowledge accumulated over several millennia, often failing to discover rules that can be found in basic grammar books.

Lightly-supervised systems aim to improve upon the accuracy of unsupervised system by using a limited amount of resources. One of such systems for fusional languages is described in the paper.

Using a reference grammar, it is relatively easy to provide information about inflectional endings, possibly organized into paradigms. In some languages, an analyzer built on such information would have an acceptable accuracy (e.g., in English, most words ending in *ed* are past/-passive verbs, and most words ending in *est* are superlative adjectives). However, in many languages, the number of homonymous endings is simply too high for such system to be useful. For example, the ending *a* has about 19 different meanings in Czech [4].

Thus our goal is to discover inflectional paradigms each with a list of words declining according to it, in other words we discover a list of paradigms and a lexicon. But we do not attempt to assign morphological categories to any of the forms. For example, given an English corpus the program should discover that *talk, talks, talking, talked* are the forms of the same word, and that *work, push, pull, miss,...* decline according to the same pattern. However, it will not label *talked* as a past tense and not even as a verb.

This kind of shallow morphological analysis has applications in information retrieval (IR), for example search engines. For the most of the queries, users aren't interested only in particular word forms they entered but also in their inflected forms. In highly inflectional languages, such as Czech, dealing with morphology in IR is a necessity. Moreover, it can also be used as a basis for a standard morphological analyzer after labeling endings with morphological tags and adding information about closed-class/irregular words.

As the basis of our system, we chose Paramor [10], an algorithm for unsupervised induction of inflection paradigms and morphemic segmentation. We extended it to handle basic phonological/graphemic alternations and to accept seeding paradigm-lexicon information.

The rest of this paper is organized as follows: First, we discuss related work on unsupervised and semi-supervised learning. Then follows a section about baseline Paramor model. After that, we motivate and describe our extension to it. Finally, we report results of experiments on Czech and Slovene.

2 Previous Work

Perhaps the best known unsupervised morphological analysers are Goldsmith's Linguistica [5] and Morfessor [1, 2, 3] family of algorithms.

Goldsmith uses minimum description length (MDL; [14]) approach to find the morphology model which allows the most compact corpus representation.

His Linguistica software returns a set of *signatures* which roughly correspond to paradigms.

Unlike Linguistica, Morfessor splits words into morphemes in a hierarchical fashion. This makes it more suitable to agglutinative languages, such as Finnish or Turkish, with a large number of morphemes per word. A probabilistic model is used to tag each morph as a prefix, suffix or stem. Kohonen et al. [9] improve the results of Morfessor by providing a small set (1000+ for English, 100+ for Finish) of correctly segmented words. While the precision slightly drops, the recall is significantly improved for both languages. Tepper and Xia [17] use handwritten rewrite rules to improve Morfessor's performance by recognising allomorphic variations.

The approaches by Yarowsky and Wicentowski [18] and Schone and Jurafsky [15] aim at combining different information sources (e.g., corpus frequencies, edit distance similarity, or context similarity) to obtain better analysis, especially for irregular inflection.

A system requiring significantly more human supervision is presented by Oflazer et al. [13]. This system takes manually entered paradigm specification as an input and generates a finite-state analyser. The user is then presented with words in a corpus which are not accepted by the analyser, but close to an accepted form. Then the user may adjust the specification and the analyser is iteratively improved.

Feldman and Hana [8, 4] build a system which relies on a manually specified list of paradigms, basic phonology and closed-class words and use a raw corpus to automatically acquire lexicon. For each form, all hypothetical lexical entries consistent with the information about the endings are created. Then competing entries are compared and only those supported by the highest number of forms are retained. Most of the remaining entries are still nonexistent; however, in the majority of cases, they licence the same inflections as the correct entries, differing only in rare inflections.

3 Paramor

Our approach builds upon Paramor [10, 11, 12], another unsupervised approach for discovery of inflectional paradigms.

Due to data sparsity, not all inflections of a word are found in a corpus. Therefore Paramor does not attempt to reconstruct full paradigms, but instead works with partial paradigms, called *schemes*. A scheme contains a set of c(andidate)-suffixes and a set of c(andidate)-stems inflecting according to this scheme. The corpus must contain the concatenation of every c-stem with every c-suffix in the same scheme. Thus, a scheme is uniquely defined by its c-suffix set. Several schemes might correspond to a single morphological paradigm, because different stems belonging to the paradigm occur in the corpus in different set of inflections.

The algorithm to acquire schemes has several steps:

1. Initialization: It first considers all possible segmentations of forms into candidate stems and endings.

2. Bottom-up Search: It builds schemes by adding endings that share a large number of associated stems.

3. Scheme clustering: Similar schemes (as measured by cosine similarity) are merged.

4. Pruning: Schemes proposing frequent morpheme boundaries not consistent with boundaries proposed by a character entropy measure are discarded.

Paramor works with types and not tokens. Thus it is not using any information about the frequency or context of forms. Below, we describe some of the steps in more detail.

3.1 Bottom-up Search

In this phase, Paramor performs a bottom-up search of the scheme lattice. It starts with schemes containing exactly one c-suffix. For each of them, Paramor ascends the lattice, adding one c-suffix at a time until a stopping criterion is met. C-suffix selected for adding is the one with the biggest c-stem ratio. (Adding a c-suffix to a scheme reduces number of the stems and the suffix reducing it the least is selected. C-stem ratio is ratio between number of stems in the candidate higher-level scheme and the current scheme.) When the highest possible c-stem ratio falls under 0.25, the search stops. It is possible to reach the same scheme from multiple searches. For example, a search starting from the scheme *(-s)* can continue by adding *(-ing)* and end by adding *(-ed)*, thus creating a scheme *(-s, -ing, -ed)*. Another search starting from *(-ed)* can continue by adding *(-s)* and then by adding *(-ing)*, creating a redundant scheme. Such duplicates are discarded.

3.2 Scheme Clustering

Resulting schemes are then subjected to agglomerative bottom-up clustering to group together schemes which are partially covering the same linguistic paradigm. For example, if the first phase generated schemes *(-s, -ing)* and *(-ing, -ed)*, the clustering phase should put them in the same scheme cluster. To determine proximity of two scheme clusters, sets of words generated by the clusters are measured by cosine similarity.[1] A scheme cluster generates a set of words which is the union of sets generated by the schemes it contains (not a Cartesian product of all stems and suffixes throughout the schemes). In order to be merged, clusters must satisfy some conditions, e.g. for any two suffixes in the cluster, there must be a stem in the cluster which can combine with both of them.

---

[1] $\mathrm{proximity}(X,Y) = \frac{|X \cap Y|}{\sqrt{|X||Y|}}$

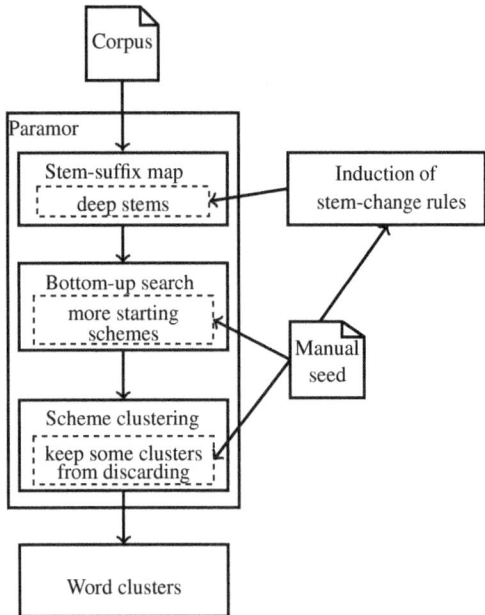

Figure 1: Altered Paramor's pipeline (our alterations are in dashed boxes and outside the Paramor box).

3.3 Pruning

After the clustering phase, there are still too many clusters remaining and pruning is necessary. In the first pruning step, clusters which generate only small number of words are discarded. Then clusters modelling morpheme boundaries inconsistent with letter entropy are dropped.

4 Our Approach

4.1 Overview

We have modified the individual steps in Paramor's pipeline in order to use (1) a manually provided seed of inflected words divided into stems and suffixes; and (2) to take into account basic allomorphy of stems. Figure 1 shows phases of Paramor on the left with dashed boxes representing our alterations.

In the bottom-up search phase and the scheme cluster filtering phase, we use manually provided examples of valid suffixes and their grouping to sub-paradigms to steer Paramor towards creating more adequate schemes and scheme clusters. The data may also contain allomorphic stems, which we use to induce simple stem rewrite rules. Using these rules, some of the allomorphic stems in the corpus can be discovered and used to find more complete schemes.

Note that the Paramor algorithm is based on several heuristics with many parameters whose values were set experimentally. We used the same settings. Moreover, when we applied similar heuristics in our modifications, we used analogical parameter values.

4.2 Scheme Seeding

The manual seed contains a simple list of inflected words with marked morpheme boundary. A simple example in English would be:

talk+0, talk+s, talk+ed, talk+ing
stop+0, stop+s, stopp+ed, stopp+ing
chat+0, chat+s, chatt+ed, chatt+ing

This can be written in an abbreviated form as:

talk, stop/stopp, chat/chatt + 0, s / ed, ing

The data are used to enhance Paramor's accuracy in discovering the correct schemes and scheme clusters in the following way:

1. In the bottom-up search, Paramor starts with single-suffix schemes. We added a 2-suffix scheme to the starting scheme set for every suffix pair from the manual data belonging to the same inflection. Note that we cannot simply add a scheme containing all the suffixes of the whole paradigm as many of the forms will not be present in the corpus.

2. Scheme clusters containing suffixes similar to some of the manually entered suffix sets are protected from the second phase of the cluster pruning. More precisely, a cluster is protected if at least half of its schemes share at least two suffixes with a particular manual suffix set.

4.3 Allomorphy

Many morphemes have several contextually dependent realizations, so-called allomorphs due to phonological/-graphemic changes or irregularities. For example, consider the declension of the Czech word *matka* 'mother' in Table 1. It exhibits stem-final consonant change (palatalisation of *k* to *c*) triggered by the dative and local singular ending, and epenthesis (insertion of *-e-*) in the bare stem genitive plural.

Case	Singular	Plural
nom	mat**k**+a	mat**k**+y
gen	mat**k**+y	mat**ek**+0
dat	mat**c**+e	mat**k**+ám
acc	mat**k**+u	mat**k**+y
voc	mat**k**+o	mat**k**+y
loc	mat**c**+e	mat**k**+ách
inst	mat**k**+ou	mat**k**+ami

Table 1: Declension of the word *matka* "mother". Changing part of the stem is in bold.

Paramor ignores allomorphy completely (and so do Linguistica and Morfessor). There are at least two reasons to handle allomorphy. First, linguistically, it makes more sense to analyze *winning* as *win+ing* than as *winn+ing* or *win+ning*. For many applications, such as information retrieval, it is helpful to know that two morphs are variants of the same morpheme. Second, ignoring allomorphy makes the data appear more complicated and noisier than they actually are. Thus, the process of learning morpheme boundaries or paradigms is harder and less successful.

This latter problem might manifests itself in Paramor's bottom-up search phase: a linguistically correct suffix triggering a stem change might be discarded, because Paramor would not consider stem allomorphs to be variants of the same stem and c-stem ratio may drop significantly. Further more, incorrect c-suffixes may be selected.

For example, suppose there are 5 English verbs in the corpus: *talk, hop, stop, knit, chat*, together with their *-s* (*talks, hops, stops, knits, chats*) and *-ing* (*talking, hopping, stopping, knitting, chatting*) forms. Let's assume we already have a scheme {*0, s*} with 5 stems. Unfortunately, a simple *ing* suffix (without stem-final consonant doubling) combines with one out the 5 stems only, therefore adding *ing* to the scheme would decrease the number of its stems to 1, leaving only *talk* in the scheme.

However, for most languages the full specification of rules constraining allomorphy is not available, or at least is not precise enough. Therefore, we automatically induce a limited number of simple rules from the seed examples and/or from the scheme clusters obtained from the previous run of algorithm. Such rules both over and undergenerate, but nevertheless they do improve the accuracy of the whole system. For languages, where formally specified allomorphic rules are available, they can be used directly along the lines of Tepper and Xia [17, 16]. For now, we consider only stem final changes, namely vowel epenthesis (e.g., *matk-a – matek-0*) and alternation of the final consonant (e.g., *matk-a – matc-e*). The extension to other processes such as root vowel change (e.g., English *foot – feet*) is quite straightforward, but we leave it for future work.

Stem change rule induction and application. Formally, the process can be described as follows. From every pair of stem allomorphs in the manual input, $s\delta_1, s\delta_2$, where s is their longest common initial substring,[2] with suffix sets f_1, f_2 we generate a rule $*\delta_1 \rightarrow *\delta_2 / (f_1, f_2)$ and also a reverse rule $*\delta_2 \rightarrow *\delta_1 / (f_2, f_1)$. Notation $*\delta_1 \rightarrow *\delta_2 / (f_1, f_2)$ means "transform a stem $x\delta_1$ into $x\delta_2$ if the following conditions hold:"

1. $x\delta_2$ is a c-stem present in the corpus.

2. C-suffix set f_1^x (from the corpus) of the c-stem $x\delta_1$ contains at least one of the suffixes from f_1 and contains no suffix from f_2.

3. C-suffix set f_2^x of the c-stem $x\delta_2$ contains at least one of the suffixes from f_2 and contains no suffix from f_1.

Induced rules are applied after the initialisation phase. So-called *deep* stems are generated from the c-stems. A deep stem is defined as a set of surface stems.

To obtain a deep stem for a c-stem t, operation of *expansion* is applied. Expansion works as a breadth-first search using a queue initialised with t and keeping track of the set D of already generated variants. While the queue is not empty, the first member is removed and its variants found by application of all the rules. (Result of applying a rule is non-empty only if the rule is applicable and its right hand side is present in the corpus.) Variants which haven't been generated so far are added to the back of the queue and to D. When the queue is emptied, D becomes the deep stem associated with t and all other members of D.

Bottom-up search and all the following phases of Paramor algorithm are then using the deep stems instead of the surface ones.

Stem change rule induction from scheme clusters. In addition to deriving allomorphic rules from the manual seed, we also use a heuristic for detecting stem allomorphy in the scheme clusters obtained from the previous run of the algorithm. Stem allomorphy increases the sparsity problem and might prevent Paramor from finding some paradigms. However, if the stem changes are systematic and frequent, Paramor does create the appropriate scheme clusters. However, it considers the changing part of the stem to be a part of suffix.

As an example, consider again the declension of the Czech word *matka* "mother" in Table 1. Paramor's scheme cluster with suffixes *ce, ek, ka, kami, kou, ku, ky, kách, kám* has correctly discovered 9 of 10 paradigm's suffixes,[3] but fused together with parts of the stem. Presence of such scheme cluster in the result is a hint that there may be a *c/k* alteration and epenthesis in the language.

First phase of the algorithm for deciding whether a scheme cluster with a c-suffix set f is interesting in this respect is following:

1. If f contains a c-suffix without a consonant, return *false*.

2. c_c = count of unique initial consonants found in c-suffixes in f.

3. If $c_c > 2$ return *false*. (Morpheme boundary probably incorrectly shifted to the left.)

4. If $c_c = 1$ and f doesn't contain any c-suffix starting with a vowel, return *false*. (No final consonant change, no epenthesis.)

5. Return *true*.

If a scheme cluster passes this test, each of its stems' subparadigms is examined. Subparadigm for stem s consists of s and f_s – all the c-suffixes from f with which s forms

---

[2]should δ_1 or δ_2 be 0, one final character is removed from s and prepended to δ_1 and δ_2

[3]Except for vocative case singular, which is rarely used.

a word in the corpus. For example, let's have a stem $s = mat$ with $f_s = \{ce, ek, ka, ku, ky\}$. Now, the morpheme boundary is shifted so that it is immediately to the right from the first consonant of the original c-suffixes. In our example, we get 3 stem variants: $matk + a, u, y$, $matc + e$, $matek + 0$. To reduce falsely detected phonological changes, we check each stem variant's suffix set whether it contains at least one of the c-suffixes that Paramor has already discovered in other scheme clusters. If the condition holds, rules the with same syntax as the manual data are created. For example, $matk / matc / matek + a, u, y / e / 0$. All generated rules are gathered in a file and can be used in the same way as the manual seed or just for the induction of phonological rules.

5   Experiments and Results

We tested our approach on Czech and Slovene lemmatised corpora. For Czech, we used two differently sized subsets of the PDT 1 corpus. The first, marked as **cz1**, contains 11k types belonging to 6k lemmas. The second, **cz2**, has 27k types and 13k lemmas and is a superset of **cz1**. The purpose of having two Czech corpora was to observe the effect of data size on performance of the algorithm. The Slovene corpus **si** is a subset of the jos100k corpus V2.0 (http://nl.ijs.si/jos/jos100k-en.html) with 27k types and 15.5k lemmas.

The manual seed consisted of inflections of 18 lemmas for Czech and inflections of 9 lemmas for Slovene. In both cases, examples of nouns, adjectives and verbs were provided. They were obtained from a basic grammar overview. For Czech, we also added information about the only two inflectional prefixes (negative prefix *ne* and superlative prefix *nej*). The decision which prefixes to consider inflectional and which not is to a certain degree an arbitrary decision (e.g., it can be argued that *ne* is a clitic and not a prefix), therefore it makes sense to provide such information manually. (Prefixes were implemented by a special form of stem transformation rules introduced in section 4.3 which create deep stems consisting of a stem with and without given prefix.)

5.1   Evaluation Method

We evaluated the experiments only on types at least 6 characters long which Paramor uses for learning. That means 8.5k types and 4500 lemmas for **cz1**, 21k types and 10k lemmas for **cz2** and 21k types and 12k lemmas for **si**.

Since corpora we used do not have morpheme boundaries marked, we could not use the same evaluation method as authors of Paramor and Morfessor – measuring the precision and recall of placing morpheme boundaries. On the other hand, corpora are lemmatised and we can evaluate whether types grouped to paradigms by the algorithm correspond to sets of types belonging to the same lemma.

We use the following terminology in this section: a *word group* is a set of words returned by our system, a *word paradigm* is a set of words from the corpus sharing the same lemma. Both word groups and word paradigms are divisions of corpus into disjoint sets of words. An *autoseed* is a seed generated by the heuristic described in Section 4.3.

Since Paramor only produces schemes and scheme clusters, we need an additional step to obtain word groups. We generated the word groups by bottom-up clustering of words using the *paradigm distance* which is designed to group together words generated by similar sets of scheme clusters. To compute paradigm distance for two words w_1, w_2, we find the set of all scheme clusters which generate w_1 and compute cosine similarity to the analogical set for w_2[4]. In the simplest case, two forms of a lemma will be generated just by one scheme cluster and therefore get distance 1. For a more complicated example, let's take two Czech words: *otrávení* "poisoned masc. anim. nom. pl." and *otrávený* "poisoned masc. anim. nom. sg.". The first one was generated by scheme clusters 33 and 41, both with *otráv* as a stem. The second word was generated by scheme cluster 41 with *otráv* as a stem and by scheme cluster 45 with *otráven* as stem. That means that only scheme cluster 41 generates both words and their paradigm distance is $\frac{1}{\sqrt{2\times2}} = 0.5$.

Precision and recall of the word groups can be computed in the following way: To compute precision, start with $p = 0$. For each word group, find a word paradigm with the largest intersection. Add the intersection size to p. Precision = p / total number of words. For computing recall, start with $r = 0$. For each word paradigm, find a word group with the largest intersection. Add the intersection size to r. Recall = r / total number of words. F1 is the standard balanced F-score.

5.2   Results

Results of the experiments are presented in Tables 2 – 4. We used the following experiment settings:

1. *no seed* – the baseline, Paramor was run without any seeding

2. *man. seed* – manual seed was used

3. *autoseed* – autoseed was used for induction of the stem change rules

4. *both seeds* – Paramor run with manual seed, stem change rules were induced from manual and autoseed.

5. *seed + pref.* – manual seed was used together with additional rules for two Czech inflectional prefixes, otherwise same as 2.

---

[4]We also have to check whether w_1 and w_2 have the same stem, so, in fact we are comparing sets of pairs ⟨scheme cluster, c-stem⟩, to make sure only words sharing c-stems are grouped together.

6. *both seeds + pref* – manual seed was used together with additional rules for two Czech inflectional prefixes, otherwise same as 4.

Experiment	Precision	Recall	F1
no seed	97.87	84.61	90.76
man. seed	97.96	87.52	92.44
autoseed	98.19	84.58	90.88
both seeds	97.96	87.52	92.44
seed + pref.	97.84	89.40	**93.43**
both seeds + pref.	97.84	89.40	**93.43**

Table 2: Results for the **cz1** corpus.

Experiment	Precision	Recall	F1
no seed	97.36	87.02	91.90
man. seed	97.04	89.30	93.01
autoseed	97.30	87.72	92.26
both seeds	96.78	89.30	92.89
seed + pref.	96.68	92.35	**94.46**
both seeds + pref.	96.31	92.49	94.36

Table 3: Results for the **cz2** corpus.

Experiment	Precision	Recall	F1
no seed	95.70	93.00	94.33
man. seed	95.62	94.44	95.02
autoseed	95.69	93.13	94.40
both seeds	95.56	94.76	**95.16**

Table 4: Results for the **si** corpus.

As can be seen from the results, the extra manual information indeed does help the accuracy of clustering words belonging to the same paradigms. What is not shown by the numbers is that more of the morpheme boundaries make linguistic sense because basic stem allomorphy is accounted for.

6 Conclusion

We have shown that providing very little of easily obtainable information can improve the result of a purely unsupervised system. In the near future, we are planning to model a wider range of allomorphic alternations, try larger (but still easy to obtain) seeds and finally test the results on more languages.

References

[1] Creutz, M., Lagus. K.: Unsupervised discovery of morphemes. In: Proceedings of the ACL-02 Workshop on Morphological and Phonological Learning, Vol. 6, MPL '02, 21–30, Stroudsburg, PA, USA, 2002, Association for Computational Linguistics

[2] Creutz, M., Lagus, K.: Inducing the morphological lexicon of a natural language from unannotated text. In: Proceedings of the International and Interdisciplinary Conference on Adaptive Knowledge Representation and Reasoning (AKRR'05), 106–113, Finland, Espoo, 2005

[3] Creutz, M., Lagus, K.: Unsupervised models for morpheme segmentation and morphology learning. ACM Trans. Speech Lang. Process. **4 (3)** (February 2007), 1–34

[4] Feldman, A., Hana, J.: A resource-light approach to morpho-syntactic tagging. Rodopi, Amsterdam/New York, NY, 2010

[5] Goldsmith, J. A.: Unsupervised learning of the morphology of a natural language. Computational Linguistics **27(2)** (2001), 153–198

[6] Hajič, J.: Morphological tagging: data vs. dictionaries. In: Proceedings of ANLP-NAACL Conference, 94–101, Seattle, Washington, USA, 2000

[7] Hajič, J.: Disambiguation of rich inflection: computational morphology of Czech. Karolinum, Charles University Press, Praha, 2004

[8] Hana, J., Feldman, A., Brew, C.: A resource-light approach to Russian morphology: Tagging Russian using Czech resources. In: Dekang Lin and Dekai Wu, (eds.), Proceedings of EMNLP 2004, 222–229, Barcelona, Spain, July 2004, Association for Computational Linguistics

[9] Kohonen, O., Virpioja, S., Lagus, K.: Semi-supervised learning of concatenative morphology. In: Proceedings of the 11th Meeting of the ACL Special Interest Group on Computational Morphology and Phonology, SIGMOR-PHON'10, 78–86, Stroudsburg, PA, USA, 2010, Association for Computational Linguistics

[10] Monson, C.: ParaMor: from paradigm structure to natural language morphology induction. PhD thesis, Language Technologies Institute, School of Computer Science, Carnegie Mellon University, 2009

[11] Monson, C., Carbonell, J., Lavie, A., Levin, L.: ParaMor: minimally supervised induction of paradigm structure and morphological analysis. In: Proceedings of Ninth Meeting of the ACL Special Interest Group in Computational Morphology and Phonology, 117–125, Prague, Czech Republic, June 2007, Association for Computational Linguistics

[12] Monson, C., Carbonell, J. G., Lavie, A., Levin, L. S.: Paramor: finding paradigms across morphology. In: Advances in Multilingual and Multimodal Information Retrieval, 8th Workshop of the Cross-Language Evaluation Forum, CLEF 2007, Budapest, Hungary, September 19-21, 2007, Revised Selected Papers, 900–907, 2007.

[13] Oflazer, K., Nirenburg, S., McShane, M.: Bootstrapping morphological analyzers by combining human elicitation and machine learning. Computational Linguistics **27(1)** (2001), 59–85

[14] Rissanen, J.: Stochastic complexity in statistical inquiry. World Scientific Publishing Co, Singapore, 1989.

[15] Schone, P., Jurafsky, D.:   Knowledge-free induction of inflectional morphologies. In: Proceedings of the North American Chapter of the Association for Computational Linguistics, 183–191, 2001.

[16] Tepper, M., Xia, F.: A hybrid approach to the induction of underlying morphology. In: Proceedings of the Third International Joint Conference on Natural Language Processing (IJCNLP-2008), Hyderabad, India, Jan 7-12, 17–24, 2008.

[17] Tepper, M., Xia, F.:   Inducing morphemes using light knowledge. ACM Trans. Asian Lang. Inf. Process. **9 (3)** (March 2010), 1–38

[18] Yarowsky, D., Wicentowski, R.:   Minimally supervised morphological analysis by multimodal alignment. In: Proceedings of the 38th Meeting of the Association for Computational Linguistics, 207–216, 2000

J. Yaghob (Ed.): ITAT 2015 pp. 73–80
Charles University in Prague, Prague, 2015

Improvements to Korektor: A Case Study with Native and Non-Native Czech

Loganathan Ramasamy[1], Alexandr Rosen[2], and Pavel Straňák[1]

[1]Institute of Formal and Applied Linguistics, Faculty of Mathematics and Physics
[2]Institute of Theoretical and Computational Linguistics, Faculty of Arts
Charles University in Prague

Abstract: We present recent developments of Korektor, a statistical spell checking system. In addition to lexicon, Korektor uses language models to find real-word errors, detectable only in context. The models and error probabilities, learned from error corpora, are also used to suggest the most likely corrections. Korektor was originally trained on a small error corpus and used language models extracted from an in-house corpus WebColl. We show two recent improvements:

- We built new language models from freely available (shuffled) versions of the Czech National Corpus and show that these perform consistently better on texts produced both by native speakers and non-native learners of Czech.

- We trained new error models on a manually annotated learner corpus and show that they perform better than the standard error model (in error detection) not only for the learners' texts, but also for our standard evaluation data of native Czech. For error correction, the standard error model outperformed non-native models in 2 out of 3 test datasets.

We discuss reasons for this not-quite-intuitive improvement. Based on these findings and on an analysis of errors in both native and learners' Czech, we propose directions for further improvements of Korektor.

1 Introduction

The idea of using the context of a misspelled word to improve the performance of a spell checker is not new [10]. Moreover, recent years have seen the advance of context-aware spell checkers such as *Google Suggest*, offering reasonable corrections of search queries.

Methods used in such spell checkers usually employ the *noisy-channel* or *window-based* approach [4]. The system described here also belongs to the *noisy-channel* class. It makes extensive use of language models based on several morphological factors, exploiting the morphological richness of the target language.

Errors detected by such advanced spell checkers have a natural overlap with those of rule-based grammar checkers – grammatical errors are also manifested as unlikely n-grams. Using language models or even complete SMT approach [8] for grammatical error correction is also becoming more common, however all the tasks and publications on grammar correction we have seen so far expect

pre-corrected text in terms of spelling. See also [15] and Table 1 in [14] for what types of errors were subject to correction at the CoNLL 2013 and 2014 Shared Tasks on English as a Second Language.

We make no such optimistic expectations. As we show in Section 2 there are many types of spelling errors both in native speakers' texts and in learner corpora. The error distributions are slightly different, though.

Richter [12] presented a robust spell checking system that includes language models for improved error detection and suggestion. To improve the suggestions further, the system employs error models trained on error corpora. In this paper we present some recent improvements to Richter et al.'s work in both respects: improved language models in Section 3 and task-dependent, adapted error models in Section 4. We apply native and non-native error models on both native and non-native datasets in Section 5. We analyze a portion of the systems output in Section 6 and provide some insight into the most problematic errors that various models make. Finally, we summarize our work and list potential scope for further improvements of Korektor components in Section 7.

2 Error Distribution for Native *vs* Non-Native Czech

Richter [11, p. 33] presents statistics of spelling errors in Czech, based on a small corpus of 9500 words, which is actually a transcript of an audio recording of a novel. The transcription was done by a native speaker. Following [1], the error analysis in Table 1 is based on the classification of errors into four basic groups: substitution, insertion, deletion/omission and swap/transposition/metathesis. Although the figures may be biased due to the small size of the corpus and the fact that it was transcribed by a single person, we still find them useful for a comparison with statistics of spelling errors made by non-native speakers.

In Table 2 the aggregate figures from Table 1 (in the last column headed by "Native") are compared with figures from an automatically corrected learner corpus ("SGT", or CzeSL-SGT) and a hand-corrected learner corpus ("MAN", or CzeSL-MAN). The taxonomy of errors is derived from a "formal error classification" used in those two corpora, described briefly in Section 4.[1] In this table we follow [3] in treating errors in diacritics as dis-

---

[1]See [7] for more details about the classification and the http://utkl.ff.cuni.cz/learncorp/ site, including all information about the corpora.

Error Type	Frequency	Percentage
Substitution	224	40.65%
– horizontally adjacent letters	142	25.77%
– vertically adjacent letters	2	0.36%
– z → s	6	1.09%
– s → z	1	0.18%
– y → i	10	1.81%
– i → y	10	1.81%
– non-adjacent vocals	13	2.36%
– diacritic confusion	21	3.81%
– other cases	19	3.45%
Insertion	235	42.65%
– horizontally adjacent letter	162	29.40%
– vertically adjacent letter	13	2.36%
– same letter as previous	14	2.54%
– other cases	46	8.35%
Deletion – other cases	58	10.53%
Swap letters	34	6.17%
TOTAL	551	100.00%

Table 1: Error types in a Czech text produced by native speakers

	SGT	MAN	PT	Native
Insertion	3.76	3.52	10.45	42.65
Omission	1.39	9.20	17.12	10.53
Substitution	31.30	37.67	12.82	36.84
Transposition	0.16	0.19	3.69	6.17
Missing diacritic	50.19	40.40	37.66	
Addition of diacritic	12.69	8.60	1.67	
Wrong diacritic	0.51	0.43	0.92	3.81

Table 2: Percentages of error types in a Czech text produced by non-native speakers, compared to Portuguese and Czech native speakers

tinct classes, adding their statistics on native Brazilian Portuguese for comparison in the "PT" column.

The high number of errors in diacritics in non-native Czech and native Portuguese in comparison with native Czech can be explained by the fact that native speakers of Czech are aware of the importance of diacritics both for distinguishing the meaning and for giving the text an appropriate status. The high number of errors in diacritics in learner texts is confirmed by results shown in Table 3, counted on the training portion of the "CzeSL-MAN" corpus by comparing the uncorrected and corrected forms, restricted to single-edit corrections.[2] The distribution is shown separately for the two annotation levels of CzeSL-MAN: somewhat simplifying, L1 is the level where non-words (forms spelled incorrectly in any context) are cor-

Error type	L1		L2	
Substitution[3]	22,695	84.36%	30,527	84.15%
– Case	1,827	8.05%	5,090	16.67%
– Diacritics	14,426	63.56%	13,367	43.79%
Insertion	1,274	4.74%	1,800	4.96%
Deletion	2,862	10.64%	3,809	10.50%
Swap	72	0.27%	143	0.39%
Total	26,903	100.00%	36,279	100.00%

Table 3: Distribution of single edit errors in the training portion of the CzeSL-MAN corpus on Levels 1 and 2

Substituting...	Frequency	Substituting...	Frequency
a for á	5255	y for ý	780
i for í	3427	á for a	695
e for ě	1284	u for ů	635
e for é	1169	y for i	482
i for y	1077	í for ý	330
í for i	1005	z for ž	297

Table 4: The top 12 most frequent substitution errors in the CzeSL corpus

rected, while L2 is the level where real-word errors are corrected (words correct out of context but incorrect in the syntactic context). For more details about CzeSL-MAN see Section 4.1.

As an illustration of the prevalence of errors in diacritics in non-native Czech, see Table 4, showing the 12 most frequent substitution errors from L1 in Table 3. There is only one error which is not an error in a diacritic (the use of the i homophone instead of y).

3   Current Improvements for Native Czech Spelling Correction

The original language model component of Korektor [12] was trained on *WebColl* – a 111 million words corpus of primarily news articles from the web. This corpus has two issues: (i) the texts are not representative and (ii) the language model from this data could not be distributed freely due to licensing issues. To obviate this, we evaluate Korektor using two new language models built from two corpora available from the *Czech National Corpus* (*CNC*): (i) SYN2005 [2] and (ii) SYN2010 [9]. Both have the size of 100 million words each and have a balanced representation of contemporary written Czech: news, fiction, professional literature etc.

We use the error model and the test data (only the *Audio* data set) described in [12]. *Audio* contains 1371 words with 218 spelling errors, of which 12 are real-word errors.

---

[2]I.e., without using the "formal error types" of [7].

[3]The two error types below are actually subtypes of the substitution error.

For the CNC corpora, we build 3rd order language models using *KenLM* [6].

The spell checker accuracy is measured in terms of standard precision and recall. The precision and recall measures are calculated at two levels: (i) error detection and (ii) error correction. These evaluation measures are similar in spirit as in [17]. For both levels, precision, recall and other related measures are calculated as: $Precision(P) = \frac{TP}{TP+FP}$, $Recall(R) = \frac{TP}{TP+FN}$, and $F-score(F1) = \frac{2*P*R}{P+R}$, where, for error detection,

- **TP** – Number of words with spelling errors that the spell checker detected correctly

- **FP** – Number of words identified as spelling errors that are not actually spelling errors

- **TN** – Number of correct words that the spell checker did not flag as having spelling errors

- **FN** – Number of words with spelling errors that the spell checker did not flag as having spelling errors

and for error correction,

- **TP** – Number of words with spelling errors for which the spell checker gave the correct suggestion

- **FP** – Number of words (with/without spelling errors) for which the spell checker made suggestions, and for those, either the suggestion is not needed (in the case of non-existing errors) or the suggestion is incorrect if indeed there was an error in the original word

- **TN** – Number of correct words that the spell checker did not flag as having spelling errors and no suggestions were made

- **FN** – Number of words with spelling errors that the spell checker did not flag as having spelling errors or did not provide any suggestions

The results for error detection and error correction are shown in Tables 5 and 6, respectively. Maximum edit distance, i.e., the number of edit operations per word is set to values from 1 to 5. In the case of error detection, the best overall performance is obtained for the SYN2005 corpus when the maximum edit distance parameter is 2, and there is no change in results for the edit distance range from 3 to 5. Of the two CNC corpora, SYN2005 consistently provides better results than SYN2010 corpus. Differences in the vocabulary could be the most likely reason.

Even in the case of error correction, the best overall performance is obtained for SYN2005 with 94.5% F1-score. We can also see that WebColl performs better in 3 out of 5 cases, but we should also note that this happens when we include top-3 suggestions in the error correction. Otherwise the SYN2005 model consistently provides better scores. We have also experimented with pruned language models and obtained similar results.

LM train data	Max. edit distance	P	R	F1
WebColl		94.7	90.8	92.7
SYN2005	1	95.7	90.8	**93.2**
SYN2010		94.7	89.9	92.2
WebColl		94.1	95.4	94.8
SYN2005	2	95.0	95.9	**95.4**
SYN2010		94.1	95.0	94.5
WebColl		94.1	95.4	94.8
SYN2005	3	95.0	95.9	**95.4**
SYN2010		94.1	95.0	94.5
WebColl		94.1	95.4	94.8
SYN2005	4	95.0	95.9	**95.4**
SYN2010		94.1	95.0	94.5
WebColl		94.1	95.4	94.8
SYN2005	5	95.0	95.9	**95.4**
SYN2010		94.1	95.0	94.5

Table 5: Error detection results with respect to different language models

4 Work in Progress for Improving Spelling Correction of Non-Native Czech

One of the main hurdle in obtaining a new error model is the availability of annotated error data for training. Many approaches are available to somehow obtain error data automatically from sources such as the web [16]. The error data obtained from the web may be good enough for handling simple typing errors, but not for the more complicated misspellings a learner/non-native speaker of a language makes. However, these approaches can be successfully used to obtain general purpose spell checkers. One resource which could be of some value to spell checking is the learner corpus. Unlike native error corpus, the learner corpus of non-native or foreign speakers tend to have more errors ranging from orthographical, morphological to real-word errors. In this work, we try to address whether error models from texts produced by native Czech speakers can be applied to errors from non-native Czech texts and vice versa. We also derive error analysis based on the results.

4.1 CzeSL — a Corpus of Czech as a Second Language

A learner corpus consists of language produced by language learners, typically learners of a second or foreign language. Deviant forms and expressions can be corrected and/or annotated by tags making the nature of the error explicit. The annotation scheme in CzeSL is based on a two-stage annotation design, consisting of three levels. The level of transcribed input (Level 0) is followed by the level of orthographical and morphological corrections (Level 1), where only forms incorrect in any context are treated. The

LM train data	Max. edit distance	top-1			top-2			top-3		
		P	R	F1	P	R	F1	P	R	F1
WebColl		85.2	89.9	87.5	90.9	90.5	90.7	93.3	90.7	92.0
SYN2005	1	87.9	90.1	**89.0**	92.3	90.5	**91.4**	93.7	90.7	**92.2**
SYN2010		86.0	89.0	87.5	91.8	89.6	90.7	92.3	89.7	91.0
WebColl		84.2	94.9	89.2	91.0	95.3	93.1	93.2	95.4	94.3
SYN2005	2	86.8	95.5	**91.0**	91.8	95.7	**93.7**	93.2	95.8	**94.5**
SYN2010		85.0	94.4	89.5	91.4	94.8	93.1	92.3	94.9	93.5
WebColl		84.2	94.9	89.2	91.0	95.3	93.1	93.2	95.4	**94.3**
SYN2005	3	86.8	95.5	**91.0**	91.4	95.7	**93.5**	92.7	95.8	94.2
SYN2010		85.0	94.4	89.5	90.9	94.8	92.8	91.8	94.8	93.3
WebColl		84.2	94.9	89.2	91.0	95.3	93.1	93.2	95.4	**94.3**
SYN2005	4	86.8	95.5	**91.0**	91.4	95.7	**93.5**	92.7	95.8	94.2
SYN2010		85.0	94.4	89.5	90.9	94.8	92.8	91.8	94.8	93.3
WebColl		84.2	94.9	89.2	91.0	95.3	93.1	93.2	95.4	**94.3**
SYN2005	5	86.8	95.5	**91.0**	91.4	95.7	**93.5**	92.7	95.8	94.2
SYN2010		85.0	94.4	89.5	90.9	94.8	92.8	91.8	94.8	93.3

Table 6: Error correction results with respect to different language models

result is a string consisting of correct Czech forms, even though the sentence may not be correct as a whole. All other types of errors are corrected at Level 2.[4]

This annotation scheme was meant to be used by human annotators. However, the size of the full corpus and the costs of its manual annotation have led us to apply automatic annotation and find ways of its improvement.

The hand-annotated part of the corpus (CzeSL-MAN) now consists of 294 thousand word tokens in 2225 short essays, originally hand-written and transcribed.[5] A part of the corpus is annotated independently by two annotators: 121 thousand word tokens in 955 texts. The authors are both foreign learners of Czech and Czech learners whose first language is the Romani ethnolect of Czech.

The entire CzeSL corpus (CzeSL-PLAIN) includes about 2 mil. word tokens. This corpus comprises transcripts of essays of foreign learners and Czech students with the Romani background, and also Czech Bachelor and Master theses written by foreigners.

The part consisting of essays of foreign learners only includes about 1.1 word tokens. It is available as the CzeSL-SGT corpus with full metadata and automatic annotation, including corrections proposed by Korektor, using the original language model trained on the WebColl corpus.[6] In the annotation Korektor detected and corrected 13.24% incorrect forms, 10.33% labeled as including a spelling error, and 2.92% an error in grammar, i.e. a 'real-word' error. Both the original, uncorrected texts and their corrected version was tagged and lemmatized, and "formal error tags," based on the comparison of the uncorrected

and corrected forms, were assigned. The share of 'out of lexicon' forms, as detected by the tagger, is slightly lower – 9.23%.

4.2 The CzeSL-MAN Error Models

We built two error models from the CzeSL-MAN corpus – one for Level 1 (L1) errors and another for Level 2 (L2) errors. As explained in Section 4.1 above, L1 errors are mainly non-word errors and L2 errors belong to real-word and grammatical errors, but still include form errors that are not corrected at L1 because the faulty form happens to be spelled as a form which would be correct in a different context. Extracting errors from the XML format used for encoding the original and the corrected text at L1 is straightforward. The only thing needed is to follow the links connecting tokens at L0 (the original tokens) and L1 (the corrected tokens) and to extract tokens for which the links are labeled as correction links. In the error extraction process, we do not extract errors that involve joining or splitting of word tokens at either level (Korektor does not handle incorrectly split or joined words at the moment).

L2 errors include not only errors identified between L1 and L2 but also those identified already between L0 and L1, if any. This is because L2 tokens are linked to L0 tokens through L1 tokens, rather than being linked directly. For example, consider a single token at Levels L0, L1 and L2: *všechy* (L0) $\xrightarrow{formSingCh,incorBase}$ *všechny* (L1) $\xrightarrow{agr}$ *všichni* (L2). The arrow stands for a link between the two levels, optionally with one or more error labels. For the L1 error extraction, the extracted pair of an incorrect token and a correct token is (*všechy*, *všechny*) with the error labels (*formSingCh*, *incorBase*), and for the L2 error extraction, the extracted error and correct token pair is

---
[4] See [5] and [13] for more details.
[5] For an overview of corpora built as a part of the CzeSL project and relevant links see http://utkl.ff.cuni.cz/learncorp/.
[6] See http://utkl.ff.cuni.cz/~rosen/public/2014-czesl-sgt-en.pdf.

Error	CzeSL-L1		CzeSL-L2	
	train	test	train	test
single-edit	73.54	72.24	67.02	69.30
multi-edit	26.46	27.76	32.98	30.70

Table 7: Percentage of single and multi edit-distance errors in the train/test of L1 and L2 errors.

(*všechy*, *všichni*) with the error labels (*formSingCh, incorBase, agr*). For the L2 errors, we project the error labels of L1 onto L2. If there is no error present or annotated between L0 and L1, then we use the error annotation between L2 and L1. The extracted incorrect token is still from L0 and the correct token from L2.

Many studies have shown that most misspellings are single-edit errors, i.e., misspelled words differ from their correct spelling by exactly one letter. This also holds for our extracted L1 and L2 errors (Table 7). We train our L1 and L2 errors on single-edit errors only, thus the models are quite similar to the native Czech error model described in [11]. The error training is based on [1]. Error probabilities are calculated for the four single-edit operations: *substitution, insertion, deletion,* and *swap*.

5 Experiments with Native and Non-Native Error Models

For the native error model (*webcoll*), we use the same model as described in [12]. For the non-native error models, we create two error models as described in Section 4.2: (i) *czesl_L1* – trained on the L1 errors (CzeSL-L1 data in Table 8) and (ii) *czesl_L2* – trained on the L2 errors (CzeSL-L2 data in Table 8). We partition the CzeSL-MAN corpus in the 9:1 proportion for training and testing.

The non-native training data include more errors than those automatically mined from web. The training of non-native error models is done on single-edit errors only (refer Table 7 for the percentage of errors used for training). For the language model, we use the best model (SYN2005) that we obtained from Section 3.

We perform evaluation on all kinds of errors in test data. We also set the maximum edit distance parameter to 2 for all our experiments. We arrived at this value based on our observation in various experiments. We run our native and non-native models on the test data described in Table 9, and their results are given in Table 10. Error correction results are shown for top-3 suggestions.

In error detection, in terms of F1-score, *czesl_L2* model posts better score than the other two models for both native and non-native data sets. When it comes to error correction, the native model *webcoll* seems to perform better in 2 out of 3 data sets, and the next better performer being the *czesl_L2* model. One has to note that, the non-native models are not tuned to any particular phenomenon such

Train data	Corpus size	#Errors
WebColl	111M	12,761
CzeSL-L1	383K	36,584
CzeSL-L2	370K	54,131

Table 8: Training data for native and non-native experiments. The errors include both single and multi-edit errors.

Test data	Corpus size	#Errors
Audio	1,371	218
CzeSL-L1	33,169	3,908
CzeSL-L2	32,597	5,217

Table 9: Test set for native and non-native experiments. The errors include both single and multi-edit errors.

as capitalization or keyboard layouts, so there is still some scope for improvements on the non-native error models. While *webcoll* and *czesl_L2* models help each other in the opposite direction, i.e., the performance of native model on the non-native data and vice versa, the *czesl_L1* model works better only on the CzesL-L1 dataset. In other words, since L1 error annotation did not involve complete correction of the test data of CzesL-MAN, they can be used, for instance, the correction of misspellings that do not involve grammar errors.

6 Discussion

We manually analyzed a part (the top 3000 tokens) of the output of Korektor for the CzeSL-L2 test data for all the three models. We broadly classify the test data as having *form* errors (occurring between the L0 and L1 level), grammar (*gram*) errors (occurring between L1 and L2) and accumulated errors (*form+gram*, where errors are present at all levels – between L0 and L1, and L1 and L2). The CzeSL-L2 test data can include any of the above types of errors. About 23% of our analyzed data include one of the above errors. More than half of the errors (around 62%) belong to the *form* errors and about 27% belong to the *gram* class. The remaining errors are the *form+gram* errors.

In the case of *form* errors, both the native (*webcoll*) and the non-native models (*czesl_L1* and *czesl_L2*) detect errors at the rate of more than 89%. Form errors may or may not be systematic and they are easily detected by all the three models. Most of the error instances in the data can be categorized under either missing/addition of diacritics, or they can occur in combination with other types of errors, for instance, *přítelkyně* was incorrectly written as *přatelkine*.

Model	Error detection									Error correction								
	Audio			CzeSL-L1			CzeSL-L2			Audio			CzeSL-L1			CzeSL-L2		
	P	R	F1	P	R	F1	P	R	F1	P	R	F1	P	R	F1	P	R	F1
webcoll	95.0	95.9	95.4	81.8	81.7	81.7	91.0	65.0	75.9	93.2	95.8	94.5	71.7	79.6	**75.4**	78.0	61.5	**68.8**
czesl_L1	95.0	96.8	**95.9**	82.2	82.2	**82.2**	91.1	64.4	75.4	93.7	96.7	**95.2**	70.2	79.8	74.7	75.5	60.0	66.8
czesl_L2	95.0	96.8	**95.9**	81.2	82.7	81.9	90.9	65.4	**76.1**	93.7	96.7	**95.2**	68.2	80.0	73.6	74.9	60.9	67.2

Table 10: Error models applied to native and non-native Czech

```
Error label: "form:formCaron0 + formSingCh
+ formY0 + incorBase + incorInfl"
Error token: přatelkine
Gold token: přítelkyně
webcoll: přatelkine
czesl_L1: <suggestions="přítelkyně|pritelkyne
|přátelíme">
czesl_L2: <suggestions="přítelkyně|pritelkyne
|přátelíme">
```

In the case of *gram* errors, most of the errors are undetected. Out of 193 *gram* errors in our analyzed data, the percentage of errors detected by the models are: *webcoll* (15.5%), *czesl_L1* (9.3%) and *czesl_L2* (15.0%). Most of the grammar errors involve agreement, dependency and lexical errors. The agreement errors are shown in Table 11. Except for a few pairs such as *jedné → jednou* (incorrect → correct), *mě → mé*, *který → kteří*, *teplí → teplý*, most of the error tokens involving agreement errors have not been recognized by any of the three models.[7]

Dependency errors (e.g. a wrongly assigned morphological case, missing a syntactic governor's valency requirement) such as $roku_{GEN} \rightarrow roce_{LOC}$ 'year', $kolej_{ACC} \rightarrow koleji_{LOC}$ 'dormitory', $roku_{SG} \rightarrow roky_{PL}$ 'year', $restauraci_{LOC} \rightarrow restaurace_{NOM}$ 'restaurant' have not been recognized by any of the models. The pair $mi_{DAT} \rightarrow mě_{ACC}$ 'me' has been successfully recognized by all the three models and the correct suggestion listed in the top:

```
Error label: "gram:dep"
Error token: mi
Gold token: mě
webcoll: <suggestions="mě|mi|ji|mu|i">
czesl_L1: <suggestions="mě|mi|ji|mu|si">
czesl_L2: <suggestions="mě|mi|ji|mu|ho">
```

For instance, the pair *ve → v* 'in' (vocalized → unvocalized) has been recognized by the *webcoll* and *czesl_L2* models, but not by the *czesl_L1* model. When it comes to grammar errors, *webcoll* and *czesl_L2* have better performance than *czesl_L1*. It was expected, because the *czesl_L1* model was not trained on grammar errors.

When the error involved a combination of *form* and *gram* errors, all the three models tend to perform better. Most of the *form+gram* errors were recognized by

incorrect usage	correct usage	category	gloss
$bavím_{SG}$	$bavíme_{PL}$	number	enjoy
byl_{SG}	$byly_{PL}$	number	was → were
byl_{SG}	$Byly_{PL}$	number	was → were
$Chci_{1ST}$	$Chce_{3RD}$	person	want → wants
$chodím_{SG}$	$chodíme_{PL}$	number	walk
$Chtěla_{FEM}$	$Chtěl_{MASC}$	gender	wanted
$dívat_{INF}$	$dívá_{3RD}$	verb form	to see → sees
$dobré_{FEM}$	$dobří_{MASC.ANIM}$	gender	good
$dobrý_{MASC}$	$dobrá_{FEM}$	gender	good
$druhý_{NOM}$	$druhého_{GEN}$	case	2nd, other
$hezké_{PL}$	$hezký_{SG}$	number	nice
je_{SG}	$jsou_{PL}$	number	is → are
$jednou_{INS}$	$jedné_{LOC}$	case	one
$jich_{GEN}$	je_{ACC}	case	them
$jsem_{SG}$	$jsme_{PL}$	number	am → are
$jsme_{PL}$	$jsem_{SG}$	number	are → am
$jsou_{PL}$	je_{SG}	number	are → is
$který_{SG}$	$kteří_{PL}$	number	which
$leželi_{MASC.ANIM}$	$ležely_{FEM}$	gender	lay
$malý_{SG}$	$malé_{PL}$	number	small
$malých_{GEN}$	$malé_{ACC}$	case	small
$mé_{ACC}$	$mí_{NOM}$	number	my
$Mě_{PERS.PRON}$	$Mé_{POSS.PRON}$	POS	me → my
$miluju_{1ST}$	$miluje_{3RD}$	person	love → loves
$mohli_{MASC.ANIM}$	$mohly_{FEM}$	gender	could
$nemocní_{PL}$	$nemocný_{SG}$	number	ill
$nich_{LOC}$	$ně_{ACC}$	case	them
$oslavili_{MASC.ANIM}$	$oslavila_{NEUT}$	gender	celebrated
$pracovní_{NOM}$	$pracovním_{INS}$	case	work-related
$pracuji_{1ST}$	$pracuje_{3RD}$	person	work → wants
$Studovali_{MASC.ANIM}$	$studovaly_{FEM}$	gender	studied
$teple_{ADV}$ cc	$teplé_{ADJ}$	POS	warmly → warm
$teplí_{PL}$	$teplý_{SG}$	number	warm
$tří_{GEN}$	$tři_{ACC}$	case	three
$tuhle_{FEM}$	$Tenhle_{MASC}$	gender	this
$typické_{FEM}$	$typická_{NEUT}$	gender	typical
$velké_{PL}$	$velký_{SG}$	number	big

Table 11: Some of the agreement errors in the analyzed portion of the CzeSL-L2 test data

all the three models: *webcoll* (85%), *czesl_L1* (86%) and *czesl_L2* (89%). For instance, the error pair **zajímavy → zajímavé* 'interesting' that was labeled at both L1 and L2 level was successfully recognized by all the models, and the correct suggestions were listed in the top. There were

---

[7]The category glosses should be taken with a grain of salt: many forms can have several interpretations. E.g. $oslavili_{MASC.ANIM} \rightarrow oslavila_{NEUT}$ 'celebrated' could also be glossed as $oslavili_{PL,MASC.ANIM} \rightarrow oslavila_{SG,FEM}$.

many errors that were successfully recognized, but the correct suggestions did not appear in top-3, such as, *nechcí → nechtěl* 'didn't want', *mym → svým* 'my', *kamarad → kamaráda* 'friend', *vzdělany → vzdělaná* 'educated'.

Based on the results in Table 10 and the manual error analysis in this section, we can make the following general observations:

- Non-native Czech models can be applied to native test data and obtain even better results than the native Czech model (Table 10).

- From the manual analysis of the test outputs of both native and non-native Czech models, the most problematic errors are the grammar errors due to missed agreement or government (valency requirements). Some of the grammar errors involve most commonly occurring Czech forms such as *jsme, byl, dobrý, je, druhý*.

- Both native and non-native error models perform well on spelling-only errors.

- The CzeSL-MAN error data include errors that involve joining/splitting of word forms that we did not handle in our experiments. We also skipped word order issues in the non-native errors which are beyond the scope of current spell checker systems.

7 Conclusions and Future Work

We have tried to improve both the language model and the error model component of Korektor, a Czech statistical spell checker. Language model improvements involved the employment of more balanced corpora from the Czech National Corpus, namely SYN2005 and SYN2010. We obtained better results for the SYN2005 corpus.

Error model improvements involved creating non-native error models from CzeSL-MAN, a hand-annotated Czech learner corpus, and a series of experiments with native and non-native Czech data sets. The state-of-the-art improvement for the native Czech data set comes from the non-native Czech models trained on L1 and L2 errors from CzeSL-MAN. Surprisingly, the native Czech model performed better for non-native Czech (L2 data) than the non-native models. This we attribute to the rich source of learner error data, since the texts come from very different texts: Czech students with Romani background, as well as learners with various proficiency levels and first languages. Another potential reason could be the untuned nature of the non-native error models that may require further improvement.

As for future work aimed at further improvements of Korektor, we plan to explore model combinations with native and non-native Czech models. We would also like to extend Korektor to cover new languages so that more analysis results could be obtained. To improve error models

further, we would like to investigate how the more complex grammar errors such as those in agreement and form errors such as joining/splitting of word forms can be modeled. Further, we would like to analyze non-native Czech models, so that Korektor can be used to annotate a large Czech learner corpus such as CzeSL-SGT more reliably.

References

[1] Church, K., Gale, W.: Probability scoring for spelling correction. Statistics and Computing **1(7)** (1991), 93–103

[2] Čermák, F., Hlaváčová, J., Hnátková, M., Jelínek, T., Kocek, J., Kopřivová, M., Křen, M., Novotná, R., Petkevič, V., Schmiedtová, V., Skoumalová, H., Spoustová, J., Šulc, M., Velíšek, Z.: SYN2005: a balanced corpus of written Czech, 2005

[3] Gimenes, P. A., Roman, N. T., Carvalho, A. M. B. R.: Spelling error patterns in Brazilian Portuguese. Computational Linguistics **41(1)** (2015), 175–183

[4] Golding, A. R., Roth, D.: A window-based approach to context-sensitive spelling correction. Machine Learning **34** (1999), 107–130 10.1023/A:1007545901558.

[5] Hana, J., Rosen, A., Škodová, S., Štindlová, B.: Error-tagged learner corpus of Czech. In: Proceedings of the Fourth Linguistic Annotation Workshop, Uppsala, Sweden, Association for Computational Linguistics, 2010

[6] Heafield, K.: KenLM: faster and smaller language model queries. In: Proceedings of the EMNLP 2011 Sixth Workshop on Statistical Machine Translation, 187–197, Edinburgh, Scotland, United Kingdom, 2011

[7] Jelínek, T., Štindlová, B., Rosen, A., Hana, J.: Combining manual and automatic annotation of a learner corpus. In: Sojka, P., Horák, A., Kopeček, I., Pala, K., (eds.), Text, Speech and Dialogue – Proceedings of the 15th International Conference TSD 2012, number 7499 in Lecture Notes in Computer Science, 127–134, Springer, 2012

[8] Junczys-Dowmunt, M., Grundkiewicz, R.: The AMU System in the CoNLL-2014 Shared Task: Grammatical error correction by data-intensive and feature-rich statistical machine translation. In: Proceedings of the Eighteenth Conference on Computational Natural Language Learning: Shared Task, 25–33, Baltimore, Maryland, Association for Computational Linguistics, 2014

[9] Křen, M., Bartoň, T., Cvrček, V., Hnátková, M., Jelínek, T., Kocek, J., Novotná, R., Petkevič, V., Procházka, P., Schmiedtová, V., Skoumalová, H.: SYN2010: a balanced corpus of written Czech, 2010

[10] Mays, E., Damerau, F. J., Mercer, R. L.: Context based spelling correction. Information Processing & Management **27 (5)** (1991), 517–522

[11] Richter, M.: An advanced spell checker of Czech. Master's Thesis, Faculty of Mathematics and Physics, Charles University, Prague, 2010

[12] Richter, M., Straňák, P., Rosen, A.: Korektor — a system for contextual spell-checking and diacritics completion. In: Proceedings of the 24th International Conference on Computational Linguistics (Coling 2012), 1019–1027, Mumbai, India, (2012), Coling 2012 Organizing Committee

[13] Rosen, A., Hana, J., Štindlová, B., Feldman, A.: Evaluating and automating the annotation of a learner corpus. Language Resources and Evaluation — Special Issue: Resources for language learning **48 (1)** (2014), 65–92

[14] Rozovskaya, A., Chang, K.-W., Sammons, M., Roth, D., Habash, N.: The Illinois-Columbia System in the CoNLL-2014 Shared Task. In: CoNLL Shared Task, 2014

[15] Rozovskaya, A., Roth, D.: Building a state-of-the-art grammatical error correction system, 2014

[16] Whitelaw, C., Hutchinson, B., Chung, G. Y., Ellis, G.: Using the web for language independent spellchecking and autocorrection. In: Proceedings of the 2009 Conference on Empirical Methods in Natural Language Processing – Volume 2, EMNLP'09, 890–899, Stroudsburg, PA, USA, Association for Computational Linguistics, 2009

[17] Wu, S.-H., Liu, C.-L., Lee, L.-H.: Chinese spelling check evaluation at SIGHAN Bake-off 2013. In: Proceedings of the Seventh SIGHAN Workshop on Chinese Language Processing, 35–42, Nagoya, Japan, Asian Federation of Natural Language Processing, 2013

J. Yaghob (Ed.): ITAT 2015 pp. 81–87
Charles University in Prague, Prague, 2015

ITAT

Spracovanie prirodzeného jazyka
pre interaktívne rečové rozhrania v slovenčine

Ján Staš, Daniel Hládek, Stanislav Ondáš, Daniel Zlacký a Jozef Juhár

Katedra elektroniky a multimediálnych telekomunikácií, Fakulta elektrotechniky a informatiky,
Technická univerzita v Košiciach, Park Komenského 13, 042 10 Košice, Slovenská republika
{jan.stas, daniel.hladek, stanislav.ondas, daniel.zlacky, jozef.juhar}@tuke.sk

Abstrakt: V príspevku sú zhrnuté priebežné výsledky aplikovaného výskumu v oblasti spracovania prirodzeného jazyka v úlohách orientovaných na výskum a vývoj modulov rečových rozhraní medzi človekom a strojom, ktorý prebieha v Laboratóriu rečových a mobilných technológií na KEMT FEI TU v Košiciach. Zahrnutie hovorenej reči, ako najprirodzenejšieho komunikačného nástroja medzi ľuďmi, má svoje nezastupiteľné miesto aj pri návrhu a vývoji interaktívnych rečových rozhraní. Pri prechode od rozpoznávania ľudskej reči k jej porozumeniu strojom je potom nevyhnutné vykonať aj dodatočnú analýzu textu po automatickom prepise. To zahŕňa aj proces transformácie textu po rozpoznaní na reprezentáciu určitého typu znalostí, ktorému dokáže stroj porozumieť. Tento zložitý proces všeobecne pozostáva z tokenizácie, automatickej korekcie a dodatočnej morfologickej, syntaktickej a sémantickej analýzy textu. Nami navrhnuté moduly a výsledky automatického spracovania textu v slovenskom jazyku budú postupne predstavené v tomto príspevku.

1 Úvod

S príchodom výpočtovej techniky sa stala potreba počítačového spracovania prirodzeného jazyka aktuálnou celosvetovou témou. Vedci sa po celom svete snažia podchytiť charakter takmer každého jazyka s cieľom zjednodušiť interakciu medzi ľuďmi a strojmi a komunikáciu medzi ľudmi samotnými. Oblasť spracovania prirodzeného jazyka zahŕňa širokú škálu disciplín, ako napr. vyhľadávanie informácií, štatistické modelovanie jazyka, strojový preklad, automatické rozpoznávanie a porozumenie reči a pod. Jednotlivé disciplíny však vo väčšine prípadov úzko súvisia, dopĺňajú sa, a pomocou nich je možné ľuďom uľahčiť prácu, štúdium, komunikáciu, či zábavu. Jednou z najaktuálnejších úloh v oblasti spracovania prirodzeného jazyka je aj automatické rozpoznávanie reči (ARR), ktorému sa v našom laboratóriu intenzívne venujeme. Vďaka viacerým zlepšeniam v oblasti automatického rozpoznávania reči v slovenčine sme schopní rozpoznať ľudskú reč s dostatočnou presnosťou v mnohých aplikačných úlohách, avšak komplexné porozumenie významu je v súčasnosti jednou z najnáročnejších úloh pri návrhu rôznych interaktívnych rečových rozhraní a to nielen v slovenčine. V tomto článku budú predstavené nami navrhnuté prístupy na spracovanie prirodzeného jazyka pre interaktívne rečové rozhrania v slovenčine.

2 Zdrojové dáta

2.1 Dolovanie textu

Rozsiahly korpus písaných textov použitý vo viacerých oblastiach spracovania prirodzeného jazyka v Laboratóriu rečových a mobilných technológií bol vytvorený pomocou nami navrhnutého systému *na dolovanie a spracovanie textových dokumentov z webových stránok* dostupných na sieti Internet s názvom webAgent [1, 2]. Systém doluje textové dáta z rôznych domén a elektronických zdrojov písaných v slovenčine a pomocou preddefinovaných pravidiel na prepis čísloviek, symbolov a skratiek ich spracúva do ich vyslovovanej podoby. Systém je navyše rozšírený o metódy tokenizácie, segmentácie na vety, metódy na kontrolu duplicity na úrovni adresy zdroja textu a obsahu dokumentu, a tiež o metódy filtrácie viet obsahujúcich veľké množstvo gramaticky nespisovných slov, číslic, akronymov, symbolov, skratiek, a iných cudzojazyčných a mimoslovníkových slov. Spracovaný text je následne rozdelený do menších celkov, t.j. podkorpusov, pomocou účinných metód na kategorizáciu textových dokumentov. Súčasný korpus písaných textov v slovenčine obsahuje približne 2,25 mld. tokenov obsiahnutých v 125 mil. vetách.

2.2 Kategorizácia textu

S narastajúcim množstvom textových dokumentov stiahnutých zo siete Internet a potrebou vytvárať čoraz presnejšie doménovo orientované interaktívne rečovozaložené systémy a rozhrania, sa vynorila otázka kategorizovať textové dáta nielen podľa adresy (URL) zdroja textu, odkiaľ daný textový dokument pochádza, ale aj na úrovni jeho obsahu. Navyše webová adresa zdroja textu nemusí byť hneď jednoznačným identifikátorom obsahu dokumentu, vychádzajúc tiež z predpokladu, že jeden dokument môže pojednávať o viacerých témach. Kategorizácia textu má preto veľký význam pri návrhu a tvorbe robustných doménovo orientovaných systémov na automatické rozpoznávanie reči, ale aj v iných úlohách využívajúcich textové dáta ako zdroj informácií, napr. pri návrhu a vývoji interaktívnych rečovo-založených rozhraní.

Narozdiel od *metód zhlukovania*, kde dokumenty s využitím štatistických prístupov spájame do určitého počtu zhlukov, v ktorých tému vopred nepoznáme, pri *kategorizácii dokumentov* sa snažíme zadeliť dokumenty do dvoch alebo viacerých tried na základe ich minimálnej vzdialenosti, resp. sémantickej podobnosti, udávajúcej prienik slov alebo celých fráz medzi dokumentami. V oboch prípadoch je nutné identifikovať tému v danom zhluku, resp. triede, a to buď pomocou prístupov založených na extrakcii kľúčových slov, pravdepodobnostných prístupov založených na výpočte podobnosti dokumentov pomocou dištančných metrík, alebo ich kombináciou.

Počiatočný výskum v oblasti kategorizácie textových dokumentov bol venovaný metódam zhlukovania pomocou iteračných algoritmov založených na *k-means* a *k-medoid zhlukovaní* a na *hierarchickom zhlukovaní* využívajúcom *aglomeračné* a *divizívne kritérium* [3]. Ako najvhodnejším prístupom sa ukázalo hierarchické zhlukovanie textu s využitím aglomeračného kritéria pri zhlukovaní článkov zo slovenskej Wikipédie. Tento spôsob zhlukovania dokumentov sme porovnali s nami navrhnutou metódou založenou na *klasifikácii textových dokumentov pomocou kľúčových fráz* využívajúcou F-skóre ako hodnotiace kritérium [4]. Nami navrhnutý spôsob klasifikácie sa výsledkami ukázal porovnateľný k hierarchickému zhlukovaniu, avšak hlavnou nevýhodou navrhnutej metódy je nutnosť mať k dispozícii zoznamy kľúčových slov, resp. fráz pre jednotlivé domény a v procese klasifikácie textu je nutné správne (v ideálnom prípade automaticky) nastaviť vhodný prah delenia. Tento typ klasifikácie sme v nasledujúcom výskume zameranom na kategorizáciu textu v úlohe robustného doménovo orientovaného modelovania jazyka rozšírili o ďalšie tri metriky určujúce vzdialenosť, resp. podobnosť medzi dokumentami, konkrétne o Bhattacharyyaov koeficient, Jaccardov index a Jensenovu-Shannonovu divergenciu. Ako najvhodnejšou mierou v úlohe klasifikácie textu sa javí použitie Jaccardovho indexu pri výpočte podobnosti dokumentov [5].

Iným spôsobom je použitie metód nekontrolovaného učenia v úlohe kategorizácie textových dokumentov. Vo viacerých súčasných výskumoch zaoberajúcich sa modelovaním jazyka sme pri kategorizácii dokumentov siahli po *latentnej Dirichletovej alokácii* (z angl. „latent Dirichlet allocation", skr. LDA). LDA je charakterizovaná ako generatívny pravdepodobnostný model, ktorý vychádza z multinomického a Dirichletovho rozdelenia pravdepodobnosti [6]. Zavedenie LDA v úlohe modelovania slovenského jazyka pri automatickom prepise diktovaných súdnych rozhodnutí prinieslo tiež výrazné zníženie perplexity modelov a miery chybovosti systému ARR [7].

3   Identifikácia tokenov a vetných hraníc

Prvým krokom v spracovaní textu je jeho príprava a prepis číslic, symbolov a skratiek do ich vyslovovanej podoby. Úlohou je pomocou sústavy pravidiel identifikovať tiež textové jednotky, ktoré sú zaujímavé z hľadiska ďalšieho spracovania, t.j. úprava na jednotný spôsob zápisu a eliminácia nepodstatných častí. Predspracovanie je tak nevyhnutným krokom pre akékoľvek ďalšie štatistické spracovanie, zvlášť v prípade textov stiahnutých z Internetu.

Najdôležitejšou časťou predspracovania textu je jeho *tokenizácia*. Jej cieľom je identifikácia jednotlivých slov a vetných hraníc, ktoré môžu slúžiť ako vstup do ďalšieho spracovania. V tomto kroku sa tiež snažíme zjednotiť spôsob zápisu číslic, diakritiky, interpunkcie, akronymov, symbolov, skratiek a iných významových jednotiek.

Tokenizácia sa zvyčajne vykonáva postupnou aplikáciou vhodne zapísaných regulárnych výrazov, ktoré obsahujú pravidlá pre identifikáciu textových jednotiek, dôležitých pre ďalšie spracovanie. Nepodstatné časti textu, ktoré nie sú pokryté pravidlami, sú z textu vynechané. Nami navrhnutý tokenizátor identifikuje tieto časti textu: diakritika, slová, akronymy, symboly, skratky, zoznamy, odseky, čísla, e-mailové adresy a adresy URL. Identifikácia vetných hraníc je ďalej vykonávaná pomocou rozlíšenia významu bodky, jej desambiguáciou. V slovenských textoch môže byť bodka súčasťou označenia číselného poradia, skratky alebo e-mailovej alebo webovej adresy.

Na začiatku procesu identifikácie významových častí je vstupný reťazec porovnaný so všetkými pravidlami v databáze. Pravidlo, ktoré vyhovuje najdlhšiemu textu, je vybraté a jeho zodpovedajúci text je prepísaný podľa požiadaviek. Tento text je potom odstránený zo vstupného reťazca. Ak nevyhovuje žiadne pravidlo, vstupný reťazec je skrátený o jeden znak a prehľadávanie bázy pravidiel pokračuje. Výsledkom tokenizácie je text, kde sú textové jednotky oddelené medzerou a vety novým riadkom.

Proces identifikácie tokenov je zvyčajne výpočtovo náročný. Pre urýchlenie sme všetky pravidlá zapísali v špeciálnom jazyku Ragel [8] a spojili do jediného stavového automatu v programovacom jazyku C, z ktorého je zvyčajným spôsobom vytvorený spustiteľný súbor resp. knižnica [9]. Podrobnejšie informácie možno nájsť v [1].

4   Anotácia textu

Tam kde to je možné, využívame pre anotáciu textu prístupy založené na štatistickom modelovaní. V trénovacej databáze sú zvyčajne tokenom priradené určité triedy alebo morfologické značky. Štatistický klasifikátor analyzuje trénovací korpus a je schopný priradiť najpravdepodobnejšiu značku aj takým kontextom, ktoré sa v trénovacej databáze nevyskytujú. Slovenčina sa vyznačuje relatívne voľným poradím slov vo vetách, vysokým počtom morfologických tvarov slov a gramatických výnimiek. Počet možných kontextov tak môže byť veľmi vysoký, a to sťažuje úlohu natrénovania čo možno najpresnejšieho štatistického klasifikátora.

4.1   Rozpoznávanie pomenovaných entít

Z dôvodu nedostatku trénovacích dát pre rozpoznávanie pomenovaných entít v súčasnosti využívame systém založený na pravidlách. Systém využíva sadu slovníkov, regulárnych výrazov a viacslovných pomenovaní, ktoré sú spojené do unifikovaného systému na automatický prepis

korpus	anotácia	TreeTagger	Dagger
Národný korpus jazyka poľského	manuálna	15,63	11,83
Český akademický korpus (v2.0)	manuálna	13,10	9,46
Slovenská časť korpusu W2C	automatická	10,30	4,47
Maďarský webový korpus	automatická	2,55	1,97

Tabuľka 1: Miera chybnej klasifikácie [v %] morfologických analyzátorov TreeTagger a Dagger

pomenovaných entít. Tento na pravidlách založený systém pracuje podobne ako tokenizátor, pomocou stavového automatu. Rozpoznané pomenované entity je možné využiť v rôznych úlohách spracovania prirodzeného jazyka v slovenčine. Vzhľadom na to, že autorom systému na rozpoznávanie pomenovaných entít nie je v súčasnosti známy žiaden iný porovnateľný nástroj vytvorený pre slovenčinu, ani databáza vhodná na testovanie, nie je preto možné tento nástroj správne ohodnotiť a vyčísliť jeho úspešnosť.

4.2 Morfologická analýza textu

Morfologické značky sú jedným z najdôležitejších príznakov v spracovaní prirodzeného jazyka. Z toho dôvodu sme *morfologický klasifikátor* Dagger [10] navrhli tak, aby bral do úvahy špecifické vlastnosti flektívnych jazykov. Klasifikátor je založený na skrytom Markovovom modeli (z angl. *„hidden Markov model"*, skr. HMM) druhého rádu a najpravdepodobnejšia postuponosť morfologických značiek je vyhľadávaná Viterbiho algoritmom.

Nami navrhnutý HMM klasifikátor Dagger pre morfologickú analýzu flektívnych jazykov sa skladá z nasledujúcich štyroch častí:

1. lexikón, ktorý navrhuje množinu možných značiek na základe slova alebo jeho koncovky;

2. model prechodov, ktorý vyjadruje pravdepodobnosť nasledujúcej značky na základe dvoch predchádzajúcich,

3. model pozorovaní, ktorý vyjadruje pravdepodobnosť slova na základe možnej značky;

4. a v prípade, že skúmané slovo sa nenachádza v trénovacej databáze, využije sa dodatočný model pozorovaní, ktorý vyjadruje pravdepodobnosť stavu na základe koncovky daného slova.

Je vhodné poznamenať, že algoritmus obsiahnutý v morfologickom analyzátore Dagger využíva vlastný algoritmus na automatickú identifikáciu koncoviek slov založený na minimálnej opisnej dĺžke.

Na natrénovanie klasifikátora pre slovenský jazyk sme využili početnosti trigramov slov z *ručne morgologicky anotovaného korpusu* r-mak-2.0 [1] a množinu *morfologických značiek* [2], získaných zo Slovenského národného korpusu Jazykovedného ústavu Ľudovíta Štúra na Slovenskej akadémii vied v Bratislave. Algoritmus sme uplatnili

najmä pri morfologickej analýze korpusu písaných textov v slovenčine a pri trénovaní štatistických modelov jazyka založených na triedach slov v systémoch na automatické rozpoznávanie plynulej reči v slovenčine [11, 12].

Presnosť klasifikácie nami navrhnutého morfologického analyzátora Dagger sme porovnali s dobre známym slovnodruhovým (z angl. *„part-of-speech"*, skr. PoS) značkovačom TreeTagger [13], ktorého algoritmus je založený na rozhodovacích stromoch. Pre porovnanie presnosti klasifikácie sme morfologické analyzátory vyhodnotili na štyroch rôznych manuálne resp. automaticky anotovaných textových korpusoch, a to na Národnom korpuse jazyka poľského [14], Českom akademickom korpuse, vo verzii 2.0 [15], Maďarskom webovom korpuse [16], a na slovenskej časti textového korpusu Web to Corpora [17][3]. V Tab. 1 sú znázornené výsledky miery chybnej klasifikácie, ktoré ukazujú, že nami navrhnutý algoritmus morfologickej analýzy dosahuje porovnateľnú presnosť s klasifikáciou obsiahnutou v nástroji TreeTagger.

4.3 Doplňovanie diakritiky

Častým javom pri komunikácii medzi ľudmi prebiehajúcej na sieti Internet je vysoký výskyt preklepov a chýbajúca diakritika. Hoci človeku to väčšinou pri porozumení správy nerobí problém, pri počítačovom spracovaní prirodzeného jazyka je potrebné nájsť vhodný spôsob pre rozlíšenie významu nejednoznačných zápisov na základe okolitého kontextu. Z toho dôvodu sme sa venovali aj problému automatického doplňovania diakritiky slov [18].

Podobne ako pri návrhu morfologického analyzátora sme pri rekonštrukcii diakritiky využili algoritmus využívajúci skrytý Markovov model. V tomto prípade je však matica prechodov tvorená trigramovým jazykovým modelom a matica pozorovaní je trénovaná pomocou algoritmu pre generovanie nesprávnych zápisov na texte, ktorý je pokladaný za správny. Úspešnosť navrhnutého systému na automatické doplňovanie diakritiky v korpusoch textov z blogov písaných v slovenskom jazyku dosahuje úroveň až 85%. Podobný nástroj na rekonštrukciu diakritiky pre slovenčinu, využívajúci štatistické modely jazyka vysokého rádu, bol vytvorený tiež tímom pracovníkov v Slovenskom národnom korpuse Jazykovedného ústavu Ľudovíta Štúra na Slovenskej akadémii vied v Bratislave[4].

---

[1] http://korpus.sk/ver_r(2d)mak.html
[2] http://korpus.juls.savba.sk/attachments/morpho/tagset-www.pdf

[3] https://lindat.mff.cuni.cz/repository/xmlui/
[4] http://korpus.juls.savba.sk/diakritik.html

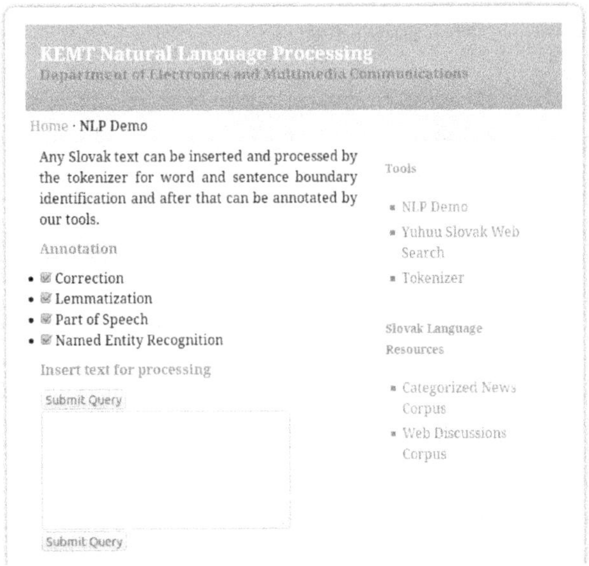

Obr. 1: Webové rozhranie k systémom na spracovanie prirodzeného jazyka na KEMT FEI TU v Košiciach

4.4 On-line webové rozhranie

Pre účely demonštrácie a testovania presnosti tokenizácie, morfologickej analýzy a automatického doplňovania diakritiky slov sme k nami navrhnutým nástrojom na počítačové spracovanie priedzeného jazyka v slovenčine vytvorili aj jednoduché on-line webové rozhranie[5], ktoré je znázornené na Obr. 1 a opísané v článku [11].

5 Aktivity v oblasti aplikovaného výskumu

5.1 Štatistické modelovanie jazyka

Konštrukcia štatistického modelu jazyka pre slovenčinu, ktorá patrí do skupiny vysoko flektívnych jazykov, je oveľa obtiažnejšia, než vytvorenie štatistického modelu pre jazyk anglický. Prvým dôvodom je *neusporiadanosť* slovenského jazyka, čo vedie k voľnejším pravidlám reťazenia slov do viet. Druhým je samotná *flektívnosť* jazyka, ktorá vytvára predpoklad pre mnohonásobne väčší slovník, než je to v prípade jazyka anglického.

Súčasný stav v oblasti štatistického modelovania jazyka v slovenčine v doposiaľ navrhnutých systémoch na interakciu človeka so strojom hovorenou rečou a automatické rozpoznávanie a prepis plynulej reči to textu sa opiera o poznatky z oblasti modelovania príbuzných jazykov, najmä jazyka českého, poľského, slovinského, srbochorvátskeho, či ruského. Čo sa týka samotného modelovania pomocou štatistických metód, hlavným predpokladom pri tvorbe kvalitného jazykového modelu je dôsledné predspracovanie textového korpusu, ktorý vstupuje do procesu trénovania. Zvýšenú pozornosť je vhodné preto venovať najmä prepisu čísloviek a skratiek, a v neposled-

nom rade aj generovaniu ohybných tvarov vlastných podstatných mien, ktoré tvoria kritickú časť pri tvorbe akéhokoľvek štatistického modelu jazyka. Taktiež slovník, ktorý vstupuje do procesu trénovania, ale aj samotného rozpoznávania reči, musí byť v podmienkach reálneho nasadenia systému ARR obmedzený čo do počtu slov. Ukázalo sa, že dobré výsledky modelovania slovenského jazyka sa dosahujú už pri veľkosti 100–150 tisíc slov pri doménovo orientovanom automatickom prepise diktovanej reči do textu a 300–400 tisíc slov v úlohách všeobecného rozpoznávania spontánnej reči [19].

V oblasti *adaptácie modelov jazyka* na vybranú tému alebo prehovor rečníka sa ukázalo, že metódy, ktoré vykazovali pozoruhodné výsledky pre štatisticky viac závislé jazyky (ako napr. angličitna), v prípade slovenčiny nebolo pozorované výrazné zlepšenie. Z toho dôvodu je použitie metódy *lineárnej interpolácie* tematicky zameraných modelov jazyka viac než postačujúce, pričom výpočet interpolačných váh by mal byť určený *minimalizáciou perplexity modelov na množine odložených dát*. Ako adaptačné dáta je vhodné použiť buď textové dáta získané z prepisov rečových nahrávok obsiahnutých v rečových databázach, alebo písané texty čo možno najviac príbuzné doméne, v ktorej rozpoznávanie reči prebieha [20].

Kvalitu jazykového modelu, ako aj úspešnosť samotného rozpoznávania reči je možné zlepšovať množstvom optimalizačných techník. Jednou z možností je modelovať vysoko frekventované javy pomocou *viacslovných výrazov*. Takéto výrazy pokrývajú zväčša kontext dvoch–troch slov a zvyčajne sú tvorené odbornými termínmi, resp. spojením predložky s podstatným, či prídavným menom. Na základe experimentov opísaných v [19] konštatujeme, že viacslovné výrazy, aj keď len v malej miere, dokážu prispieť k zlepšeniu presnosti rozpoznávania plynulej reči, a to najmä na začiatku rečového prejavu pri rozpoznávaní krátkych jednoslabičných slov a nadobúdajú na význame aj v čiastkových úlohách pri reprezentácii viacslovných pomenovaní v jazykovom modeli, a tým prispievajú aj k postspracovaniu dát po rozpoznaní systémom ARR.

Ďalšou možnosťou je modelovať málopočetné javy v jazyku pomocou *morfémových modelov*. Delením singletónov a javov s veľmi malým výskytom vo vybranom jazyku na subslovné jednotky (koreň a koncovku), je možné štatisticky pokryť aj také javy, ktoré sa priamo v jazykovom modeli nevyskytujú. Výsledky modelovania slovenského jazyka pomocou morfémových modelov ukazujú výraznú redukciu počtu mimoslovníkových tvarov a perplexity modelov približne o jednu tretinu [21].

Naopak javy, ktoré sa v danom jazyku menia dynamicky a počet všetkých možných tvarov ohybných slov nie je v jazyku limitovaný, je vhodné modelovať pomocou *modelov založených na triedach slov*. Medzi takéto javy možno zahrnúť najmä vlastné podstatné mená, ako sú krstné mená, priezviská, geografické názvy alebo číslovky. Experimentálne výsledky modelovania slovenského jazyka pomocou slovných tried odvodených od koncoviek slov ukazujú mierne zlepšenie presnosti rozpoznávania

---

[5]http://nlp.web.tuke.sk/

reči oproti štandardným modelom približne o 5% v relatívnej miere a sú určitým kompromisom medzi modelmi založenými na slovných druhoch a štandardnými modelmi, čo do počtu slovných tried, percenta mimoslovníkových slov, či perplexity jazykového modelu [22].

Pri tvorbe jazykových modelov použiteľných v systémoch na automatický prepis spontánnej reči je často nevyhnutné sa vysporiadať aj s rôznymi mimorečovými prejavmi, ktoré pochádzajú priamo do rečníka. Tie sú spôsobené zlou výslovnosťou, nevhodnou artikuláciou a nedokonalosťou rečového prejavu. *Modely vložených páuz a dysfluentných javov* sa preto snažia podchytiť a zahrnúť do jazykového modelu rôzne suprasegmentálne javy obsiahnuté v rečovom prejave, ako napr. zaváhanie, prolongovanie a opakovanie slov resp. fráz, skomolenie slov, či časté dýchanie. Tieto javy vo vysokej miere vplývajú aj na celkovú chybovosť systému ARR. Bolo dokázané, že vhodným výberom a správnou reprezentáciou vybraných typov vložených páuz a dysfluentných javov v slovníku výslovnosti a v modeli jazyka je možné dosiahnuť zlepšenie presnosti rozpoznávania reči relatívne až do 10%.

V neposlednom rade, kvalitu jazykového modelu je možné zvýšiť aj na úrovni *rozširovania štatistík* pomocou internetových vyhľadávačov [7], prekladom slov alebo slovných párov z príbuzných jazykov a pod.

Oblasť štatistického modelovania slovenského jazyka má za sebou krátku minulosť a donedávna jej nebola venovaná taká pozornosť ako napr. v susednej Českej republike. Z toho dôvodu bolo nevyhnutné pri tvorbe štatistických modelov slovenského jazyka, ako aj pri samotnom spracovaní textových dát, ktoré sa používajú najmä v procese ich trénovania a adaptácie, podrobne naštudovať aj oblasť komputačnej lingvistiky, a vytvoriť tak rad programových nástrojov na počítačové spracovanie slovenského jazyka. Absencia dostupnosti niektorých kľúčových nástrojov slúžiacich najmä k morfologickej či syntaktickej analýze tiež obmedzili použitie jazykových modelov založených na triedach slov v plnom rozsahu, aj keď prvé kroky v tejto oblasti už boli uskutočnené. Napriek týmto obmedzeniam, modely slovenského jazyka dosahujú vysokú úspešnosť na úrovni 84–95% v reálnych aplikáciách systému ARR, ktorých výsledky sú zhrnuté v Tab. 2, od jednoduchých hlasových rozhraní slúžiacich na ovládanie robotických systémov, cez jednoduché rečové dialógové manažéry poskytujúce hlasové, interaktívne, či multimodálne služby, až ku komplexným diktačným a transkripčným systémom, ktoré pracujú s veľmi veľkými slovníkmi, nezávisle od rečníka, dokážu sa adaptovať na vybranú tému, či konkretného rečníka, sú robustné a prebiehajú v reálnom čase [7, 12, 20, 23].

Vďaka narastajúcemu záujmu o interaktívne rečové technológie v slovenskom jazyku sa ďalšie smerovanie v tejto oblasti uberá cestou využitia doménovo orientovaných modelov jazyka pri tvorbe diktačných systémov aj pre takú oblasť ako je medicína, ďalej systémov na automatický prepis akademických prednášok, spravodajských relácií, športových prenosov, televíznych, rozhlasových, či

parlamentných debát, resp. obchodných rokovaní v malých konferenčných miestnostiach s viacerými účastníkmi, a to s využitím robustných algoritmov na adaptáciu jazykových modelov na vybranú tému alebo rečníka [20], ale aj ďalších aplikácií určených pre robotické systémy.

úloha	Acc* [%]
hlasové ovládanie ramena robota (SNR=10dB)	90,19
automatický prepis diktátu z oblasti súdnictva	95,11
elektronické služby Generálnej prokuratúry	94,65
prepis reálnych parlamentných debát	90,93
prepis spravodajských relácií	84,33

* Presnosť (z angl. „*accuracy*", Acc) automatického prepisu vychádza z miery chybovosti slov (z angl. „*word error rate*"), definovanej ako minimálna vzdialenosť medzi referenčnou sekvenciou slov a automatickým prepisom po rozpoznaní.

Tabuľka 2: Celková presnosť jazykových modelov v doposiaľ navrhnutých systémoch na automatické rozpoznávanie a prepis reči do textu v slovenskom jazyku

5.2 Porozumenie prirodzeného jazyka

K tomu, aby bolo možné obohatiť počítačové systémy o schopnosť skutočného *porozumenia prirodzenej reči a jazyka*, je potrebné realizovať proces *sémantickej analýzy*, a získaný význam tak vhodným spôsobom zachytiť a zosúladiť s databázou znalostí. Je možné konštatovať, že ide o neľahkú úlohu, vzhľadom na komplexnosť slovenského jazyka, jej sémantiky a ostatných faktorov, ktoré súvisia s porozumením. Aj keď sémantická analýza nie je hlavným zameraním nášho laboratória, pre využitie systému ARR v aplikáciách interakcie človeka so strojom hovorenou rečou, je nevyhnutné sa vysporiadať s interpretáciou vyjadrení tiež v prirodzenom jazyku.

V systéme IRKR [24] sme na tento účel implementovali v jednotke riadenia dialógu podporu jazyka pre sémantickú analýzu – W3C SISR, ktorý umožňuje vložiť interpretačné inštrukcie priamo do *deterministických gramatík*, napísaných podľa W3C odporúčania SRGS. V danom prípade sa jednalo iba o limitované porozumenie, ktoré bolo zamerané skôr na naplnenie doménovo špecifických sémantických slotov hodnotami získanými z rečového prejavu používateľa. Takéto riešenie bolo v tom čase pomerne komfortné a postačujúce pre celý rad rečových aplikácií a rozhraní, ktoré poskytujú tzv. rečovo-založené dialógové systémy.

Pri riešení projektu zameraného na implementáciu hlasového ovládania do robotickej platformy, kde kvôli kompaktnosti a rýchlosti systému na limitovanom hardvéri nebolo možné použiť rečové gramatiky, keďže komplexnosť riadiacich povelov bola omnoho väčšia, implementovali sme pre tento účel tzv. „*keyword-spotting*" techniku sémantickej analýzy. Vytvorili sme viacero doménovošpecifických sémantických slotov, ktoré zachytávali preddefinované slová z rečového dopytu používateľa [23].

Spolu s návrhom a vývojom systémov ARR, nielen pre rozpoznávanie jednoduchých povelov a fráz, ale aj pre diktovanú a spontánnu reč, vzrastali nároky na ich interpretáciu a predchádzajúce prístupy pre ne neboli použiteľné. Pre potreby rozpoznávania plynulej reči sa namiesto deterministických gramatík začali vo veľkom využívať štatistické modely jazyka, nakoľko plynulá reč poskytuje podstatne väčšiu výrazovú variabilitu.

Zvlášť dôležitou sa sémantická analýza a interpretácia ukázala pri experimentovaní s *virtuálnym konverzačným agentom*, ktorý má ľudský zjav [25]. Pri takomto druhu komunikácie má človek tendenciu očakávať od virtuálneho konverzačného agenta podobné výrazové prostriedky ako majú ľudia, predovšetkým v oblasti komunikačných schopností a porozumenia, ktoré spolu úzko súvisia. Ďalším špecifikom je, že systémy s hlasovým rozhraním sa stávajú viac doménovo nezávislými, teda umožňujú dialógovú interakciu v rámci množstva tém (ako napr. Apple SIRI[6] a pod.), čo posúva interpretáciu významu ďalej, od relatívne „bezpečných" doménových sémantických slotov k viac všeobecným sémantickým roliam.

T. E. Payne v [26] definuje sémantické roly nasledovne: *„Sémantická rola predstavuje základný vzťah, ktorý daná entita má k hlavnému slovesu vo vete."* Ďalej vysvetľuje, že: *„Sémantická rola je aktuálna rola, ktorú participant hrá v nejakej reálnej alebo imaginárnej situácii, bez ohľadu na lingvistickú realizáciu danej situácie."*

Aj keď je teória sémantických rolí a s nimi súvisiacich valenčných rámcov slovies pomerne dobre rozpracovaná pre rôzne jazyky, neexistujú však žiadne systémy na *automatické určovanie sémantických rolí* (z angl. *„automatic semantic roles labeling"*, skr. ASRL) v slovenčine. Za veľmi dôležitú prácu v oblasti automatického určovania sémantických rolí pre slovenčinu možno považovať prácu E. Paleša, ktorý detailne opísal proces porozumenia prirodzeného jazyka na jednotlivých vrstvách a vyvinul prvý systém SAPFO – Parafrázovač slovenčiny, ktorého súčasťou bol aj modul pre určovanie sémantických rolí [27]. Tento systém však nie je podľa našich vedomostí voľne dostupný. Navyše sa jedná o deterministický systém, ktorý ako konštatuje M. Laclavík [28], nie je možné uspokojivo skonštruovať pre analýzu slovenského jazyka, z dôvodu veľkého množstva výnimiek. Z tohto pohľadu sú štatistické metódy jednoznačne lepšou voľbou.

Štatistické metódy pre systémy ASRL využívajú tzv. štatistické modelovanie typické pre rôzne úlohy v oblasti spracovania prirodzeného jazyka. Pre natrénovanie štatistických modelov je potrebná textová databáza anotovaná na úrovni sémantických rolí, ktorá v prípade slovenčiny doposiaľ prakticky neexistovala. Označenie vetných participantov pomocou sémantických rolí je náročná úloha a vyžaduje tiež dobré lingvistické znalosti.

Vzhľadom na neexistenciu databázy pre slovenčinu anotovanej na úrovni sémantických rolí, sme sa rozhodli vytvoriť aj takýto druh korpusu. Korpus SEMIENKO [29]

aktuálne obsahuje tristo viet v slovenskom jazyku anotovaných podľa nami *modifikovanej dvojúrovňovej schémy na označovanie sémantických rolí*, prevzatej z anotačnej schémy podľa E. Paleša [27] a upravenej pre potreby automatickej sémantickej analýzy. Sémantická anotácia korpusu SEMIENKO je ilustrovaná na nasledujúcom príklade:

$$\text{AGS}|\text{KOG}^{[Ján]} \quad \text{VRB}^{[spoznal]} \quad \text{PAC}|\text{FEN}^{[Máriu]} .$$

Úloha automatického určovania sémantických rolí všeobecne pozostáva z dvoch základných častí, a to z rozdelenia viet na vetné participanty a následného priradenia sémantických rolí daným participantom. Pre klasifikáciu vetných participantov sme experimentálne vyskúšali dve techniky. Prvá metóda modeluje jednotlivé pravdepodobnosti nepriamo pomocou *n*-gramových modelov [30], pričom účinnosť klasifikácie na danom korpuse dosahuje úspešnosť na úrovni 48%. Druhá metóda využíva modifikovaný HMM klasifikátor, obsiahnutý v nástroji Dagger [10], ktorý v procese prehľadávania výstupnej sekvencie implementuje Viterbiho dekódovanie. Úspešnosť tohto typu klasifikácie v súčasnosti dosahuje úroveň až 56%, čo je vzhľadom na veľkosť trénovacej množiny adekvátne. Na základe predbežných výsledkov sémantickej analýzy v slovenskom jazyku môžeme konštatovať, že pre ďalšie zlepšenie je nevyhnutné významne rozšíriť manuálne anotovaný korpus, čo je však neľahká a veľmi pracná úloha.

6   Záver

V tomto príspevku boli predstavené úlohy z oblasti spracovania a modelovania slovenského jazyka, ktorým sa v Laboratóriu rečových a mobilných technológií na KEMT FEI TU v Košiciach v súčasnosti intenzívne venujeme. Je možné konštatovať, že úspešnosť nami navrhnutých algoritmov stále dobieha úroveň svetových výskumov, avšak súčasné výsledky je možné už teraz aplikovať v rôznych systémoch na rozpoznávanie a porozumenie reči, ale aj v iných systémoch intrakcie človeka so strojom hovorenou rečou, ktoré na našom pracovisku vyvíjame.

Poďakovanie

Táto práca vznikla realizáciou projektu Univerzitný vedecký park TECHNICOM pre inovačné aplikácie s podporou znalostných technológií (kód ITMS: 26220220182) vďaka podpore operačného programu Výskum a vývoj spolufinancovaného zo zdrojov Európskeho fondu regionálneho rozvoja (25%) a výskumných projektov: Výskum a vývoj modulov pre jazykovo-adaptívne multimodálne rozhrania na základe Zmluvy č. SK-HU-2013-0015 podporujúcej spoluprácu medzi organizáciami v Slovenskej republike a v Maďarsku (50%), a Slovník viacslovných pomenovaní (lexikografický, lexikologický a komparatívny výskum) v rámci projektu APVV-0342-11 (25%), realizovaných vďaka podpore Agentúry na podporu výskumu a vývoja financovanej z prostriedkov Ministerstva školstva, vedy, výskumu a športu Slovenskej republiky.

---

[6]https://www.apple.com/ios/siri/

Literatúra

[1] Hládek, D., Staš, J.: Text mining and processing for corpora creation in Slovak language. Journal of Computer Science and Control Systems. **3, 1** (2010) 65–68

[2] Hládek, D., Staš, J., Juhár, J.: Building organized text corpora for speech technologies in the Slovak language. Jazykovedné štúdie XXXI: Rozvoj jazykových technológií a zdrojov na Slovensku a vo svete (10 rokov Slovenského národného korpusu). **31** (2014) 173–181

[3] Zlacký, D., Staš, J., Juhár, J., Čižmár, A.: Slovak text document clustering. Acta Electrotechnica et Informatica. **13, 2** (2013) 3–7

[4] Zlacký, D., Staš, J., Čižmár, A.: Supervised text document clustering algorithm with keywords in Slovak. In Proc. of the 7th Int. Workshop on Multimedia and Signal Processing, Redžúr 2013. Smolenice, Slovakia (2013) 31–34

[5] Staš, J., Juhár, J., Hládek, D.: Classification of heterogeneous text data for robust domain-specific language modeling. EURASIP Journal on Audio, Speech and Music Processing. **2014, 14** (2014) 1–12

[6] Zlacký, D., Staš, J., Juhár, J., Čižmár, A.: Text categorization with latent Dirichlet allocation. Journal of Electrical and Electronics Engineering. **7, 1** (2014) 161–164

[7] Staš, J., Hládek, D., Juhár, J.: Recent advances in the statistical modeling of the Slovak language. In Proc. of the 56th Int. Symp. ELMAR 2014. Zadar, Croatia (2014) 39–42

[8] Thurson, A.: Parsing computer languages with an automaton compiled from a single regular expression. In Implementation and Application of Automata: Proc. of the 18th Intl. Conf. CIAA 2013, Halifax, NS, Canada. Ibarra, O. H., Yen, H. Ch. (Eds.). LNCS **4094**. Springer Berlin Heidelberg (2006) 285–286

[9] Ćavar, D., Jazbec, I.-P., Stojanov, T.: CroMo - Morphological analysis for standard Croatian and its synchronic and diachronic dialects and variants. In Proc. of the 8th Int. Conf. on Finite-State Methods and Natural Language Processing, FSMNLP 2009. Pretoria, South Africa (2009) 183–190

[10] Hládek, D., Staš, J., Juhár, J.: Dagger: The Slovak morphological classifier. In Proc. of the 54th Int. Symp. ELMAR 2012. Zadar, Croatia (2012) 195–198

[11] Hládek, D., Ondáš, S., Staš, J.: Online natural language processing of the Slovak language. In Proc. of the 5th IEEE Int. Conf. on Cognitive Infocommunications, CogInfoCom 2014. Vietri sul Mare, Italy (2014) 315–316

[12] Rusko, M., Juhár, J., Trnka, M., Staš, J., Darjaa, S., Hládek, D., Sabo, R., Pleva, M., Ritomský, M., Ondáš, S.: Recent advances in the Slovak dictation system for judicial domain. In Proc. of the 6th Language and Technology Conference: Human Language Technologies as a Challenge for Computer Science and Linguistics, LTC 2013, Poznań, Poland (2013) 555–560

[13] Schmid, H.: Probabilistic part-of-speech tagging using decision trees. In Proc. of Int. Conf. on New Methods in Language Processing. Manchester, UK (1994) 44–49

[14] Przepiórkowski, A., Górski, R. L., Łaziński, Pęzik, P.: Recent developments in the National corpus of Polish. In Proc. of the 7th Int. Conf. on Language Resources and Evaluation, LREC 2010. Valletta, Malta (2010) 994–997

[15] Hladká, B., Hajič, J., Hana, J., Hlaváčová, J., Mírovský, J., Raab, J.: The Czech academic corpus 2.0 guide. The Prague Bulletin of Mathematical Linguistics. **89** (2008) 41–96

[16] Halácsy, P., Kornai, A., Németh, L., Rung, A., Szakadát, I., Trón, V.: Creating open language resources for Hungarian. In Proc. of the 4th Int. Conf. on Language Resources and Evaluation, LREC 2004. Lisbon, Portugal (2004)

[17] Majliš, M.: W2C – Web to Corpus – Corpora. LINDAT/C-LARIN Digital Library at Institute of Formal and Applied Linguistics, UFAL, Charles University in Prague, Czech Republic (2011)

[18] Hládek, D., Staš, J., Juhár, J.: Unsupervised spelling correction for the Slovak text. Advances in Electrical and Electronics Engineering. **11, 5** (2013) 392–397

[19] Juhár, J., Staš, J., Hládek, D.: Recent progress in development of language model for Slovak large vocabulary continuous speech recognition. In New Technologies - Trends, Innovations and Research, Volosencu, C. (Ed.). InTech Open Access, Rijeka, Croatia (2012) 261–276

[20] Staš, J., Hládek, D., Juhár, J.: Language model speaker adaptation for transcription of Slovak parliament proceedings. In Proc. of the 17th Int. Conf. on Speech and Computer, SPECOM 2015, Athens, Greece (2015) to be published

[21] Staš, J., Hládek, D., Juhár, J., Zlacký, D.: Analysis of morph-based language modeling and speech recognition in Slovak. Advances in Electrical and Electronic Engineering. **10, 4** (2012) 291–296

[22] Staš, J., Hládek, D., Juhár, J.: Morphologically motivated language modeling for Slovak continuous speech recognition. Journal of Electrical and Electronics Engineering. **5, 1** (2012) 233–236

[23] Ondáš, S., Juhár, J., Holcer, R.: Methodology for training small domain-specific language models and its application in service robot speech interface. Journal of Electrical and Electronics Engineering. **7, 1** (2014) 107–110

[24] Ondáš, S., Juhár, J.: Development and evaluation of the spoken dialogue system based on the W3C recommendations. In Product and Services; from R&D to Final Solutions, Fuerstner, I. (Ed.). Scyio, Rijeka, Croatia (2010) 315–330

[25] Ondáš, S., Juhár, J., Trnka, M.: SIMONA – The Slovak embodied conversational agent. Intelligent Decision Technologies. **8, 4** (2014) 277–288

[26] Payne, T. E.: Describing morphosyntax: A guide for field linguists. Cambridge University Press, Cambridge (1997)

[27] Páleš, E.: SAPFO - Parafrázovač slovenčiny. Veda. Bratislava, Slovenská republika (1994)

[28] Laclavík, M., Ciglan, M., Krajči, S., Hluchý, L., Furdík, K.: Dostupné zdroje a výzvy pre počítačové spracovanie informačných zdrojov v slovenskom jazyku. In Proc. of the 1st Workshop on Intelligent and Knowledge Oriented Technologies, WIKT 2006. Bratislava, Slovakia (2006) 92–98

[29] Staš, J., Hládek, D., Ondáš, S., Juhár, J.: On building the Slovak example-based meaning corpus. In Proc. of the 8th Int. Conf. on NLP, Corpus Linguistics, Lexicography, Slovko 2015. Bratislava, Slovakia (2015) to be published

[30] Ondáš, S., Hládek, D., Juhár, J.: Semantic roles labeling system for Slovak sentences. In Proc. of the 5th IEEE Int. Conf. on Cognitive Infocommunications, CogInfoCom 2014. Vietri sul Mare, Italy (2014) 161–166

J. Yaghob (Ed.): ITAT 2015 pp. 88–94
Charles University in Prague, Prague, 2015

Giving a Sense: A Pilot Study in Concept Annotation from Multiple Resources

Roman Sudarikov and Ondřej Bojar

Charles University in Prague
Faculty of Mathematics and Physics
Institute of Formal and Applied Linguistics
Malostranské náměstí 25, 11800 Praha 1, Czech Republic
http://ufal.mff.cuni.cz/
{sudarikov,bojar}@ufal.mff.cuni.cz

Abstract: We present a pilot study in web-based annotation of words with senses coming from several knowledge bases and sense inventories. The study is the first step in a planned larger annotation of "grounding" and should allow us to select a subset of these "dictionaries" that seem to cover any given text reasonably well and show an acceptable level of inter-annotator agreement.

Keywords: word-sense disambiguation, entity linking, linked data

1 Introduction

Annotated resources are very important for training, tuning or evaluating many NLP tasks. Equipped with experience in treebanking, we now move to resources for word sense disambiguation (WSD) and entity linking (EL). By EL, we mean the task of attaching a unique ID from some database to occurrences of (named) entities in text [1]. Both entity linking and word-sense disambiguation have been extensively studied, see for example [2–4]. Although only a few researches consider several knowledge bases and sense inventories at once [1, 5], the convergence between these two task is apparent, for example, the 2015 SemEval Task 13 promoted research in the direction of joint word sense and named entity disambiguation [6].

We understand the terms *ontology*, *knowledge base* and *sense inventory* in the following way:

- *Ontology* is a formal representation of a domain of knowledge. It is an abstract entity: it defines the vocabulary for a domain and the relations between concepts, but an ontology says nothing about how that knowledge is stored (as physical file, in a database, or in some other form), or indeed how the knowledge can be accessed.
- *Knowledge base* is a database, a repository of information that can be accessed and manipulated in some predefined fashion. Knowledge is stored in knowledge base according to an ontology.
- *Sense inventory* is a database, often build based on a corpus, and providing clustered *senses* for the words or expressions in the corpus.

However, we recognize the blending of *knowledge bases* and *sense inventories*, so we will use very generic terms *dictionary* or *resource* interchangeably for either of them.

In this pilot study, we examine several such *dictionaries* in terms of their coverage and annotator agreement. Unlike other works on "grounding", which try to link only the most important words in the sentence [7, 8], we aim at *complete* coverage of a given text, i.e. all content words or multi-word expressions regardless their part of speech or role in the sentence. Some of the examined resources have a clear bias towards some parts of speech, for example, valency dictionaries cover only verbs. We nevertheless ask our annotators to annotate even across parts of speech if the matching POS is not included in the resource. For instance, verbs can get nominal entries in Wikipedia and nouns get verb frames.[1]

In Section 2, we describe the sense inventories included in our experiment. Section 3 provides a unifying view on these sources and introduces our annotation interface. We conducted two experiments with English and Czech texts using the interface, slightly adapting interface for the second run. Details are in Section 4 and Section 5.

2 Resources Included

Sense inventories and knowledge bases are plentiful and they differ in many aspects including the domain coverage, level of detail, frequency of update, integration of other resources and ways of accessing them. Some of them implement Resource Description Framework, the metadata data model designed by W3C for the better data representation in Semantic Web, while others are simply collections of links in the web.

We selected the following subset of general *resources* for our experiment:

BabelNet [10] is a multilingual knowledge base, which combines several knowledge resources including Wikipedia, Wordnet, OmegaWiki and Wiktionary. The sources are automatically merged and accessible via offline Java API or online REST API. An added benefit is the multilinguality of BabelNet: the same resource can be used for genuine (as opposed to cross-lingual) annotation for both languages of our interest, English and Czech.

---

[1]The conversion of nouns to predicates whenever possible is explicitly demanded in some frameworks, e.g. in Abstract Meaning Representation (AMR, [9]).

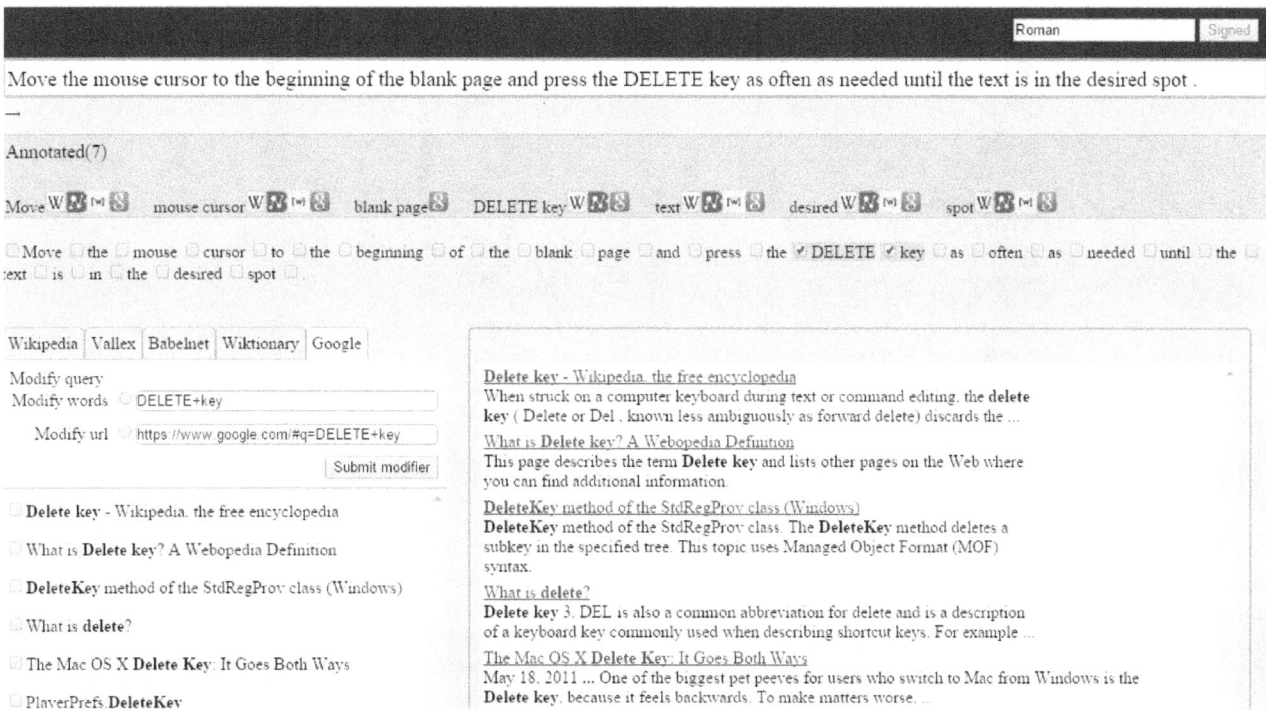

Figure 1: Annotation interface, annotating the words "DELETE key" in the sentence "Move the mouse cursor..." with Google Search "senses".

The main limitation is that BabelNet is not updated continuously, so we also added both live Wikipedia and Wiktionary as separate sources. BabelNet provides information about nouns, verbs, adjectives and adverbs, but as stated above, we are interested also in cross-POS annotation.

Wikipedia[2] is currently the biggest online encyclopedia with live updates from (hundreds of) thousands of contributors so it can cover new concepts very quickly. Wikipedia tries to nest all possible concepts as nouns. For example, `en.wikipedia.org/wiki/funny` redirects to the page "Humour".

Wiktionary[3] is a companion to Wikipedia that covers all parts of speech. It includes multilingual thesaurus, phrase books, language statistics. Each word in Wiktionary can have etymology, pronunciation, sample quotations, synonyms, antonyms and translations, for better understanding of the word.

PDT-VALLEX and EngVallex (Valency lexicons for Czech and English): Valency or subcategorization lexicons formally capture verb valency frames, i.e. their syntactic neighborhood in the sentence [11, 12]. We use the valency lexicons for Czech and English in their offline XML form as distributed with the tree editor TrEd 2.0[4].

Google Search[5] (GS): From our preliminary experiments, we had the impression that no resource covers all

expressions seen in our data, but searching the web provides some explanation almost always. We thus include the top ten results returned by Google Search as a special kind of *dictionary*, where the "concept" is a query string and each result is considered to be its' "sense".

Aside from coverage and frequency of updates, another reason to include GS is that it provides "senses" at a very different level of granularity than others. For instance, the whole Wiktionary page can appear as one of the options in GS "senses". It will also often be a very sensible choice, despite it actually covers several different meanings of the word.

We find the task of matching senses coming from different ontologies and providing a different angle of view or granularity very interesting. The current experiments serve as a basis for its further investigation.

3 Annotation Interface

To provide a unified view on the various resources, we use the terms *query*, *selection list* and *selection*. Given an expression in a text, which can be a word or a phrase, even a non-continuous one, and a resource which should be used to annotate it, the system construct a query. Querying the resource, we get a selection list, i.e. a list of possible senses.

The process of extracting the selection list depends on the resource. It is straightforward for Google Search (each result becomes an option) and complicated for Wiktionary,

---

[2]`http://wikipedia.org`
[3]`http://wiktionary.org`
[4]`http://ufal.mff.cuni.cz/tred/`
[5]`http://google.com`

Source	Total	Whole page	Bad List	None	One or more senses selected			
					1	2	3	4 or more
Babelnet	28	-	1	3	23	1	0	0
Google Search	71	-	1	9	36	15	5	5
CS Vallex	38	-	0	2	29	6	1	0
EN Vallex	19	-	1	0	18	0	0	0
CS Wikipedia	38	-	9	12	15	1	0	1
EN Wikipedia	114	-	26	16	63	3	0	6
CS Wiktionary	21	-	1	3	7	4	5	1
EN Wiktionary	21	-	0	0	18	2	1	0
Babelnet	98	24	0	10	54	6	2	2
Google Search	93	0	0	26	19	16	11	21
EN Vallex	15	4	0	3	6	2	0	0
EN Wikipedia	103	23	7	36	35	2	0	0
EN Wiktionary	98	17	23	4	40	9	2	3

Table 1: Selection statistics, the first (upper part) and second (lower part) annotation experiments

see Section 3.1 below. In principle and to include any conceivable resource, even field-specific or ad hoc ones, the annotator should be free to *select the selection list* prior to the annotation.

Our annotation interface allows to overwrite the query for cases where the automatic construction does not lead to a satisfactory selection list.

Finally, the annotator is presented with the selection list to make his choice (or multiple choices). Overall, the annotator picks one of these options:

Whole Page means that the current URL is already a good description of the sense and no selection list is available on the page. The annotators were asked to change the query and rather obtain a selection list (e.g. a disambiguation page in Wikipedia) whenever possible.

Bad List means that the extraction of selection list failed to provide correct senses. The annotators were supposed to try changing the query to obtain a usable list and resort to the "Bad List" option only if inevitable.

None indicates that the selection list is correct but that it lacks the relevant sense.

One or more senses selected is the desired annotation: The list, for the particular pair of *selected word(s)* and *selected resource*, was correct and the annotator was able to find the relevant sense(s) in the list.

Our annotation interface (Figure 1) shows the input sentence, tabs for individual sense inventories, the selection list from the current resource and also the complete page where the selection list comes from. The procedure is straightforward: (1) select one or more words in the sentence using checkboxes, (2) select a resource (we asked our annotators to use them all, one by one), (3) check if the selection list is OK and modify the query if needed, (4) make the annotation choice by marking one or more of the checkboxes in the selection list, and (5) save the annotation.

3.1 Queries and Selection Lists for Individual Resources

This is how we construct queries and extract selection lists for each of our *dictionaries* given one or more words from the annotated sentence:

BabelNet We search BabelNet for the lemma of the selected word (or the phrase of lemmas if more words are selected). The selection list is the list of all obtained BabelNet IDs.

Google Search We search for the lemmas of the selected words and return the snippets of the top ten results. The selection list is the list of snippets' titles.

Wikipedia We search for the disambiguation page for the selected words and, if not found, we search for the page with the title matching the lemmas of the selected words. The selection list for disambiguation pages is constructed by fetching hyperlinks appearing within listings nested in particular HTML blocks. For other pages we fetch links from the Table of Contents and the first hyperlink from each listing item.

Wiktionary We search for the page with the title equal to the lemmas of the selected words. The selection list is created using the same heuristics as for Wikipedia.

Vallex We scan the XML file and return all the frames belonging to the verb with the lemma matching the selected word's lemma.

4 First Experiment

The first experiment was held in March 2014. The 7 participating annotators (none of whom had any experience in annotation tasks) were asked to annotate the sentences from PCEDT 2.0 [6] with Czech and English sources: Wikipedia and Wiktionary for both languages, BabelNet,

---

[6]`http://ufal.mff.cuni.cz/pcedt2.0/en/index.html`

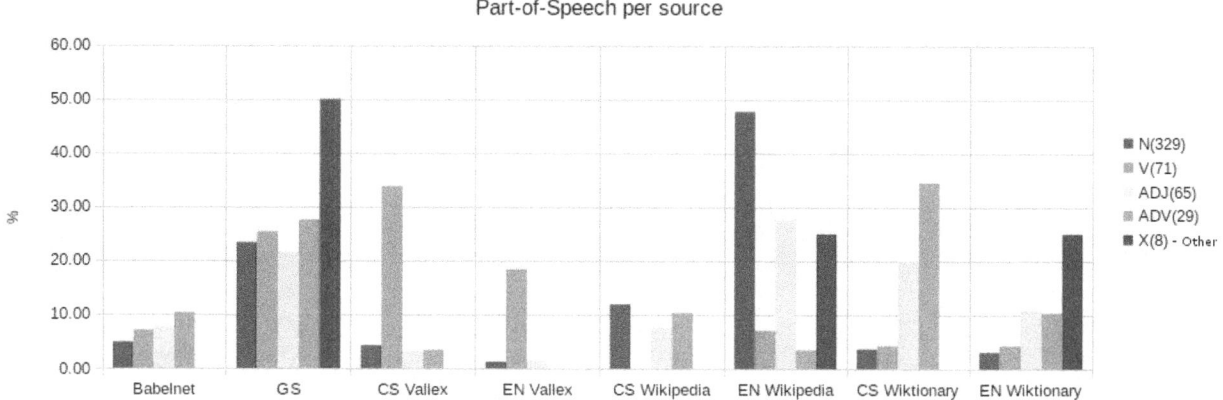

Figure 2: Annotations from a given dictionary in the first experiment broken down by part of speech of the annotated words.

Google Search, and the Czech and English Vallexes. Each annotator was given a set of sentences in English or Czech and they were asked to annotate as many words or phrases in each sentence as possible, with as many reasonable meanings as they can. We required the annotators to annotate across parts of speech if possible (for instance to annotate the noun "teacher" with the corresponding verb "to teach"). This requirement appeared because we wanted to evaluate the possibility of using more abstract *senses* as used, for instance, in works with AMR.

4.1 Gathered Annotations

In total, we collected 507 annotations for 158 units. 75 of these units had more than one annotation.

The upper part of Table 1 provides details on how often each of the annotation options was picked for a given source in the first annotation experiment. Note that in the first experiment, we did not offer the "Whole Page" option.

We see that the sources exhibit slightly different patterns of use. Wikipedia has lots of "Bad List" options selected due to the issue described in Section 4.2. GS is the most ambiguous resource, the user has picked two or more sense in about one half of GS annotations. The highest number of "Bad Lists" was received by the English Wikipedia (18 out of 40).

Figure 2 shows the distribution of different POS per source. Google Search seems to be the most versatile resource, covering all parts of speech well. The relatively low use of BabelNet was due to the web API usage limit. Vallexes work well for verbs but cross-POS annotation is only an exception. Wikipedia and Wiktionary are indeed somewhat complementary in covered POSes.

4.2 Bad List vs. None Issue

The "Bad List" annotations should be used in two cases: (1) when the system fails to extract the selection list from

Source	Annotations	2-IAA	Annotations	2-IAA
Babelnet	29	0.69	29	0.69
GS	120	0.24	120	0.24
CS Vallex	46	0.58	46	0.58
EN Vallex	19	1.00	19	1.00
CS Wikipedia	47	0.32	43	0.35
EN Wikipedia	183	0.05	181	0.10
CS Wiktionary	38	0.29	38	0.35
EN Wiktionary	25	0	25	0
Total:	507	0.21	501	0.24

Table 2: Inter-annotator agreement in the first experiment, before (left) and after (right) the "Bad List" fix.

a good page, and (2) when the whole page is wrong, for example when the system shows the Wikipedia page "South Africa" for the word "south". "None" was meant for correct selection lists (matching domain, reasonable options) but the right option missing. The guidelines for the first experiment were not very clear on this so some annotators marked problems with selection list as "Bad List" and some used the label "None".

Manual revision revealed that only 10 out of 40 "Bad List" annotations were indeed "Bad List" in one of the two meanings described above. The right hand part of Table 2 shows IAA after changing wrongly annotated "Bad Lists" into "None".

4.3 Inter-Annotator Agreement

Inter-annotator agreement is a measure of how well two annotators can make the same annotation decision for a certain item. In our case it is measured as the percentage of cases when a pair (2-IAA) of annotators agree on the (set of) senses for a given annotation unit. The measurement was made pairwise for all the annotations, which had more that one annotator. The results are presented in Table 2, before and after fixing the "Bad List" issue.

In general, the IAA estimates should be treated with caution. Many units were assigned only to a single anno-

Part-of-speech per source, second experiment

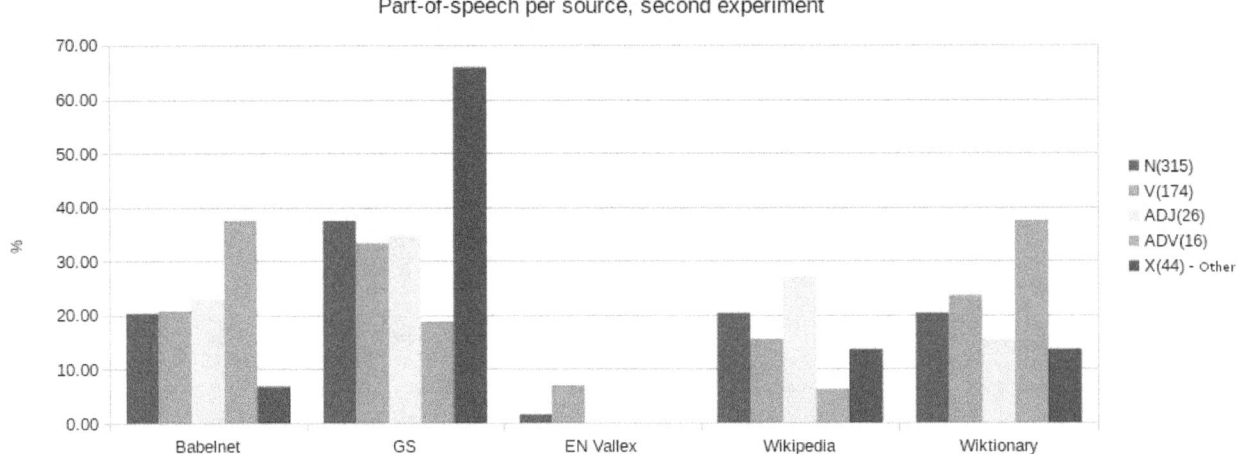

Figure 3: Annotations from a given dictionary in the second experiment broken down by part of speech of the annotated words.

tator, so they weren't taken into account while computing IAA.

The extremely low IAA for English Wikipedia was caused by the following issue. For several units, one annotator tried to select all the *senses* to show that the whole page can be used, while others have picked one or only a few *senses*. We resolved the issue by introducing a new option "Whole page" in the second experiment.

Interestingly, we see a negative correlation (Pearson correlation coefficient of -0.37) between the number of units annotated for a given source and the 2-IAA.

We also report Cohen's kappa [13], which reflects the agreement when disregarding agreement by chance. In our setting, we estimate the agreement by chance as one over the length of the selection list plus two (for "None" and "Bad List"). This is a conservative estimate, in principle the annotators were allowed to select any subset of selection list. We compute kappa using $K = \frac{P_a - P_e}{1 - P_e}$, where P_a was the total 2-IAA and P_e was the arithmetical average of agreements by chance for each annotation. Kappa for the first experiment was 0.13.

To assess the level of uncertainty for the estimates, we use bootstrap resampling with 1000 resamples, which gives us IAA of 0.25 ± 0.1 and kappa of 0.135 ± 0.115 for 95% of samples.

5 Second Experiment

The second experiment was held in March 2015 with another group of 6 annotators. One of the annotators had experience in annotating tasks, while others had no such experience. The setting of the experiment was slightly different. The annotators were asked to annotate only English sentences from QTLeap project[7] using BabelNet, Google

Search, English Wikipedia, English Wiktionary and ENG-VALLEX. The guidelines were refined, asking the annotators to mark the largest possible span for each concept in the sentence, e.g. to annotate "mouse cursor" jointly as one concept and not separately as "computer pointing device" for the word "mouse" and "graphic representation of computer mouse on the screen" for the word "cursor". The option "Whole page" was newly introduced to help users indicate that the whole page can be used as a sense.

5.1 Gathered Annotations

We collected 570 annotations for 35 words, 32 of which had annotations from more than one annotator. The number of units here is lower that in the first experiment, because all our annotators used the same sentences. Also, for the second experiment we required the annotators to use all the resources for each unit, so we have more results per unit.

During the second experiment, the system processed 147 unique (in terms of *selected word(s)* and *selected resource*) queries. All the resources got nearly equal number of queries (about 30), except for Vallex, which got only 10 queries. The annotators changed the queries 59 times, but this also includes cases, when Wikipedia used its own inner redirects, which our system did not distinguish from users' changes. BabelNet was changed 9 times, Google Search – 2, Vallex – 8, Wikipedia – 21 and Wiktionary – 19. Based on these numbers, GS may seem more reliable but it is not necessary true. One reason is that some of part of the changes for Wikipedia was made automatically by Wikipedia itself. The other argument is that users could limit their effort and after examining the first 10 GS results for the query they just picked "Bad List" option and moved on, not trying to change the query.

The POS per source distribution (see Figure 3) for the second experiment is similar to the first one, except for the

---

[7]http://qtleap.eu/

| | Content words | | Annotators | | | | | |
Source	Attempted	Labeled	A1	A2	A3	A4	A5	A6
Babelnet	100%	**91%**	53%	20%	67%	66%	**79%**	40%
GS	100%	85%	50%	13%	53%	46%	**76%**	20%
Vallex	32%	26%	7%	6%	10%	**40%**	0%	0%
Wikipedia	100%	58%	39%	20%	35%	**53%**	50%	26%
Wiktionary	100%	88%	53%	20%	32%	40%	**76%**	26%
Total *content words*	34	34	28	15	28	15	34	15

Table 3: Coverage per *content word* (second experiment). The left part reports the union across annotators, the right part reports the percentage of content words receiving a valid label (Labeled) for each annotator separately.

Source	Annotations number	2-IAA
Babelnet	114	0.49
GS	217	0.45
Vallex	17	0.60
Wikipedia	105	0.61
Wiktionary	117	0.28
Total:	570	0.46

Table 4: Inter-annotator agreement, second experiment

BabelNet, which did not reach any technical limit this time and was therefore used more often across all POSes.

5.2 Coverage

In Table 3, we show the coverage of *content words* in the second experiment. By *content words* we mean all the words in the sentence, except for auxiliary verbs, punctuation, articles and prepositions. The instructions asked to annotate all content words. Each annotator completed a different number of sentences, so the number of words annotated differs. The column *Content words Attempted* shows the total number of words with some annotation at all, while *Labeled* are words which received some sense, not just "None" or "Bad List". Both numbers are taken from the union over all annotators. Babelnet get the best coverage in terms of *Labeled* annotations. The right hand side of the table shows how many words each annotator has *labeled*. Since the union is considerably higher than the most productive annotator, we need to ask an important question: How many annotators do we need to have a perfect coverage of the sentence.

5.3 Inter-Annotator Agreement

Results presented in Table 4 are overall better than in the first experiment. The kappa was computed as in Section 4.3 with the only one difference: we added 3 instead of 2 options when estimating the local probability of the agreement by chance (for the new "Whole Page" option). Kappa for the second experiment was 0.40. Bootstrapping showed IAA 0.39 ± 0.055 and kappa 0.32 ± 0.06 for 95% central resamples. Again, the 2-IAA is negatively correlated with the number of units annotated (Pearson correlation coefficient -0.22).

6 Discussion

Comparing first and second experiment, one can see, that we managed to improve IAA by expanding the set of available options and refining the instructions, but IAA is still not satisfactory.

For resources where IAA reaches 60% (Vallex and Wikipedia), the coverage is rather low, 26% and 58%. BabelNet gives the best coverage but suffers in IAA. Google Search seems an interesting option for its versatility across parts of speech, on par with established knowledge bases like BabelNet in terms of inter-annotator agreement but with much more ambiguous "senses". The cross-POS annotation does not seem very effective in practice, but a more thorough analysis is desirable.

7 Comparison with Other Annotation Tools

Several automatic systems for sense annotation are available. Our dataset could be used to compare them empirically on the annotations from the respective repository used by each of the tools. For now we provide only an illustrative comparison of these three systems: TAGME[8], DBpedia Spotlight[9], and Babelfy[10]

Figure 4 provides an example of our manually collected annotations for the sentence "Move the mouse cursor to the beginning of the blank page and press the DELETE key as often as needed until the text is in the desired spot.".

For this sentence, the TAGME system with default settings returned three entities ("mouse cursor", "DELETE key" and "text"). DBpedia Spotlight with default settings (confidence level = 0.5) returned one entity ("mouse"). Babelfy showed the best result among these systems in terms of coverage, failing to recognize only the verb "move" and adverbs "often" and "until", but it also provided several false meanings for found entities.

8 Conclusion

In this paper, we examined how different dictionaries can be used for entity linking and word sense disambiguation.

---

[8] http://tagme.di.unipi.it/
[9] http://dbpedia-spotlight.github.io/demo/
[10] http://babelfy.org/

	BabelNet	Wikipedia	TAGME	Spotlight	Babelfy
Move	bn:00087012v,bn:00090948v bn:00056033n, bn:00056155n	Motion_(physics)	-	-	-
mouse	*bn:00021487n*,bn:00090942v *bn:00024529n*	mouse_(disambiguation), mouse_cursor	Mouse_(computing)	Mouse_(computing)	*bn:00024529n*,*bn:00021487n*
cursor	*bn:00024529n*	**mouse_cursor**, cursor_(disambiguation)	**mouse_cursor**	-	*bn:00024529n*
beginning	bn:00009632n,bn:00009633n bn:00009634n,bn:00009635n	beginning, beginning_(disambiguation)	-	-	bn:00083340v
blank	*bn:00098524a*	blank_page_(disambiguation)	-	-	bn:01161190n,*bn:00098524a*
page	*bn:00060158n*	blank_page_(disambiguation)	-	-	bn:01161190n,*bn:00060158n*
press	bn:00091988v,bn:00091986v	press_(disambiguation)	-	-	bn:00046094n
DELETE	*bn:01208543n*	**Delete_key**, DELETE	**Delete_key**	-	*bn:01208543n*, bn:00045088n
key	*bn:01208543n*, bn:00048996n	**Delete_key**, key_(disambiguation)	**Delete_key**	-	*bn:01208543n*, bn:00048985n
often	bn:00114048r, bn:00115452r bn:00116418r	often	-	-	-
needed	bn:00107194a	Need_(disambiguation)	-	-	bn:00082822v
until	-	until	-	-	-
text	*bn:00076732n*	text_(disambiguation)	Plain_text	-	*bn:00076732n*
desired	bn:00100580a, bn:00026550n bn:00100607a	Desire_(disambiguation), desired	-	-	bn:00086682v
spot	*bn:00062699n*	spot_(disambiguation)	-	-	*bn:00062699n*

Figure 4: Our BabelNet and Wikipedia manual annotations and outputs of three automatic sense taggers for the sentence "Move the mouse cursor to the beginning of the blank page and press the DELETE key as often as needed until the text is in the desired spot." Overlap indicated by italics (BabelNet and Babelfy) and bold (Wikipedia and TAGME).

In our unifying view based on finding the best "selection list" and selecting one or more senses from it, we tested standard inventories like BabelNet or Wikipedia, but also Google Search.

We proposed and refined annotation guidelines in two consecutive experiments, reaching average inter-annotator agreement of about 46%, with Wikipedia and Vallex up to 60%. Higher agreement seems to go together with lower coverage, but further investigation is needed for confirmation and to find the best balance of granularity, coverage and versatility among existing sources.

Acknowledgements

This research was supported by the grants FP7-ICT-2013-10-610516 (QTLeap). This research was partially supported by SVV project number 260 224. This work has been using language resources developed, stored and distributed by the LINDAT/CLARIN project of the Ministry of Education, Youth and Sports of the Czech Republic (project LM2010013).

References

[1] Demartini, G., et al.: Zencrowd: leveraging probabilistic reasoning and crowdsourcing techniques for large-scale entity linking. In: Proceedings of the 21st international conference on World Wide Web, ACM (2012) 469–478

[2] Bennett, P.N., et al.: Report on the sixth workshop on exploiting semantic annotations in information retrieval (ESAIR'13). In: ACM SIGIR Forum. Volume 48., ACM (2014) 13–20

[3] Ratinov, L., et al.: Local and global algorithms for disambiguation to wikipedia. In: Proc. of ACL/HLT, Volume 1. (2011) 1375–1384

[4] Navigli, R.: Word sense disambiguation: A survey. ACM Comput. Surv. **41**(2) (February 2009) 10:1–10:69

[5] Pereira, B.: Entity linking with multiple knowledge bases: An ontology modularization approach. In: The Semantic Web–ISWC 2014. Springer (2014) 513–520

[6] Moro, A., Navigli, R.: SemEval-2015 Task 13: Multilingual All-Words Sense Disambiguation and Entity Linking. In: Proc. of SemEval-2015. (2015) In press.

[7] Ferragina, P., Scaiella, U.: Tagme: on-the-fly annotation of short text fragments (by wikipedia entities). In: Proc. of CIKM, ACM (2010) 1625–1628

[8] Zhang, L., Rettinger, A., Färber, M., Tadić, M.: A comparative evaluation of cross-lingual text annotation techniques. In: Information Access Evaluation. Multilinguality, Multimodality, and Visualization. Springer (2013) 124–135

[9] Banarescu, L., , et al.: Abstract Meaning Representation for Sembanking (2013)

[10] Navigli, R., Ponzetto, S.P.: BabelNet: The automatic construction, evaluation and application of a wide-coverage multilingual semantic network. Artificial Intelligence **193** (2012) 217–250

[11] Žabokrtský, Z., Lopatková, M.: Valency information in VALLEX 2.0: Logical structure of the lexicon. The Prague Bulletin of Mathematical Linguistics (87) (2007) 41–60

[12] Lopatková, M., Žabokrtský, Z., Ketnerová, V.: Valenční slovník českých sloves. (2008)

[13] Cohen, J.: A Coefficient of Agreement for Nominal Scales. Educational and Psychological Measurement **20**(1) (1960)

J. Yaghob (Ed.): ITAT 2015 pp. 95–99
Charles University in Prague, Prague, 2015

ITAT

Czech Aspect-Based Sentiment Analysis:
A New Dataset and Preliminary Results

Aleš Tamchyna, Ondřej Fiala, Kateřina Veselovská

Charles University in Prague, Faculty of Mathematics and Physics
Institute of Formal and Applied Linguistics
Malostranské náměstí 25, Prague, Czech Republic
{tamchyna,fiala,veselovska}@ufal.mff.cuni.cz

Abstract: This work focuses on aspect-based sentiment analysis, a relatively recent task in natural language processing. We present a new dataset for Czech aspect-based sentiment analysis which consists of segments from user reviews of IT products. We also describe our work in progress on the task of aspect term extraction. We believe that this area can be of interest to other workshop participants and that this paper can inspire a fruitful discussion on the topic with researchers from related fields.

1 Introduction

Sentiment analysis (or opinion mining) is a field related to natural language processing (NLP) which studies how people express emotions (or opinions, sentiments, evaluations) in language and which develops methods to automatically identify such opinions.

The most typical task of sentiment analysis is to look at some short text (a sentence, paragraph, short review) and determine its *polarity* – positive, negative or neutral.

Aspect-based sentiment analysis (ABSA) refers to discovering aspects (aspect terms, opinion targets) in text and classifying their polarity. The prototypical scenario are product reviews: we assume that products have several aspects (such as size or battery life for cellphones) and we attempt to identify users' opinions on these individual aspects.

This is a more fine-grained approach than the standard formulation of sentiment analysis where the goal would be to classify the polarity of entire sentences (or even whole reviews) without regard for internal structure.

Recently, ABSA has been gaining researchers' interest, as evidenced e.g. by the two consecutive shared tasks organized within SemEval in 2014 and 2015 [7, 6].

ABSA can be roughly divided into two subtasks: (i) identification of aspects (or aspect term extraction) in text, i.e. marking (occurrences of) words which are evaluated; (ii) polarity classification, i.e. deciding whether the opinions about the identified words are positive, negative or neutral.

In this work, we introduce a new Czech dataset of product reviews annotated for ABSA and describe a preliminary method of aspect term identification which combines a rule-based approach and machine learning.

2 Dataset of IT Product Reviews

We downloaded a number of user product reviews which are publicly available on the website of an established Czech online shop with electronic devices. Each review consists of negative and positive aspects of the product. This setting pushes the customer to rate its important characteristics.

The dataset consists of two parts: (i) random short segments and (ii) longest reviews. The difference in length is reflected also in the use of language.

The first part of this dataset contains 1000 positive and 1000 negative reviews which were selected from source data and their targets were manually tagged. These targets were either aspects of the evaluated product or some general attributes (e.g. price, ease of use). The polarity of each aspect is based on whether the user submitted the segment as negative or positive. These short reviews often contain only the aspect without any evaluative phrase.

The second part of dataset consists of the longest reviews. We chose 100 of them for each polarity. These reviews represent more usual text and they tend to keep proper sentence structure. The longest review has 7057 characters.

The whole dataset provides a consistent view of language used in the on-line environment preserving both specific word forms and language structures. There is also a large amount of domain specific slang due to the origin of the text.

Dataset part	#targets	#reviews	Avg. length
Random, positive	640	1000	34.17
Random, negative	508	1000	39.72
Longest, positive	484	100	953.35
Longest, negative	353	100	855.04

Table 1: Statistics of the annotated data.

The data was annotated by a single annotator. The basic instruction was to mark all aspects or general characteristics of the product. The span of the annotated term should be as small as possible (often a single noun). For evaluation, the span can be expanded e.g. to the immediate dependency subtree of the target. Any part of speech can be marked; e.g. both "funkčnost" ("functionality") and "funkční" ("functional") should be marked.

Figure 1: Dependency tree for the sentence *"The fried rice is amazing."* Morphological tags (such as *NN* for nouns) and analytical functions (e.g. *Sb* for sentence subject) are shown in the parse tree. The positive evaluative word "amazing" triggers a rule which marks "rice" as a possible aspect.

The whole dataset contains 1985 target tags; 1124 of these are positive and 861 are negative. Detailed target statistics are shown in Table 1.

The dataset is freely available for download at the following URL:

```
http://hdl.handle.net/11234/1-1507.
```

3 Pipeline

Our work is inspired by the pipeline of [15]. We run morphological analysis and tagging on the data to identify the parts of speech of words and their morphological features (e.g. case or gender for Czech). We also obtain dependency parses of the sentences. Then, we use several handcrafted rules based on syntax to mark the likely aspects in the data. Figure 1 shows a sample dependency parse tree and rule application.

Unlike [15], the core of our approach is a machine-learning model and the outputs of the rules only serve as additional "hints" (features) to help the model identify aspects.

3.1 Syntactic Rules

We use the same rules as [15], Table 2 contains their description. Here, we categorize the rules somewhat differently, their types correspond to the actual features presented to the model.

The rules are designed for *opinion target identification*, i.e. discovering targets of evaluative statements.[1] They are based on syntactic relations with evaluative words, i.e.

words listed in a subjectivity lexicon for the given language.

In the example in Figure 1, the rule `vbnm_sb_adj` is triggered because "amazing" is an evaluative word and it is a predicate adjective – the word "rice", as the subject of this syntactic construction, is then marked as a likely aspect term.

Originally, the rules were written for English. Their adaptation to Czech proved very simple. We modified expressions which involved morphological tags to work with the Czech positional tagset [1]. Some of the rules included lexical items, such as the lemma *"be"* for identifying the linking verbs of predicate nominals. Simple translation of these few words to Czech sufficed in such cases.

3.2 Model

We chose linear-chain conditional random fields (CRFs) for our work [2]. In this model, aspect identification is viewed as a sequence labeling task. The input $\mathbf{x}$ are words in the sentence and the output is a labeling $\mathbf{y}$ of the same length: each word is marked as either the beginning of an aspect (B), inside an aspect (I) or outside an aspect (O).[2]

A linear-chain CRF is a statistical model. It is related to hidden Markov models (HMMs), however it is a discriminative model, not a generative one – it directly models the conditional probability of the labeling $P(\mathbf{y}|\mathbf{x})$. Linear-chain CRFs assume that the probability of the current label (B, I or O) only depends on the previous label and on the input words $\mathbf{x}$.

Formally, a linear-chain CRF is the following conditional probability distribution:

$$P(\mathbf{y}|\mathbf{x}) = \frac{1}{Z(\mathbf{x})} \exp\{\sum_{t=1}^{T}\sum_{k=1}^{K} \lambda_k f_k(y_t, y_{t-1}, t, \mathbf{x})\} \quad (1)$$

Roughly speaking, $P(\mathbf{y}|\mathbf{x})$ is the score of the sentence labeling $\mathbf{y}$, exponentiated and normalized.

The score of $\mathbf{y}$ corresponds to the sum of scores for labels y_t at each position $t \in \{1, \ldots, T\}$ in the sentence. The score at position t is the product between the values of feature functions $f_k(y_t, y_{t-1}, t, \mathbf{x})$ and their associated weights λ_k, which are estimated in the learning stage.

Feature functions can look at the current label y_t, the previous label y_{t-1} and the whole input sentence $\mathbf{x}$ (which is constant).

$Z(\mathbf{x})$ is the normalization function which sums over all possible label sequences:

$$Z(\mathbf{x}) = \sum_{\mathbf{y}'} \exp\{\sum_{t=1}^{T}\sum_{k=1}^{K} \lambda_k f_k(y_t', y_{t-1}', t, \mathbf{x})\} \quad (2)$$

To train the model, we require training data, i.e. sentences with the labeling already assigned by a human annotator. During CRF learning, the weights λ_k are optimized to maximize the likelihood of the observed labeling

---

[1] The underlying assumption of this approach is that opinion targets tend to be the sought-after aspects.

[2] This "BIO" labeling scheme is common for CRFs. In practice, it brings us a consistent slight improvement as opposed to using only binary classification (inside vs. outside an aspect).

ID	Description	Example
adverb	Actor or patient of a verb with a subjective adverb.	*The pizza tastes so good.*
but_opposite	Words coordinated with an aspect with "but".	*The food is outstanding, but everything else sucks.*
coord	Words coordinated with an aspect are also aspects.	*The excellent mussels, goat cheese and salad.*
sub_adj	Nouns modified by subjective adjectives.	*A very capable kitchen.*
subj_of_pat	Subject of a clause with a subjective patient.	*The bagel have an outstanding taste.*
verb_actant_pat	Patient of a transitive evaluative verb.	*I liked the beer selection.*
verb_actant_act	Actor of an intransitive evaluative verb.	*Their wine sucks.*
vbnm_patn	Predicative nominal (patient).	*Our favourite meal is the sausage.*
vbnm_sb_adj	Subject of predicative adjectives.	*The fried rice is amazing.*

Table 2: List of syntactic rules.

in the dataset. Gradient-based optimization techniques are usually applied for learning.

At prediction time, the weights λ_k are fixed and we are looking for such a labeling $\hat{\mathbf{y}}$ which is the most probable according to the model, i.e.:

$$\hat{\mathbf{y}} = \arg\max_{\mathbf{y}} P(\mathbf{y}|\mathbf{x}) \tag{3}$$

$\hat{\mathbf{y}}$ can be found efficiently using a variant of the Viterbi algorithm (dynamic programming). In our work, we use the CRF++ toolkit[3] both for training and prediction.

3.3 Feature Set

We now describe the various feature sets evaluated in this work.

Surface features. We use the surface forms of the current word, two preceding and two following words as separate features. Additionally, we extract all (four) bigrams and (three) trigrams of surface forms from this window. We also use the CRF++ *bigram* feature template without any arguments; this simply produces the concatenation of the previous and current label (y_{t-1}, y_t).

Morpho-syntactic features. We extract unigrams, bigrams and trigrams from a limited context window (identical to the above) around the current token but instead of surface forms, we look at:

- lemma,

- morphological tag,

- analytical function.

Analytical functions are assigned by the dependency parser and their values include "Sb" for subject, "Pred" for predicate etc.

Sublex features. We mark all words in the data whose lemma is found in the subjectivity lexicon. For each token in the window of size 4 around the current token (included), we extract a feature indicating whether it was marked as subjective. We also concatenate these indicator features with the surface form of the current token.

Rule features. Finally, for each type of rule, we extract features for the current token, the preceding and the following token, indicating whether the rule marked that token. Again, these features have two versions: one standalone and one concatenated with the surface form of the current token.

4 Experiments

We analyze our data using Treex [8], a modular NLP toolkit. Sentences are first tokenized and tagged using Morphodita [12]. Then we obtain their dependency parses using the MST parser [4]. We use Czech SubLex [14] is our subjectivity lexicon both for the CRF sublex features and for the rules. The rules are implemented as blocks within the Treex platform.

4.1 Results

Table 3 shows the obtained precision (P), recall (R) and f-measure (F1) for both parts of the data set. The results in all cases were acquired using 5-fold cross-validation on the training data.

Random segments. The baseline (surface-only) features achieve the best precision but the recall is very low.

Morpho-syntactic features lower the precision by a significant margin but push recall considerably. As the review data come from the "wild", they are quite noisy; many segments are written without punctuation, reducing the benefit of using morphological analysis, let alone dependency parsing.[4]

Often, the segments are rather short, such as "Rychlé *dodání*" ("fast *delivery*") or "*Fotky* fakt parádní." ("*Photos* really awesome."). This also considerably limits the benefit that a parser can bring – there is a major domain mismatch both in the text topic and types of sentences between the parser's training data and this dataset, so we cannot expect parsing accuracy to be high.

Most of the improvement from adding morpho-syntactic features thus probably comes from the availability of word lemmas – this allows the CRF to learn which

---

[3] http://taku910.github.io/crfpp/

[4] This issue could perhaps be addressed by using a spell-checker, we leave that to future work.

Feature set	Random segments (2000)			Longest reviews (200)		
	P	**R**	**F1**	**P**	**R**	**F1**
surface	**85.22**	36.85	51.45	47.18	8.05	13.76
+morpho-syntactic	75.88	54.17	63.21	40.17	**23.08**	29.31
+sublex	78.19	55.09	64.64	**58.74**	18.99	28.70
+rules	76.54	**57.69**	**65.79**	51.74	21.39	**30.27**

Table 3: Precision, recall and f-measure obtained using various feature sets on the two parts of the dataset.

words are frequently marked as aspects in this domain and to generalize this information beyond their current inflected form.

Adding the information from the sentiment lexicon further improves performance, though not as much as we would expect. We could possibly further increase its impact through more careful feature engineering – so far, the features only capture whether a subjective term is present in a small linear context. For example, the lemma of the evaluative word could be included in the feature.[5]

Finally, adding the output of syntactic rules further improves the results. Due to the uncommon syntactic structure of the segments, most rules were not active very often, so the space for improvement is quite limited. Yet the results show that when the rules do trigger, their output can be a useful signal for the CRF.

The observed improvement in recall at the slight expense of precision is in line with the results of [15] where the system based on the same rules achieved high recall and rather low precision.

Long reviews. It is immediately apparent that the long reviews are a much more difficult dataset than review segments – the best f-measure achieved on the short segments is 65.79 while here it is only 30.27. This can be explained by the lower density of aspect terms compared to random review segments and a much higher sentence length – after sentence segmentation, the average sentence length is over 29 words, compared to only 6 words for the random segments.

When using only the baseline features, the recall is extremely low. Adding morpho-syntactic features has a similar effect as for the random segments – precision is lowered but recall nearly triples.

Interestingly, adding features from the subjectivity lexicon changes the picture considerably. This feature set obtains the highest precision but recall is lower compared to both +morpho-syntactic and +rules. It may be that due to the high sentence length, sublex features help identify aspects within the short window but their presence pushes the model to ignore the more distant ones. A more thorough manual evaluation would be required to confirm this.

Finally, the addition of syntactic rules leads to the highest f-measure, even though neither recall nor precision are the best. In this dataset, possibly again thanks to the length

of sentences, the rules are trigged much more often than for the random segments. Rule features can therefore have a more prominent effect on the model.

5 Related Work

In terms of using rules for ABSA, our work is inspired by [15]. Such rules can also be used iteratively to expand both the aspects and evaluative terms using the double propagation algorithm [10]. Other methods of discovering opinion targets are described, inter alia, in [3, 9, 5]. Linear-chain CRFs have been applied in sentiment analysis and they are also well suited for ABSA, they were used e.g. by the winning submission by [13] to the SemEval 2014 Task 4.

For Czech, a dataset for ABSA was published by [11]. This dataset is in the domain of restaurant reviews and closely follows the methodology of [7]. Our work focuses on reviews of IT products, naturally complementing this dataset. It should further support research in this area and enable researchers to evaluate their approaches on diverse domains.

6 Conclusion

We have presented a new dataset for ABSA in the Czech language and we have described a baseline system for the subtask of aspect term extraction.

The dataset consists of segments from user reviews of IT products with the annotation of aspects and their polarity.

The system for aspect term extraction is based on linear-chain CRFs and uses a number of surface and linguistically-informed features. On top of these features, we have shown that task-specific syntactic rules can provide useful input to the model.

Utility of the syntactic rules could be further evaluated on other domains (such as the Czech restaurant reviews) or languages (e.g. using the official SemEval data sets) and the impact of individual rules could be thoroughly analyzed across these data sets.

---

[5]CRF++ feature templates do not offer a simple way to achieve this without also generating a large number of uninformative feature types.

Acknowledgements

This research was supported by the grant GA15-06894S of the Grant Agency of the Czech Republic and by the SVV project number 260 224. This work has been using language resources developed, stored and distributed by the LINDAT/CLARIN project of the Ministry of Education, Youth and Sports of the Czech Republic (project LM2010013).

References

[1] Hajič, J., Vidová-Hladká, B.: Tagging inflective languages: Prediction of morphological categories for a rich, structured tagset. In: Proceedings of the COLING — ACL Conference, 483–490, 1998

[2] Lafferty, J. D., McCallum, A., Pereira, F. C. N.: Conditional random fields: Probabilistic models for segmenting and labeling sequence data. In: Proceedings of the Eighteenth International Conference on Machine Learning, ICML'01, 282–289, San Francisco, CA, USA, 2001, Morgan Kaufmann Publishers Inc.

[3] Liu, B.: Web data mining: exploring hyperlinks, contents, and usage data (Data-centric systems and applications). Springer-Verlag New York, Inc., Secaucus, NJ, USA, 2006

[4] McDonald, R., Pereira, F., Ribarov, K., Hajič. J.: Nonprojective dependency parsing using spanning tree algorithms. In: Proceedings of the conference on Human Language Technology and Empirical Methods in Natural Language Processing, 523–530, 2005

[5] Qiaozhu Mei, Xu Ling, Matthew Wondra, Hang Su, and ChengXiang Zhai. Topic sentiment mixture: Modeling facets and opinions in weblogs. In *Proceedings of the 16th International Conference on World Wide Web*, WWW '07, pages 171–180, New York, NY, USA, 2007. ACM.

[6] Pontiki, M., Galanis, D., Papageorgiou, H., Manandhar, S., Androutsopoulos, I.: SemEval-2015 task 12: Aspect based sentiment analysis. In: Proceedings of the 9th International Workshop on Semantic Evaluation (SemEval 2015), 486–495, Denver, Colorado, June 2015, Association for Computational Linguistics

[7] Pontiki, M., Galanis, D., Pavlopoulos, J., Papageorgiou, H., Androutsopoulos, I., Manandhar, S.:. SemEval-2014 task 4: Aspect based sentiment analysis. In: Proceedings of the 8th International Workshop on Semantic Evaluation (SemEval 2014), 27–35, Dublin, Ireland, August 2014, Association for Computational Linguistics and Dublin City University

[8] Popel, M., Žabokrtský, Z.: TectoMT: modular NLP framework. In: Hrafn Loftsson, Eirikur Rögnvaldsson, and Sigrun Helgadottir, (eds.), IceTAL 2010, volume 6233 of Lecture Notes in Computer Science, 293–304, Iceland Centre for Language Technology (ICLT), Springer, 2010

[9] Popescu, A. -M., Etzioni, O.: Extracting product features and opinions from reviews. In: Proceedings of the Conference on Human Language Technology and Empirical Methods in Natural Language Processing, HLT'05, 339–346, Stroudsburg, PA, USA, 2005

[10] Qiu, G., Liu, B., Bu, J., Chen, C.: Opinion word expansion and target extraction through double propagation. Computational Linguistics 37 (1) (March 2011), 9–27

[11] Steinberger, J., Brychcín, T., Konkol, M.: Aspect-level sentiment analysis in Czech. In: Proceedings of the 5th Workshop on Computational Approaches to Subjectivity, Sentiment and Social Media Analysis, Baltimore, USA, June 2014, Association for Computational Linguistics

[12] Straková, J., Straka, M., Hajič, J.: Open-source tools for morphology, lemmatization, POS tagging and named entity recognition. In: Proceedings of 52nd Annual Meeting of the Association for Computational Linguistics: System Demonstrations, 13–18, Baltimore, Maryland, June 2014, Association for Computational Linguistics

[13] Toh Z., Wang, W.: DLIREC: Aspect term extraction and term polarity classification system. In: Proceedings of the 8th International Workshop on Semantic Evaluation (SemEval 2014), 235–240, Dublin, Ireland, August 2014, Association for Computational Linguistics and Dublin City University

[14] Veselovská, K., Bojar, O.: Czech SubLex 1.0, 2013

[15] Veselovská, K., Tamchyna, A.: ÚFAL: Using hand-crafted rules in aspect based sentiment analysis on parsed data. In: Proceedings of the Eighth International Workshop on Semantic Evaluation (SemEval 2014), 694–698, Dublin, Ireland, 2014, Dublin City University, Dublin City University

Implementation of a search engine for DeriNet

Jonáš Vidra

Institute of Formal and Applied Linguistics, Faculty of Mathematics and Physics, Charles University in Prague
vidra.jonas@seznam.cz,
WWW home page: http://jonys.cz/

Abstract: DeriNet, a new database of lexical derivates, is being developed at the Institute of Formal and Applied Linguistics. Since it is a wordnet containing large amounts of trees, it is hard to visualize, search and extend without specialized tools. This paper describes a program which has been developed to display the trees and enable finding certain types of errors, and its query language. The application is web-based for ease of use and access. The language has been inspired by CQL, the language used for searching the Czech National Corpus, and uses similar keywords and syntax extended to facilitate searching in tree structures.

1 Introduction

1.1 DeriNet

DeriNet[1] is a database of lexical derivates.[5] It contains lexemes structured into oriented trees that express derivational relations – the nodes are lexemes and edges are child→parent relations. Each lexeme represents one lemma with its meaning (word sense) partially disambiguated – several lexemes may represent the same lemma in different senses, if they differ in their derivational history – that is, they're derived from different words[2] or they have different descendants[3].

The database is stored in a tab-separated-values file using a simple format; see table 1 for an excerpt. Each line contains one lexeme. The meanings of the columns are: ID is a nonnegative integer which serves as a unique identifier; lemma is the base form of a word; techlemma is the lemma plus additional annotation in the style of the m-layer of the Prague Dependency Treebank[2]. POS is part of speech – A for adjectives, D for adverbs, N for nouns, V for verbs. Parent is the ID of the direct derivational parent. Every lexeme has at most one parent, which simplifies the structure, although it doesn't allow representing compounding. Version 0.9, released in December 2014, contains 305,781 lexemes sampled from the Czech National Corpus and 117,327 oriented edges.

---

[1]http://ufal.mff.cuni.cz/derinet
[2]such as "sladit" (to sweeten / to harmonize) from "sladký" (sweet) or "ladit" (to tune)
[3]as in the case of "hnát" (vulgar form of limb / to drive) whose child is either "hnátek" (diminutive from hnát) or "hnaný" (the chased one)

Table 1: Excerpt from the DeriNet database showing four lexemes.

ID	lemma	techlemma	POS	parent
177751	nosní	nosní	A	177725
177752	nosník	nosník	N	177753
177753	nosný	nosný	A	
177754	nosně	nosně_^(*1ý)	D	177751

1.2 Requirements

Since DeriNet is a graph database contained in a TSV file, it's very hard to visualize the structure without a specialized tool. Therefore, we've decided to create a specialized application which would use a simple language for searching. Ideally, this application should be accessible and usable by both skilled linguists and random visitors. The requirements were to support:

- Searching for lemmas, techlemmas and parts-of-speech using regular expressions.

- Querying of ancestor-descendant and sibling-sibling relations.

- Displaying the trees. The largest trees in DeriNet 0.9 contain 31 lexemes, the deepest are 7 levels deep, but most of them are smaller. This means that the visualization can be straightforward.

2 Related work

We've decided to survey the available query languages and either choose a suitable one or create a custom domain-specific language while drawing inspiration from existing sources. Our criteria were

1. support for searching within tree structures: specifying criteria for ancestors, descendants and siblings,

2. familiarity within our target group of Czech linguists – nonprogrammers; or at least a shallow learning curve,

3. easy and effective writing of new queries, and

4. simple implementation.

The first two points ruled out relational database query languages such as SQL, since they have no provisions for searching in trees. We've examined XPath, PML-TQ and CQL in more detail.

XPath[1] is a long-established language designed specifically for tree querying and it has an existing implementation in Javascript[6]. DeriNet has an XML version which could be used as a backing database with just a few changes. Unfortunately, testing has shown that existing web browsers cannot handle XML files the size of DeriNet. Loading the database easily took several minutes on weaker machines and any XPath query took tens of seconds to process. The latter could be overcome by using an external XPath library such as Wicked Good XPath, but the problem with representing the XML in a browser remains.

PML-TQ (Prague Markup Language – Tree Query[4]) is a language used for querying the Prague Dependency Treebank. It has many options, but it is also very complex and hard to implement. Therefore, it could only be reasonably used with the existing implementation as a server-side backend. Writing queries by hand is not very convenient, they are mostly edited using an offline application which we wouldn't be able to replicate in a web application. There is a web interface[4], but it doesn't offer the interactive editor. Also, many of its features would be unused in our simple system.

CQL (Corpus Query Language[3]) is a language used for searching the Czech National Corpus[5]. It only supports searching in sentences, i.e. linearly ordered sequences of nodes, but the syntax is simple and easy to modify and expand. It could be easily extended to handle basic tree queries as well. Also, its syntax supports using bare keywords as a query, while XPath and PML-TQ require a richer construction even for the simplest cases.

3 Query language

We've decided to go for maximum simplicity and familiarity and base our language on CQL. We wrote our own implementation using Jacob[6], an alternative of Flex and Bison for JavaScript, which creates an LALR parser from an EBNF grammar.

The simplest query is the attribute expression, which describes attributes of a single node in the database. Its results are all nodes that match the attributes. It has the following general form:

`[attribute="value" …]`

Multiple attributes can be specified at once – the query then matches iff all of them match. We currently support the following comparison operators: "="

for RegExp comparison, "==" for string comparison and "!=" and "!==" as their negations. The following attributes may be used: "lemma", "techlemma", "pos", "id" and "parent". There is a special syntax for matching nodes without a parent: `parent=="-1"` matches all nodes *without* a parent. This special case is guaranteed not to clash with any actual IDs, since normal IDs are nonnegative integers.

Examples:

`[lemma="stroj"]`

Matches all nodes whose lemma contains "stroj".

`[lemma="^stroj$"]`

Matches all nodes whose lemma is exactly "stroj".

`[lemma=="stroj"]`

The same as the previous one.

`[]`

Matches any node.

`[id=="12345"]`

Matches nodes whose ID is exactly "12345".

`[pos=="A" lemma="ký$"]`

Matches adjectives ending in "ký".

`[parent=="-1"]`

Matches all nodes without a parent.

Attribute expressions also have shorthand forms: By selecting the appropriate "default attribute", you can write just `"stroj"` instead of `[defaultattribute="stroj"]` and `stroj` instead of `[defaultattribute=="stroj"]`.

Attribute expressions can be combined to form a tree expression. The standard CQL was designed for querying sentences, which contain totally ordered nodes and so the only higher-order expression it supports is a linear sequence. However, nodes in our database are only partially ordered, so we had to define an extension which would allow querying of non-linearly ordered sets of nodes. To do this, we've introduced parentheses. The sequence of attribute expressions encodes parent→child relation, with multiple children enclosed in a parenthesis and separated by commas: `car carský (carsky, carismus, carství)`

4 Implementation

The application is a web-based program written in JavaScript and runs entirely client-side; the server is only used for storing the databases and the program itself. After the page loads, a background task downloads and parses the selected database. You can change the database used for searching by selecting a new one from a drop-down menu.

After you enter a search query into the text field, your query is processed and the results are displayed as SVG trees. Possible larger amounts of results are split into several pages. You can view the techlemma and part-of-speech of a single node by hovering

---

[4]https://lindat.mff.cuni.cz/services/pmltq/

[5]https://kontext.korpus.cz/first_form?
queryselector=cqlrow

[6]https://github.com/Canna71/Jacob

over it with the pointer, or, alternatively, you can enable the detailed display for all nodes by selecting the „Show additional information" checkbox. The results can be exported to the same tab-separated-values format that DeriNet uses for storage and stored to a file.

A screenshot of the application is included in figure 1 and a running version of the search engine is available at `http://jonys.cz/derinet/search/`.

Figure 1: Screenshot of the application.

5 Detecting errors

The main reason for creating this application was enabling a quick way of searching for certain common errors that were present in the then-current version of DeriNet. DeriNet is built from the source lexicon using a series of scripts based on handwritten rules. Many of the rules have the form of pattern+exceptions, for example the pattern "A-ačný N-ačnost" which connects (among others) the noun "lačnost" to the adjective "lačný", or "jednoznačnost" to "jednoznačný", has an exception of "senzačnost", which is derived from "senzační" instead. These rules often have overlooked exceptions.

In addition, there are several systematic errors in 0.9 that have been discovered by this search tool. For example:

- `[techlemma="_\^.*\(.*\*[0-9].*\)" parent="-1"]`
  Many techlemmas contain derivational information, which is not parsed correctly in version 0.9.

- `"á$" "ý$"`
  Matches female and male variants of names that have the derivation order reversed. The female variant ending in "á" should be derived from the male one.

- `[lemma="(ovat|it)$" pos=="A"]`
  These words are not adjectives, but errors in the source lexicon. These endings are both typical for verbs.

- `[lemma!="(í|ý|ost)$"] "ost$"`
  Bad derivations. Words ending in "ost" that aren't derived by prefixation should be nearly always derived from adjectives ending in "í" or "ý".

6 Problems

6.1 Performance

An important issue is performance. A loaded webpage with DeriNet Search can easily consume over 100 MiB of RAM and complex queries (especially queries with many siblings with few constraints on a single level, such as `[] ([],[],[],[],[])`) may take over a minute to complete. However, although the searching isn't instantaneous, for simpler queries it is reasonably fast. Even a dated laptop (CPU: Intel Atom Z520, 1.33 GHz) is able to complete simple searches in under a second.

The design of the application is well suited to parallelization, but JavaScript has only limited provisions for parallel execution. Its threading model is based on message-passing with no shared data structures, so speedup can only be achieved at the expense of using more memory, because each thread either needs its own copy of the whole database, or we have to split the DB into clusters (connected components) and allocate these clusters to search threads. The latter design would highly complicate displaying the results. This means that if we want to avoid complexity and needless copying of data, we are limited to single threaded execution only.

However, if there are multiple search windows open, each one runs in its own thread. It would be possible to create a global execution thread or a pool with user-determined amount of threads, delegate all work to it and keep only a single database copy in memory, but not all browsers support these techniques and we've decided not to complicate the design for small gains.

6.2 Compatibility

Apart from large memory requirements, which may be a problem on mobile devices, the application requires support for ECMAScript 5.1, HTML5, CSS3 and SVG. All modern desktop browsers (less than three years old) should pass these requirements, with the exception of Internet Explorer before version 11, but support on browsers for mobile devices is poor. Tested browsers include Firefox 31, Konqueror 4.14 and Chrome 36. The application is only partially usable in Opera 12.16, which doesn't support the history manipulation method needed to switch between pages of results and so only the first page is ever shown.

6.3 Scalability

Another set of problems was uncovered when testing the application with the newest version of DeriNet, 0.10, which is as of 2015-06-14 not yet publicly available. Version 0.10 is four times larger than 0.9 (45 MiB uncompressed and 9.6 MiB when compressed with gzip, as opposed to 12 MiB uncompressed and 3 MiB compressed) and downloading a database of this size is problematic for users without a broadband network connection.

Search times have also increased, because DeriNet 0.10 contains 968,929 lexemes – over three times more than 0.9. To test the scalability of our application with increasing database size, we've prepared six progressively smaller datasets by cutting 500,000, 250,000, 100,000, 50,000 and 10,000 lexemes with the lowest IDs out of DeriNet 0.10. We've then measured the time needed to parse a query, search the database and draw the resulting SVG. Each measurement was repeated five times, the results averaged and the error expressed using standard deviation.

Three queries were selected for testing: `Absolón`, which produces a single result; `[]`, which matches all nodes; and `[pos!=="A"]` (`[pos="[ADNV]$"]`,

`[] []`, `[]`) as a test of more complex queries. Figure 2 shows that the time needed for parsing and display is approximately constant and that the searching time quickly dominates. Figure 3 indicates that the query `[pos!=="A"]` (`[pos="[ADNV]$"]`, `[] []`, `[]`) scales supralinearly, but figure 4 shows that for the other queries the algorithm is, in fact, linear. The explanation for why the complex query behaves differently lies in the structure of the database, as evidenced by the number of results. The database is scanned from the beginning to the end without using any acceleration technique such as an index, and the beginning (low IDs) contains lemmas that start with capital letters. These are typically names which contain fewer derivational links than the rest of the database. This causes the complex query to abort quickly, just after discovering that the node doesn't have enough children to satisfy the requirements.

7 Future work

Due to the problems encountered with DeriNet 0.10, we plan on switching to server-side searching. This is feasible with few changes thanks to Node.js, which allows running arbitrary JavaScript code on the server. Doing so would allow us to support even primitive browsers on devices with little processing power – the demands for modern JavaScript features could be dropped; support for SVG would be sufficient.

Another needed change is optimization of the search routines. Queries such as `[]` (`[]`, `[]`, `[]`, `[]`, `[]`) (find all trees with at least five siblings on a single level) take a very large amount of time, sometimes even minutes. The current algorithm goes through all the possible combinations, which is not needed, since many of the results will be duplicates of each other. Stopping at the first match is not a solution, though, because advanced features of CQL such as "within" and "containing" clauses rely on this behavior to work.

The third possible area of improvement is introducing new language features.

- A way of marking "no more nodes here". We have the `[parent=="-1"]` syntax for finding nodes that don't have a parent, but there should be a better way of specifying this, that would also allow searching for trees that are smaller than a certain limit, as opposed to larger than a limit. The latter can be done using `[]` (`[pos="A"]`, `[]`, `[]`) (find all adjectives with more than two siblings).

- Support for iteration: "*", "+", "?", "{x,y}". The last type partially satisfies (and depends on) the previous point.

- Support for "within" and "containing".

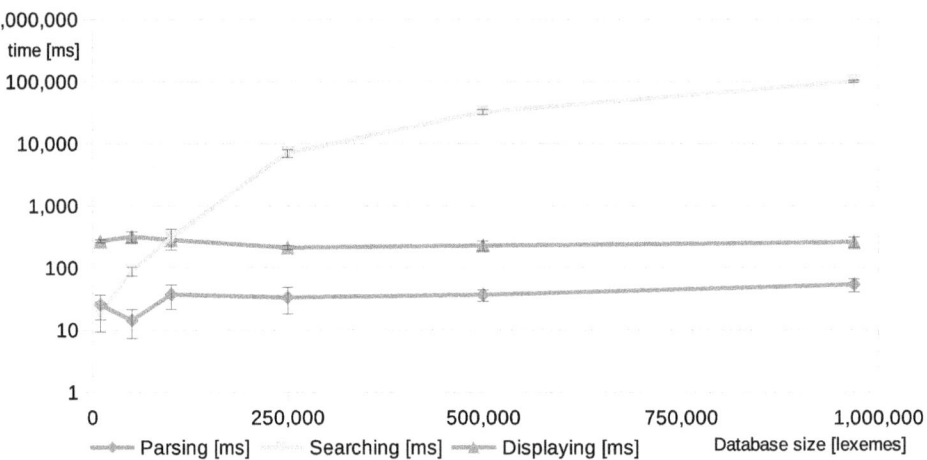

Figure 2: A graph showing the increase of time needed to process query [pos!=="A"] ([pos="[ADNV]$"], []
[], []) with increasing database size. The error bars represent one standard deviation over five trials.

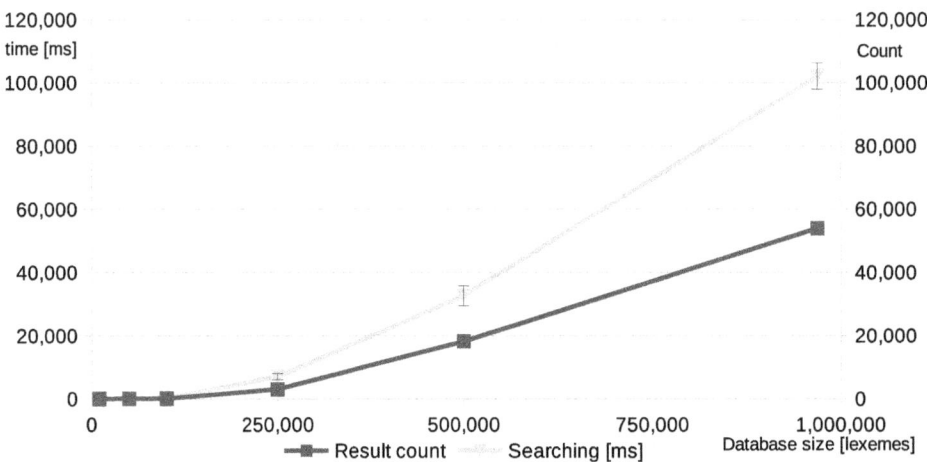

Figure 3: Graph showing that when processing [pos!=="A"] ([pos="[ADNV]$"], [] [], []), the time needed
increases supralinearly. The number of results has been included for comparison. The error bars represent one
standard deviation over five trials.

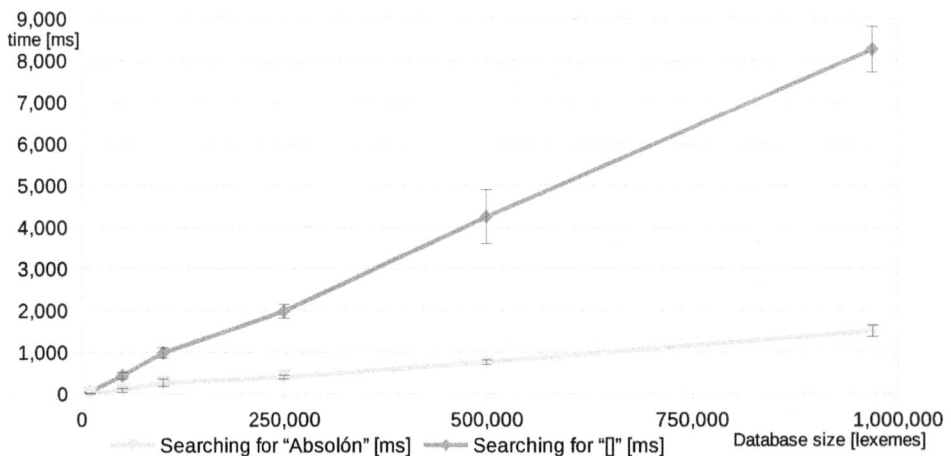

Figure 4: Graph showing that processing Absolón and [] scales linearly with increasing database size. The
error bars represent one standard deviation over five trials.

The application itself could be improved, too. The following would be useful:

- User-friendlier handling of erroneous input. When you enter `[lemma!="ický$"` `[lemma="ičnost$"]`, instead of "Unexpected token lsbra "[" at (0:16)" the parser should probably report "Missing close-bracket before position 16" and show the query with the offending part highlighted in red. Unfortunately, Jacob, the tool we use for building the parser, doesn't have any support for returning structured error information and printing useful messages.

- A method of exporting the SVG to a file that produces a useful image. Currently, the SVG images are styled using CSS, which gives us a flexible way of coloring the nodes and showing/hiding technical information. This method doesn't work outside of browsers, though, which means that the exported SVG is black-and-white only and, unless the user had the "Show additional information" checkbox checked, contains overlapping text.

- A more compact style of drawing trees. Even in the current database there are some trees that are too wide to fit on screen without horizontal scrolling. Trees in DeriNet 0.10 are even larger, which makes this a big problem.

8 Conclusion

The stated goals were to build an application that would enable users to easily search DeriNet and help its creators discover both systematic and random errors. The utility fulfilled our expectations with the second part – error searching – since several mistakes and faults have been detected with its help. The query language is only very basic so far, but extensions are possible.

Acknowledgements

This work has been using language resources developed and/or stored and/or distributed by the LINDAT/CLARIN project of the Ministry of Education of the Czech Republic (project LM2010013).

References

[1] James Clark, Steve DeRose, et al. *XML path language (XPath) version 1.0.* 1999.

[2] Jan Hajič. *Disambiguation of rich inflection: computational morphology of Czech.* Karolinum, 2004.

[3] Miloš Jakubíček et al. "Fast Syntactic Searching in Very Large Corpora for Many Languages." In: *PACLIC.* Vol. 24. 2010, pp. 741–747.

[4] Petr Pajas and Jan Štěpánek. "System for querying syntactically annotated corpora". In: *Proceedings of the ACL-IJCNLP 2009 Software Demonstrations.* Association for Computational Linguistics. 2009, pp. 33–36.

[5] Magda Ševčíková and Zdeněk Žabokrtský. "Word-Formation Network for Czech". In: *Proceedings of the Ninth International Conference on Language Resources and Evaluation (LREC'14).* Ed. by Nicoletta Calzolari (Conference Chair) et al. Reykjavik, Iceland: European Language Resources Association (ELRA), May 26–31, 2014. ISBN: 978-2-9517408-8-4.

[6] Ray Whitmer et al. "Document Object Model (DOM) Level 3 XPath Specification". In: *W3C, http://www.w3.org/TR/DOM-Level-3-XPath* (2004).

Workshop on Computational Intelligence and Data Mining (WCIDM 2015)

Workshop Program Committee

Jose Luis Balcazar, Technical University of Catalonia, Barcelona
Petr Berka, University of Economics, Prague
Hans Engler, University of Georgetown
Jan Faigl, Czech Technical University, Prague
Xuhui Fan, University of Technology, Sydney
Pitoyo Hartono, University of Chukyo
Martin Holeňa, Czech Academy of Sciences, Prague
Ján Hric, Charles University, Prague
Robert John, University of Nottingham
Jan Kalina, Czech Academy of Sciences, Prague
Jirí Kléma, Czech Technical University, Prague
Pavel Kordík, Czech Technical University, Prague
Tomas Krilavicius, Vytautas Magnus University, Kaunas
Jirí Kubalík, Czech Technical University, Prague
Vera Kurková, Czech Academy of Sciences, Prague
Stéphane Lallich, University of Lyon
Philippe Lenca, Telecom Bretagne, Brest
Arnaud Martin, University of Rennes
Mirko Navara, Czech Technical University, Prague
Engelbert Memphu Nguifo, Blaise Pascal University, Clermont-Ferrand
Tomáš Pevný, Czech Technical University, Prague
Petr Pošík, Czech Technical University, Prague
Jan Rauch, University of Economics, Prague
Patrick Siarry, University of Paris-East
Marta Vomlelová, Charles University, Prague
Filip Železný, Czech Technical University, Prague

J. Yaghob (Ed.): ITAT 2015 pp. 108–114
Charles University in Prague, Prague, 2015

Using Multi-Objective Optimization for the Selection of Ensemble Members

Tomáš Bartoň[1,2] Pavel Kordík[1]

[1] Faculty of Information Technology
Czech Technical University in Prague, Thakurova 9, Prague 6, Czech Republic
http://www.fit.cvut.cz
tomas.barton@fit.cvut.cz
[2] Institute of Molecular Genetics of the ASCR, v. v. i.
Vídeňská 1083, 142 20 Prague 4, Czech Republic

Abstract: In this paper we propose a clustering process which uses a multi-objective evolution to select a set of diverse clusterings. The selected clusterings are then combined using a consensus method. This approach is compared to a clustering process where no selection is applied. We show that careful selection of input ensemble members can improve the overall quality of the final clustering. Our algorithm provides more stable clustering results and in many cases overcomes the limitations of base algorithms.

1 Introduction

Clustering is a popular technique that can be used to reveal patterns in data or as a preprocessing step in data analysis. Clustering tends to group similar objects into groups that are called clusters and to place dissimilar objects into different clusters. Its application domains include data mining, information retrieval, bioinformatics, image processing and many others.

Many clustering algorithms have been introduced so far, an extensive overview of clustering techniques can be found in [1]. Since we have the No Free Lunch Theorem [36], which states that there is no single, supreme algorithm that would optimally run on all datasets, it is complicated for a user to decide which algorithm to choose.

Ensemble methods have been successfully applied in supervised learning [25] and similar attempts appeared in the unsupervised learning area [32], [14], [34]. Cluster ensemble methods should eliminate the drawbacks of individual methods by combining multiple solutions into one final clustering.

2 Cluster Ensembles

The main objective of clustering ensembles is to combine multiple clusterings into one, preferably high-quality solution.

Let $X = \{x_1, x_2, \ldots, x_n\}$ be a set of n data points, where each $x_i \in X$ is represented by a vector of d attributes. A cluster ensemble is defined as $\Pi = \{\pi_1, \pi_2, \ldots, \pi_m\}$ with m clusterings. Each base clustering π_i consists of a set of clusters $\pi_i = \{C_1^i, C_2^i, \ldots, C_{k_i}^i\}$, such that $\cup_{j=1}^{k_i} C_j^i = X$, where k_i is the number of clusters in a given ensemble member (it does not have to be the same for all members).

The problem is how to obtain a final clustering $\pi^* = \{C_1^*, C_2^*, \ldots, C_K^*\}$, where K is the number of clusters in the result and π^* summarizes the information from ensemble Π.

The process consists of two major steps. Firstly, we need to generate a set of clustering solutions and secondly we need ti combine the information from these solutions. A typical result of the ensemble process is a single clustering. It has been shown in supervised ensembles that the best results are achieved when using a set of predictors whose errors are dissimilar [25]. Thus it is desirable to introduce diversity between ensemble members.

2.1 Ensemble Generation Strategies

Several approaches have been used to initialize clustering solutions in order to create an ensemble.

- **Homogeneous ensembles**: Base clusterings are created using repeated runs of a single clustering algorithm. This is quite a popular approach, repeated runs of k-means with random center initialization have been used in [14]. When using k-means the number of clusters is typically fixed to $\lceil \sqrt{n} \rceil$, where n is the size of the dataset [22].

- **Varying k**: Repeated runs of k-means with random initialization and k [15], a golden standard is using k in the range from 2 to $\sqrt{n}$.

- **Random subspacing** An ensemble is created from base clusterings that use different initial data. This could be achieved by projecting data onto different subspaces [13], [16] choosing different subsets of features [32], [37], or using data sampling techniques [10].

- **Heterogeneous ensembles**: Diversity of solutions is introduced by applying different algorithms on the same dataset.

2.2 Consensus Functions

Having multiple clusterings in an ensemble, many functions have been proposed to derive a final clustering. When only one solution is considered as the result, it is usually referred as a consensus function, unlike meta clustering where the output is a set of multiple clusterings [6].

There are several approaches as to how to represent information contained in base clusterings, some use matrices while others use graph representation.

- **Pairwise similarities**: A pairwise similarity matrix is created and afterwards a clustering algorithm (e.g. hierarchical agglomerative clustering) is applied to group together items that were most frequently together in the same cluster in all the base clusterings [14]. the Cluster-based Similarity Partitioning Algorithm (CSPA) from Strehl and Ghosh [32] uses METIS [24] for partitioning a similarity matrix into k components.

- **Feature-based approach**: The ensemble problem is formulated as categorical data clustering. For each data point an m-dimensional vector containing labels in base clusterings is created. The goal is to find a partition π^* which summarizes the information gathered from the partitions Π [28], [33], [4].

- **Graph based**: Many methods use graph representation for capturing relationships between base clusterings. Strehl and Ghosh [32] also proposed the HyperGraph-Partitioning Algorithm (HGPA), where vertices correspond to data points and a hyperedge represents clusters. Another approach chooses CO-MUSA [27] which increases the weight of the edge for each occurrence of data pairs in the same cluster. Afterwards the nodes are sorted by the attachment score, which is defined as the ratio between the sum of the node's weights and its number of incident edges. The nodes with the highest attachment score are then used as a foundation for new clusters. This approach is relatively fast to compute, however it might fail to capture complex relationships between very diverse clusterings.

3   Multi-Objective Clustering

Multi-objective clustering usually optimizes two objective functions. Using more than a few objectives is not usual because the whole process of optimization becomes less effective.

The first multi-objective evolutionary clustering algorithm was introduced in 2004 by Handl and Knowles [18] and is called VIENNA (the Voronoi Initialized Evolutionary Nearest-Neighbour Algorithm).

Subsequently, in 2007 Handl and Knowles published a Pareto-based multi-objective evolutionary clustering algorithm called MOCK [19] (Multi-Objective Clustering with automatic K-determination). Each individual in MOCK is represented as a directed graph which is then translated into a clustering. The genotype is encoded as an array of integers whose length is same as the number of instances

in the dataset. Each number is a pointer to another instance (an edge in the graph), since it is connected to the instance at a given index. This easily enables the application of mutation and crossover operations.

As a Multi-Objective Evolutionary Algorithm (MOEA), MOCK employs the Pareto Envelope-based Selection Algorithm version 2 (PESA-II) [7], which keeps two populations, an internal population of fixed size and a larger external population which is exploited to explore good solutions. Two complementary objectives, *deviation* and *connectivity*, are used as objectives in the evolutionary process.

A clear disadvantage of MOCK is its computation complexity, which is a typical characteristic of evolutionary algorithms. Nevertheless, the computation time spent on MOCK should result in high-quality solutions. Faceli et al. [12] reported that for some high-dimensional data it is not guaranteed that the algorithm will complete, unless the control front distribution has been adjusted for the given data set.

Faceli et al. [12] combined a multi-objective approach to clustering with ensemble methods and the resulting algorithm is called MOCLE (Muli-Objective Clustering Ensemble Algorithm). The objectives used in the MOCLE algorithm are the same as those used in MOCK [19]: deviation and connectivity. Unlike MOCK, in this case the evolutionary algorithm used is NSGA-II [9].

4   Our Approach

There are many ways to produce a final consensus, nonetheless in this work we focus on the selection of high-quality and diverse clusterings.

In order to optimize ensemble member selection, we apply a multi-objective optimization. The whole process is shown in Fig. 1. In our previous work [3] we have shown that introducing multiple objectives into the clustering process can improve the quality of clustering, specifically using the Akaike information criterion (AIC) [2] (or BIC [31]) with another criterion leads to better results. The second criterion is typically based on computing a ratio between cluster compactness and the sum of distances between cluster centres.

AIC is typically used in supervised learning when trying to estimate model error. Essentially it attempts to estimate the optimism of the model and then add it to the error [20]:

$$f_{AIC} = -2 \cdot \frac{\log\left(L_n(k)\right)}{n} + 2 \cdot \frac{k}{n} \qquad (1)$$

where $L_n(k)$ is the maximum likelihood of a model with k parameters based on a sample of size n.

The SD index was introduced in 2000 by Halkidi et al. [17] and it is based on concepts of *average cluster scattering* and *total separation of clusters* which were previously used by Rezaee et al. [29] for evaluation of fuzzy clusterings.

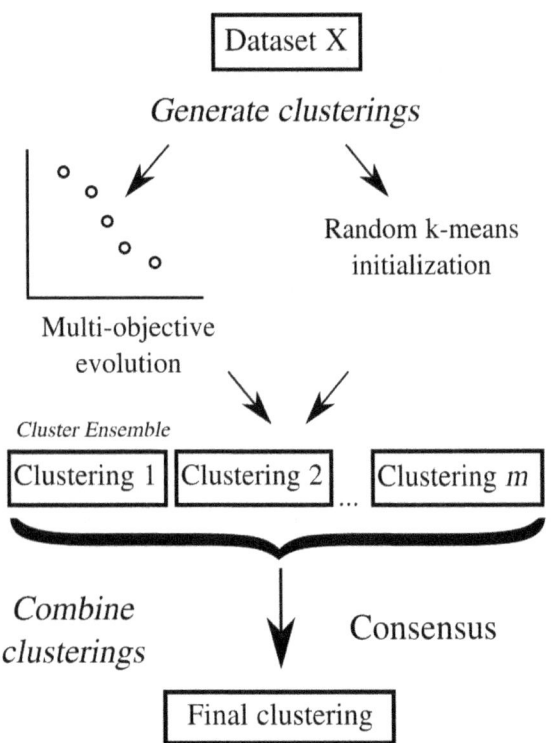

$$Generate\ clusterings$$

Random k-means
initialization

Multi-objective
evolution

Cluster Ensemble

| Clustering 1 | Clustering 2 | ... | Clustering m |

*Combine
clusterings*   Consensus

Final clustering

Figure 1: Cluster Ensemble Process: Firstly using the same input dataset we generate multiple clusterings, then select m best clusterings for the ensemble and combine them into a single solution.

The average scattering is defined as:

$$Scatt(k) = \frac{1}{k} \sum_{i=1}^{k} \frac{\|\sigma(\bar{\mathbf{c}}_i)\|}{\|\sigma(X)\|} \qquad (2)$$

where $\|\mathbf{x}\|$ is the norm of a vector,
$\bar{\mathbf{c}}_i$ is the centroid of the i-th cluster,
$\sigma(X)$ is the variance of the input dataset.

$\sigma(X) \in \mathbb{R}^m$ with m being the number of dataset dimensions. Variance for a dimension d is defined as:

$$\sigma^d = \frac{1}{n} \sum_{i=1}^{n} \left(x_i^d - \bar{x}^d \right)^2$$

$$\|\sigma(X)\| = \sqrt{\sum_{i=1}^{m} (\sigma^d)^2}$$

The total separation is given by:

$$Dis(k) = \frac{D_{max}}{D_{min}} \sum_{i=1}^{k} \left(\sum_{j=1}^{k} \|\bar{\mathbf{c}}_i - \bar{\mathbf{c}}_j\| \right)^{-1} \qquad (3)$$

where D_{max} is the maximum distance and D_{min} is the minimum distance between cluster centers ($\bar{c}_i$) and k is the number of clusters.

$$D_{max} = \max_{\substack{i,j \in \{1,...,k\} \\ i \neq j}} (\|\bar{\mathbf{c}}_i - \bar{\mathbf{c}}_j\|) \qquad (4)$$

$$D_{min} = \min_{\substack{i,j \in \{1,...,k\} \\ i \neq j}} (\|\bar{\mathbf{c}}_i - \bar{\mathbf{c}}_j\|) \qquad (5)$$

Then we can define the SD validity index as follows:

$$f_{SD} = \alpha \cdot Scat(k) + Dis(k) \qquad (6)$$

where α should be a weighting factor equal to $Dis(c_{max})$ with c_{max} being the maximum number of clusters [17]. This makes perfect sense for fuzzy clustering (as was proposed in [29]), however it is rather unclear how to compute c_{max} in the case of crisp clustering, when $c_{max} \gg k$ without running another clustering with c_{max} as the requested number of clusters. Nonetheless, [17] mentions that "SD proposes an optimal number of clusters almost irrespective of c_{max}, the maximum number of clusters", thus we consider the special case where $c_{max} = k$:

$$f_{SD} = Dis(k) \cdot Scat(k) + Dis(k) \qquad (7)$$
$$= Dis(k) \cdot (Scat(k) + 1) \qquad (8)$$

The idea of minimizing inner cluster distances and maximizing distances between cluster centres it not really new. In fact most of the clustering objective functions use this idea. Instead of the SD index we could have used C-index [21], Calinski-Harabasz Index [5], Davies-Bouldin [8] or Dunn's index [11], just to name a few. Each of these indexes might perform better on some dataset, however it is out of scope of this paper to compare which combination would be better. We considered

As an evolutionary algorithm we use NSGA-II [9] while using only the mutation operator and a relatively small population (see Table 2). Each individual in the population contains a configuration of a clustering algorithm. For this experiment we used only the basic k-means [26] algorithm with random initialization. The number of clusters k is the only parameter that we change during the evolutionary process. For k we used a constraint in order to keep the number of clusters within reasonable boundaries. For all test datasets we used the interval $\langle 2, \sqrt{n} \rangle$.

In order to evaluate the effect of Pareto optimal solutions selection for the ensemble bagging process, we compare two strategies for generating base clusterings. The first one generates 10 k-means clusterings with random initialization with k within the interval $\langle 2, \sqrt{n} \rangle$. The second method uses multi-objective evolution of clusterings where each individual represents a k-means clustering.

As a consensus function we use either a graph based COMUSA algorithm or a hierarchical agglomerative clustering of the co-association matrix. Both approaches have their drawbacks, the COMUSA algorithm is locally consistent, however fails to integrate the information contained in different base clusterings, thus final clustering does not overcome limitations of base clustering algorithms. The latter could be very slow on larger datasets (complexity $O(n^3)$) and might produce noisy clustering in case of very diverse base clusterings.

Dataset	d	n	classes	source
aggregation	2	788	7	[15]
atom	3	800	2	[35]
chainlink	3	1000	2	[35]
complex8	2	2551	8	[30]
complex9	2	3031	9	[30]
diamond9	2	3000	9	[30]
jain	2	373	2	[23]
long1	2	1000	2	[18]
longsquare	2	900	6	[18]
lsun	2	400	3	[35]
target	2	770	6	[35]
tetra	3	400	4	[35]
triangle1	2	1000	4	[18]

Table 1: Datasets used for experiments.

Parameter	Setting
Number of generations	5
Population size	10
Mutation probability	30%
Mutation	Polynomial mutation

Table 2: Parameter settings for NSGA-II evolution

Dataset	C-RAND	C-MO	k-means
aggregation	0.75	**0.84**	**0.84**
atom	0.39	**0.61**	0.28
chainlink	0.07	**0.50**	0.07
complex8	0.59	**0.64**	0.59
complex9	0.64	**0.66**	0.65
diamond9	**1.00**	0.87	0.97
jain	0.36	**0.51**	0.37
long1	0.00	**0.49**	0.03
longsquare	0.83	**0.86**	0.80
lsun	0.54	**0.69**	0.54
target	0.67	0.57	**0.69**
tetra	**1.00**	0.86	0.99
triangle1	**0.95**	0.78	0.94

Table 3: NMI_{sqrt} score from various artificial dataset clusterings. C-RAND uses bagging of 10 k-means runs with a fixed value k that corresponds to the number of classes in the given dataset. The same approach is used in the case of k-means, but without bagging. C-MO is a NSGA-II based algorithm with criteria AIC and SD index; the number of clusters is randomized during evolution within the interval 2 to $\sqrt{n}$. Both C-RAND and C-MO use the COMUSA algorithm to form final (consensus) partitioning. The NMI values are averages from 30 independent runs.

To evaluate clustering quality we use NMI[1] (Normalized Mutual Information) [32]. NMI has proved to be a better criterion than the Adjusted Rand Index (ARI) for the evaluation of datasets with apriori known labels. However it does not capture noise in clusters (which does not occur in the case of traditional k-means), which might be introduced by some consensus methods. Another limitation of these evaluation metrics is apparent from Figure 6, where clustering with obvious 50% assignment error gets lowest possible score. Both indexes do not penalize correctly higher number of clusters, thus algorithms producing many small clusters will be preferred by these objectives.

5 Results

Most of the datasets used in these experiments come from the Fundamental Clustering Problems Suite [35]. We have intentionally chosen low dimensional data in order to be able to visually evaluate the results. An overview of the datasets used can be found in Table 1.

As objectives during the multi-objective evolution we used AIC and SD index in all cases. The advantage is combination of two measures that are based on very different

grounds. We tried using Deviation and Connectivity, as it was proposed by Handl [19] (and later used by [12]), but for all tested datasets, our approach was better or at least comparable with these objectives.

In the case of the COMUSA approach, the multi-objective selection either improved the final NMI score or it was comparable with the random selection. There were only a few exceptions, especially in case of compact datasets like the 9 diamond dataset which contains a specific regular pattern that is unlikely to appear in any real-world dataset (see Table 3). The number of clusters produces by COMUSA approach was usually higher than the correct number of clusters and clustering contains many nonsense clusters. Despite these facts the NMI score does not penalize these properties appropriately.

Agglomerative clustering of the co-association matrix despite its high computational requirement provide pretty good results. Nonetheless in the case of the *chainlink* dataset there is over 50% improvement in NMI. It is important to note, that for HAC-RAND and HAC-MO we did not provide the information about the correct number of clusters. Still these methods manages to estimate number of clusters or at least provide result that is very close to the number of supervised classes.

---

[1]There are several versions of this evaluation method, sometimes it is referred as NMI_{sqrt}

Dataset	HAC-RAND	HAC-MO
aggregation	**0.74**	0.71
atom	0.44	**0.68**
chainlink	0.47	**0.99**
complex8	**0.72**	0.71
complex9	**0.71**	0.70
diamond9	**0.83**	0.75
jain	**0.61**	0.60
long1	0.73	**0.81**
longsquare	0.73	**0.89**
lsun	0.66	**0.71**
target	0.40	0.40
tetra	0.75	**0.99**
triangle1	0.78	**0.89**

Table 4: NMI_{sqrt} score from the clustering of various artificial datasets. As a consensus method, hierarchical agglomerative clustering (with complete linkage) of co-association matrix is used. In the case of *HAC-RAND* we run 10 independent k-means clusterings with a random number of clusters (between 2 and $\sqrt{n}$), then form a co-association matrix and finally run agglomerative clustering of the matrix. *HAC-RAND* works in very similar manner, but instead of the first step a multi-objective evolution of k-means is performed. 10 dominating solutions are selected and the rest of the algorithm is the same.

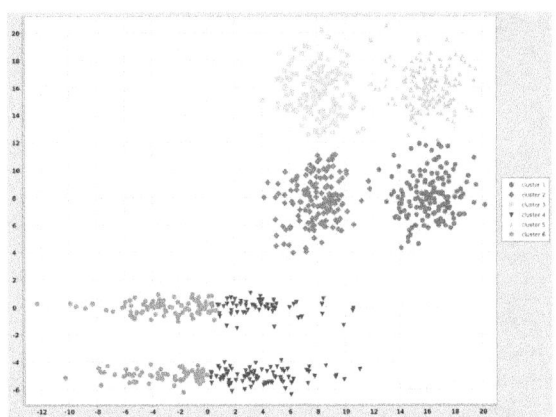

Figure 2: Typical clustering *longsquare* dataset using k-means ($k = 6$). K-Means algorithm fails to reveal non-spherical clusters (Ajusted Rand Index $= 0.94$, NMI $= 0.85$).

6   Conclusion

During our experiments we have shown that careful selection of clusterings for the ensemble process can significantly improve overall clustering quality for non-trivial datasets (measured by NMI).

It is interesting that non-spherical clusters could be discovered by consensus function when agglomerative hierarchical clustering is used (compare Fig. 2 and 3).

Using a multi-objective optimization for clustering se-

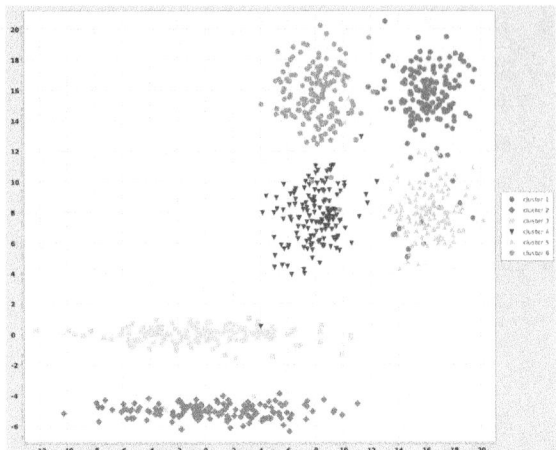

Figure 3: Consensus of 10 independent k-means runs on the *longsquare* dataset using $k = 6$. The co-association matrix obtained was clustered using hierarchical clustering (average linkage). The resulting clustering is qualitatively better than a single k-means run, however it contains noise in all clusters (Ajusted Rand Index $= 0.99$, NMI $= 0.95$).

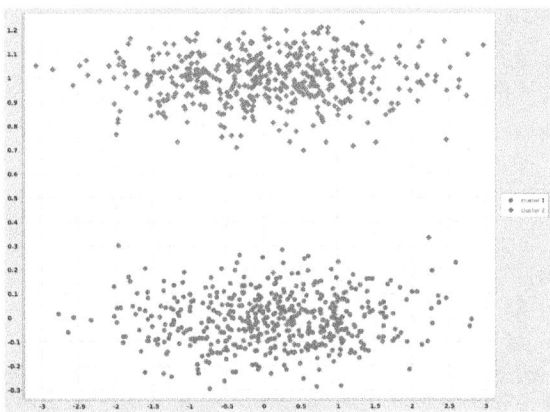

Figure 4: Consensus of 10 independent k-means runs on the *long1* dataset using $k = 6$. On many datasets setting the correct k does not improve the quality of the resulting clustering. The co-association matrix obtained was clustered using hierarchical clustering (average linkage).

lection improved the overall quality of clustering results, however ensemble methods might produce noisy clustering with a higher evaluation score. Noisy clusterings are hard to measure with current evaluation metrics, therefore it might be beneficial to include an unsupervised score in the results. In further research we would like to examine the process of selecting optimal objectives for each dataset.

Acknowledgement

The research reported in this paper was supported by the Czech Science Foundation (GAČR) grant 13-17187S.

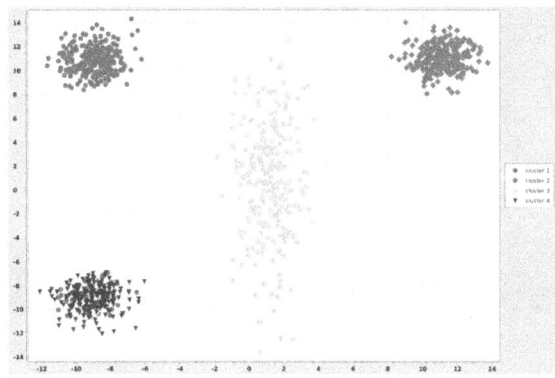

Figure 5: Consensus of 10 independent k-means runs on the *triangle1* dataset using $k = 6$. In this case the consensus produces a nonsense cluster that mixes items from 3 other clusters together.

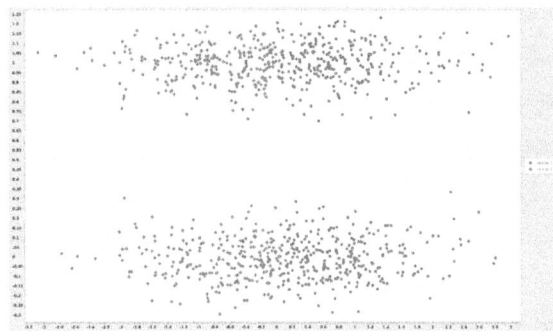

Figure 6: Single k-means runs on the *long1* dataset using $k = 2$. K-Means algorithm fails to reveal non-spherical clusters. Both supervised indexes Ajusted Rand Index and NMI assigns this clustering score 0.0, even though there is 50% assignment error.

References

[1] Aggarwal, C. C., Reddy, C. K., (Eds.): Data clustering: algorithms and applications. CRC Press, 2014

[2] Akaike, H.: A new look at the statistical model identification. IEEE Transactions on Automatic Control **19** (6 December 1974), 716–723

[3] Bartoň, T., Kordík, P.: Evaluation of relative indexes for multi-objective clustering. In: Hybrid Artificial Intelligent Systems, E. Onieva, I. Santos, E. Osaba, H. Quintián, and E. Corchado, (Eds.), vol. 9121 of Lecture Notes in Computer Science. Springer International Publishing, 2015, 465–476

[4] Boulis, C., Ostendorf, M.: Combining multiple clustering systems. In: PKDD (2004), J.-F. Boulicaut, F. Esposito, F. Giannotti, and D. Pedreschi, (Eds.), vol. 3202 of Lecture Notes in Computer Science, Springer, 63–74

[5] Caliński, T., Harabasz, J.: A dendrite method for cluster analysis. Communications in Statistics-theory and Methods **3** (1) (1974), 1–27

[6] Caruana, R., Elhawary, M., Nguyen, N., Smith, C.: Meta clustering. In: Proceedings of the Sixth International Conference on Data Mining (Washington, DC, USA, 2006), ICDM'06, IEEE Computer Society, 107–118

[7] Corne, D., Jerram, N., Knowles, J., Oates, M.: PESA-II: region-based selection in evolutionary multiobjective optimization. In: Proceedings of the Genetic and Evolutionary Computation Conference (GECCO-2001) (2001)

[8] Davies, D. L., Bouldin, D. W.: A cluster separation measure. IEEE Trans. Pattern Anal. Mach. Intell. **1** (2) (1979), 224–227

[9] Deb, K., Pratap, A., Agarwal, S., Meyarivan, T.: A fast and elitist multiobjective genetic algorithm: NSGA-II. IEEE Transactions on Evolutionary Computation **6** (2) (2002), 182–197

[10] Dudoit, S., Fridlyand, J.: Bagging to improve the accuracy of a clustering procedure. Bioinformatics **19** (9) (2003), 1090–1099

[11] Dunn, J. C.: Well-separated clusters and optimal fuzzy partitions. Journal of Cybernetics **4** (1) (1974), 95–104

[12] Faceli, K., de Souto, M. C. P., de Araujo, D. S. A., de Carvalho, A. C. P. L. F.: Multi-objective clustering ensemble for gene expression data analysis. Neurocomputing **72 (13-15)** (2009), 2763–2774

[13] Fern, X. Z., Brodley, C. E.: Random projection for high dimensional data clustering: a cluster ensemble approach. In: ICML (2003), T. Fawcett and N. Mishra, (Eds.), AAAI Press, 186–193

[14] Fred, A. L. N., Jain, A. K.: Combining multiple clusterings using evidence accumulation. IEEE Trans. Pattern Anal. Mach. Intell. **27** (6) (2005), 835–850

[15] Gionis, A., Mannila, H., Tsaparas, P.: Clustering aggregation. TKDD **1** (1) (2007)

[16] Gullo, F., Domeniconi, C., Tagarelli, A.: Projective clustering ensembles. Data Min. Knowl. Discov. **26** (3) (2013), 452–511

[17] Halkidi, M., Vazirgiannis, M., Batistakis, Y.: Quality scheme assessment in the clustering process. In: PKDD (2000), D. A. Zighed, H. J. Komorowski, and J. M. Zytkow, (Eds.), vol. 1910 of Lecture Notes in Computer Science, Springer, 265–276

[18] Handl, J., Knowles, J.: Evolutionary multiobjective clustering. Parallel Problem Solving from Nature – PPSN VIII (2004), 1081–1091

[19] Handl, J., Knowles, J.: An evolutionary approach to multiobjective clustering. Evolutionary Computation, IEEE Transactions on **11** (1) (2007), 56–76

[20] Hastie, T., Tibshirani, R., Friedman, J., Corporation, E.: The elements of statistical learning. Springer, Dordrecht, 2009

[21] Hubert, L., Levin, J.: A general statistical framework for assessing categorical clustering in free recall. Psychological Bulletin **83** (6) (1976), 1072

[22] Iam-on, N., Boongoen, T.: Improved link-based cluster ensembles. In: IJCNN (2012), IEEE, 1–8.

[23] Jain, A. K., Law, M. H. C.: Data clustering: a user's dilemma. In: PReMI (2005), S. K. Pal, S. Bandyopadhyay, and S. Biswas, (Eds.), vol. 3776 of Lecture Notes in Computer Science, Springer, 1–10

[24] Karypis, G., Kumar, V.: A fast and high quality multi-level scheme for partitioning irregular graphs. SIAM J. Sci. Comput. **20** (December 1998), 359–392

[25] Kittler, J., Hatef, M., Duin, R. P. W.: Combining classifiers. In: Proceedings of the Sixth International Conference on Pattern Recognition (Silver Spring, MD, 1996), IEEE Computer Society Press, 897–901

[26] MacQueen, J. B.: Some methods for classification and analysis of multivariate observations. In: Proc. of the Fifth Berkeley Symposium on Mathematical Statistics and Probability (1967), L. M. L. Cam and J. Neyman, (Eds.), vol. 1, University of California Press, 281–297

[27] Mimaroglu, S., Erdil, E.: Obtaining better quality final clustering by merging a collection of clusterings. Bioinformatics **26 (20)** (2010), 2645–2646

[28] Nguyen, N., Caruana, R.: Consensus clusterings. In: ICDM (2007), IEEE Computer Society, 607–612

[29] Rezaee, M. R., Lelieveldt, B., Reiber, J.: A new cluster validity index for the fuzzy c-mean. Pattern Recognition Letters **19 (3–4)** (1998), 237–246

[30] Salvador, S., Chan, P.: Determining the number of clusters/segments in hierarchical clustering/segmentation algorithms. In: ICTAI (2004), IEEE Computer Society, 576–584

[31] Schwarz, G.: Estimating the dimension of a model. Annals of Statistics **6** (1978), 461–464

[32] Strehl, A., Ghosh, J.: Cluster ensembles – a knowledge reuse framework for combining multiple partitions. Journal on Machine Learning Research (JMLR) 3 (December 2002), 583–617

[33] Topchy, A. P., Jain, A. K., Punch, W. F.: Clustering ensembles: models of consensus and weak partitions. IEEE Trans. Pattern Anal. Mach. Intell. **27 (12)** (2005), 1866–1881

[34] Tumer, K., Agogino, A. K.: Ensemble clustering with voting active clusters. Pattern Recognition Letters **29 (14)** (2008), 1947–1953

[35] Ultsch, A., Mörchen, F.: ESOM-Maps: tools for clustering, visualization, and classification with Emergent SOM. Technical Report No. 46, Dept. of Mathematics and Computer Science, University of Marburg, Germany (2005)

[36] Wolpert, D. H., Macready, W. G.: No free lunch theorem for optimization. IEEE Transactions on Evolutionary Computation **1 (1)** (1997)

[37] Yu, Z., Wong, H. -S., Wang, H. -Q.: Graph-based consensus clustering for class discovery from gene expression data. Bioinformatics **23 (21)** (2007), 2888–2896

J. Yaghob (Ed.): ITAT 2015 pp. 115–120
Charles University in Prague, Prague, 2015

An Optimisation Strategy for the Catalytic Transformation of Bioethanol into Olefins Using Computational Intelligence

Gorka Sorrosal[1], Cristina Martin, Cruz E. Borges, Ana M. Macarulla, and Ainhoa Alonso-Vicario

Deusto Institute of Technology – DeustoTech Energy,
University of Deusto, Avda. Universidades 24, 48007 Bilbao, Spain,
gsorrosal@deusto.es,
http://energia.deusto.es

Abstract: This paper presents a strategy for the optimisation of the operational conditions of the catalytic transformation of Bioethanol into Olefins (BTO) process. The variables to optimise are the main operating variables of the process (temperature, space-time and water content in the feed), and the objective function is to maximise the total production of olefins. The proposed strategy is based on evolutionary algorithms guided by surrogate models used to simulate the process behaviour under different experimental conditions. This paper compares the optimisation results of the BTO process obtained using an existing mechanistic model with those obtained with a surrogate model. The results suggest that the proposed methodology achieves similar results than those using mechanistic models but 43 times faster. This is a preliminary study where only constant set points have been tested; further research will include dynamic optimisation of the operational conditions by testing expected dynamic trajectories for each operating variable.

1 Introduction

Nowadays, we are becoming aware that crude oil is a finite source of energy and raw material. Therefore, our society begins to impulse the sustainable development using alternative sources of energy and raw materials, such as coal or biomass. In nature, being 170 billions tones of biomass annually produced, only the 3-4 % is exploited. Thus, there is a huge quantity of biomass available for its valorization as raw material to obtain biofuels and other chemical products [6]. Consequently, the scientific developments in this field are very important in order to advance in a future post-petroleum society and to reduce our dependency from the petroleum and its derivatives. The development of new tools to study the optimal operation of biomass transformation processes for a future scaling up to industrial level is a new interesting research line.

An important biomass transformation process is the Bioethanol-To-Olefins (BTO) process. The use of biomass as raw material has a great interest as an alternative to the petrochemistry for the production of light olefins like ethylene and propylene. The use of the computational intelligence in this research field can improve the optimisation procedures and allow faster developments in the design and optimal operation of the production processes.

The optimal control laws of biorefinery production processes are mainly unknown. Therefore, it is necessary to test several operational conditions over the whole operational range to study the influence of each manipulated variable on the final production objectives.

One of the key points for the implementation of the BTO process is to perform an advanced control strategy by adjusting the operating variables to maintain product quality while extending the lifespan of the catalyst. Due to the influence of multiple variables simultaneously over the reaction kinetics and the catalyst deactivation, it is necessary to develop advanced optimisation strategies of the operational conditions that guarantee specific production objectives without exceeding the operation limits to avoid an irreversible deactivation of the catalyst.

Therefore, the search of the parameters and operational conditions that allow to reach those production objectives give rise to optimisation problems in which the calculation of an analytic solution can present some difficulties with conventional search techniques [19]. While these techniques require characteristics of the process or from the optimisation problem, such as gradients, Hessians or linearities, to calculate the next points; there are stochastic search techniques as the Evolutionary Algorithms (EA) that solve the optimisation problems only with stochastic rules.

In the field of chemical engineering, there are several applications where these algorithms have been employed for the design, optimisation and optimal control of chemical reactors and plants [8, 1], such as in fermentation processes with fed-batch reactors [19] or to optimise the operational conditions of industrial scale reactor [18] or chemical plants [16].

To perform a dynamic optimisation of a process or reactor, one of the problems is that each cost function evaluation may require from minutes to hours of calculation time, and when using EA hundreds of evaluations are generally needed [14]. Therefore, as knowledge models use to be non lineal (and hence difficult to be solved, even numerically), surrogate models are commonly used to simulate the real process during its optimisation [13, 10]. Thanks to their characteristics, Artificial Neural Networks (ANN) are being increasingly used as a modelling technique for process simulation using evolutionary optimisation techniques [2, 5].

ANN attempt to mimic the structural principles of the

biologic brains to learn the existing relations inside input-output datasets. They have been able to successfully model any type of complex models and chemical reactors, such as batch reactors [7, 12], laboratory or industrial scale reactors [15] or even catalytic reactions as in the case of the BTO process [11].

The present work has the objective of performing an optimisation of the Bioethanol-To-Olefins (BTO) process in order to maximise the olefins total production while extending the catalyst lifespan. The BTO process, as other biomass transformation processes, uses a specific catalyst to stimulate the formation of a specific product at the reactor output. This catalyst is deactivated with time depending on the operational conditions. Therefore, in the design and optimisation of new catalytic transformation processes, such as the BTO process, there are two aspects to take into account in order to maximise the production of the desired product: the composition of the catalyst and the operational conditions. In the first case, different approaches based on soft computing techniques have been proposed to optimise the catalyst composition [17]. For the operational conditions, we can find some studies for different processes but without catalyst deactivation [16] or only in a discrete way [9]. In this work, we proposed a strategy to study the optimal dynamic operational conditions (with a specific catalyst composition) to achieve the optimal production results, taking into account the catalyst deactivation to design an optimal operation policy that not only maximise the production objectives but it is also able to counteract the catalyst deactivation.

This paper presents the preliminary results for the optimisation of constant operation set points. The optimisation objectives are the operating conditions that govern the BTO process. The optimisation process has been implemented using computational intelligence algorithms. An EA explores the possible optimal solutions guided by a surrogate model of the process based on ANN that is integrated in its evaluation function.

The paper is organized as follows: Section 2 presents the BTO process. Next, Section 3 describes the proposed methodology for the optimisation of the BTO process. Section 4 presents the obtained results. The conclusions and future works are finally summarized in Section 5.

2 BTO Process

The BTO process consists in the catalytic transformation of bioethanol into olefins over an acid catalyst. This is a key process in the concept of sustainable refinery, incorporating biomass or derivatives as an alternative feedstock to petroleum. This process uses a very selective catalyst to maximise the olefins conversion rate (X_O). However, this catalyst is deactivated due to the accumulation of coke. Therefore, when the catalyst reaches a minimum activity value, it is necessary to stop the production step and carry out the catalyst regeneration phase. In addition, the regeneration does not achieve the total catalyst recovery, so the

number of possible cycles of production-regeneration is also limited. Depending on the operational conditions, both the production and the catalyst deactivation reversibility will be affected.

Therefore it is necessary to explore experimentally those operational conditions that fulfill the desired objectives, which would be very costly and complicated. The use of EA and surrogate models to explore those operational conditions is much more practical and reasonable.

The following variables are the main operating variables that govern the behaviour of the BTO process:

- **Operating variables**:

  T: reaction temperature (K).

  X_w: mass fraction of water based on the equivalent mass of ethylene in the reactor feed $\left(g_{water}g^{-1}\right)$.

  WF_{EO}^{-1}: space-time $\left(g_{catalyst}h^{-1}\left(g_{ethanol}\right)\right)$.

- **Activity level**:

  a: catalyst activity. The activity has been considered as a disturbance that quantifies the rate of catalyst deactivation by coke.

In previous works, experimental runs of this process were carried out in an automated device equipped with an isothermal laboratory scale fixed bed reactor connected on-line to a gas chromatograph and a micro-GC for the analysis of the reaction product (see Figure 1). Details about the reaction equipment, catalyst preparation and experimental methodology can be found in previous works [3, 4].

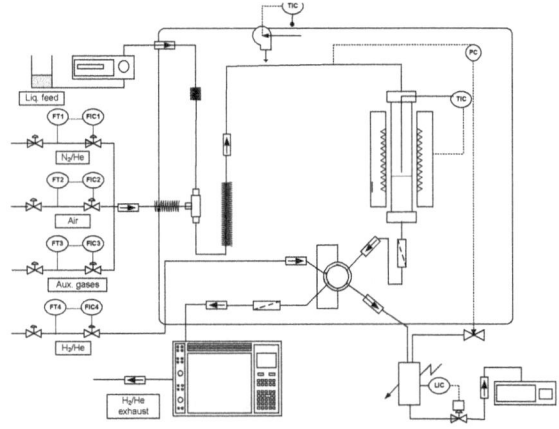

Figure 1: Diagram of the reaction equipment used to obtain the experimental data.

3 Methodology

This section presents the proposed dynamic optimisation strategy using Soft Computing techniques. An EA guided by an ANN based surrogate model of the process is proposed. The surrogate models are used in the evaluation operator to simulate the process behaviour under each operational condition proposed by the EA. The results will be validated with an existing mechanistic model (MECH) [3, 4], since the experimental validation is not technical or economically viable.

The EA and the BTO process models (both the MECH and the surrogate) have been implemented using the programming package MATLAB^{TM} (version 8.0, 2012b, Mathworks Company). Simulation and optimisation have been carried out in an PC with an Intel® Core^{TM} i5-2467M CPU at 1.6 GHz and 4.0 GB of physical memory (RAM).

3.1 Surrogate Model

Chemical knowledge models are generally computationally very demanding to be used in evolutionary approaches [14]. Thus, in this work, an ANN based surrogate model is proposed to dynamically simulate the process behaviour.

A nonlinear autoregressive with exogeneous inputs (NARX) neural network topology has been selected to model the BTO process. The developed ANN model estimates the olefins conversion rate at the reactor output (X_O) using as inputs the previously mentioned process operating variables (T, X_W, WF_{EO}^{-1}), the catalyst activity level (a) and the previously estimated output values in a recursive loop ($\hat{X}_O$).

An iterative methodology modifying the number of layers and neurons in each layer has been carried out in order to select the neural model structure that better fits the process using the Leave-One-Out Cross-Validation (LOOCV) technique. This technique consists of setting aside a set of experiments (representing unique operational conditions) from the model training phase and only using them for the validation phase. The process is repeated until every single set of experiments is used in the validation stage. These type of techniques test the generalization capability of a model structure.

Consequently, the available data are divided into training, validation and test datasets. The training and validation datasets are used in the model structure selection. The first ones are used to train several models with different structures. Aspects such as the number of hidden layers, the neurons in each layer or the connections between neurons are iteratively modified. The performance of each model is tested and compared following the LOOCV procedure.

Once the neural model structure is selected, the final model is trained with all the available data except from the test dataset which is used to validate the fitted model. The Levenberg-Marquardt Algorithm has been used for the training and the main comparison criterion has been the Root Mean Squared Error (RMSE) of the model (Equation (1)).

$$RMSE := \sqrt{\frac{1}{n}\sum_{i=1}^{n}(X_O - \hat{X}_O)^2}. \qquad (1)$$

A first ANN model, trained only with experimental data, was able to describe correctly the process behaviour for the experimentally tested operating conditions (see Section 4). However, for optimisation purposes, it is necessary to test several operating conditions where limited experimental data are available. In particular, the dataset does not contain any experiment describing the dynamic behaviour of X_W and WF_{EO}^{-1}. To provide the ANN with the required information about the process behaviour in the whole operating range, some experiments were simulated with the MECH model and introduced in the training dataset.

3.2 Optimisation Problem

The aim of this optimisation problem is to obtain the operational conditions that achieve the best production results per amount of catalyst needed. Due to the catalyst deactivation, is important to bear in mind the deactivation rate, keeping in the whole simulation time the catalyst activity and the olefins production rate over the established minimum value of 0.10. Therefore, the Equation (2) defines the objective function in order to maximise the total production of olefins.

$$\max_{T, X_w, WF_{EO}^{-1}} \frac{\int_0^\tau X_O\left(T, X_w, WF_{EO}^{-1}\right) dt}{WF_{EO}^{-1}}. \qquad (2)$$

Please note that there are two stopping criteria (τ constant in the upper bound of the integral):

- The catalyst involves a major process cost. Therefore, to maximise the production per amount of catalyst we have set a lower bound on the activity in order to reduce the total production costs. This bound has been set to 0.10 ($a < 0.10$).

- In order to maintain the production, it is needed that the olefins conversion rate does not decrease below the 0.10 ($X_O < 0.10$).

If any of the above criteria are passed over, we consider that the production has reached its maximum span and should be stopped to proceed with a catalyst regeneration phase.

The operating variables to optimise are bounded based on the physical-chemical properties of the process. In fact, for the temperature (T), $573 K$ is the inferior bound where the complete dehydration of the ethanol happens and $673 K$ is the upper bound to avoid the irreversible deactivation of the catalyst. The variable X_W will range between $[0.0821, 4.8889]$ and WF_{EO}^{-1} will range between $[0.068, 1.525]$ respectively. Please note that those intervals have been chosen as are the ones used to adjust the mechanistic model [3, 4].

As previously stated, in order to solve the optimisation problem an EA has been used. In particular we have implemented a Genetic Algorithm (GA). Table 1 summarizes the principal parameters of the EA used. The surrogate

Parameter	Value
Individuals	Vector of real numbers
Population	50 individuals
Generations	200 generations
Fitness Operator	Olefins Production
Genesis Operator	Random (uniform) initialization
Selection Operator	4-Tournament
Crossover Operator	Arithmetic crossover
Mutation Operator	Adaptive mutation
Elitism	True

Table 1: Main parameters of the GA used to optimise the operational conditions.

model previously developed will be used now to simulate the temporal behaviour of the process under the operating constant set points provided by the GA.

4   Results

In this section the results obtained for the modelling and optimisation of the BTO process are shown. Following the modelling procedure mentioned above, an ANN based surrogate model, with the topology showed in the Figure 2, has been implemented.

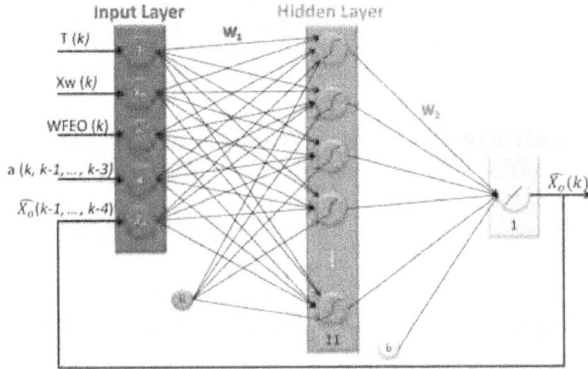

Figure 2: Feed-Forward ANN topology for the BTO process model.

Table 2 shows the mean estimation errors of the mechanistic and surrogate models for the experimental data. The discrepancy between both models is rather low, but the surrogate model simulates the process behaviour 43 times faster.

Model	RMSE ($g_o g^{-1}$)
MECH	0.0323
ANN	0.0387

Table 2: Root mean square error of the mechanistic (MECH) and surrogate (ANN) model to experimental data.

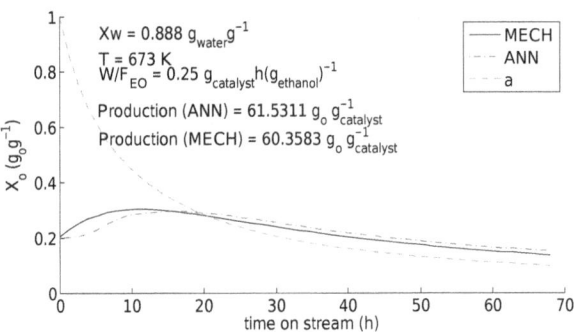

Figure 3: Comparison of the estimation generated by the MECH and ANN models for a test experiment.

MECH ($g_o g_{catalyst}^{-1}$)		ANN ($g_o g_{catalyst}^{-1}$)	
$\mu_{Production}$	$\sigma_{Production}$	$\mu_{Production}$	$\sigma_{Production}$
70.2204	0.6385	63.7376	1.6946

Table 3: Comparison of the mean production results for the "optimal" operational conditions when repeating several times the evolutionary optimisation using the mechanistic (MECH) and surrogate (ANN) models.

Figure 3 represents the estimates of the process behaviour calculated with both models for a test operational conditions that were excluded for the training procedure. The ANN model has been validated using the testing dataset (see Section 3), obtaining a root mean squared error of 0.0323 $g_o g^{-1}$ for the whole test experiments. These results show the capacity of the ANN to properly assimilate and reproduce the BTO process dynamics in the same way as the mechanistic model.

Once the surrogate model has been validated, the evolutionary optimisation has been carried out. Table 3 shows the mean production results obtained with the "optimal" operational conditions generated by the EA. Being the optimisation procedure stochastic, it has been launched several times to guarantee its convergence to a local optimum. Notice the small deviations repeating all the procedure 50 and 100 times for the mechanistic and surrogate approaches respectively. Although the standard deviation using the surrogate model doubles the deviation that results from using the mechanistic model, in both cases they are still very small. So we can conclude that the proposed approach is converging to a local optimum.

Finally, Figure 4 shows the behaviour of the process under the optimum solution provided by the evolutionary optimisation using the surrogate model. The maximum production has been 68.05 $g_o g_{catalyst}^{-1}$. Please note that the maximum production using the surrogate model only differs a 5.67 % over the optimisation carried out using the mechanistic model, but with much less computation cost. This maximum is reached by operating the reactor at $645 K$ with a high content of water in the feed ($X_W = 4.875 g_{water} g^{-1}$)

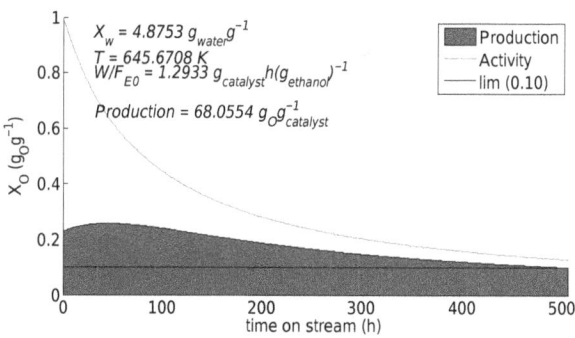

Figure 4: BTO process performance under the best solution provided by the evolutionary optimisation guided by the surrogate model.

and a space-time of $1.293 g_o h (g_{ethanol})^{-1}$. This water quantity attenuates the catalyst deactivation, which allows the extension of the production phase.

5 Conclusions

An optimisation strategy for the catalytic transformation of Bioethanol-To-Olefins (BTO) process based on computational intelligence has been presented. The proposed evolutionary optimisation guided by an Artificial Neural Networks based surrogate models has been able to optimise the process obtaining similar results than those obtained when using a mechanistic model but 43 times faster. A clear optimum has been defined. The optimum operational conditions have shown to be, temperature: $645 K$; water quantity in the feed: $4.875 g_{water} g^{-1}$; space-time: $1.293 g_o h (g_{ethanol})^{-1}$. These operational conditions allow to extend the catalyst lifespan maximising the production phase and hence the total production.

The presented results are a preliminary study on the optimisation of the BTO process looking only for constant set points that maximise the production objectives. This has been the first step for a more complex dynamic optimisation of the process, in which the optimal dynamic trajectory for each operating variable will be described. In this line, several trajectory types will be defined for each of the operating conditions to study the process dynamic behaviour.

Acknowledgement

The research team would like to thank the financial support of the Basque Government by the research project BIOTRANS (PI2013-40).

References

[1] Angira, R., Babu, B.V.: Optimization of process synthesis and design problems: A modified differential evolution approach. Chem. Eng. Sci. **61(14)** (2006) 4707–4721

[2] Farshad, F., Iravaninia, M., Kasiri, N., Mohammadi, T., Ivakpour. J.: Separation of toluene/n-heptane mixtures experimental, modeling and optimization. Chem. Eng. J. **173(1)** (2011) 11–18

[3] Gayubo, A. G., Alonso, A., Valle, B., Aguayo, A. T., Bilbao., J.: Kinetic model for the transformation of bioethanol into olefins over a HZSM-5 zeolite treated with alkali. Ind. Eng. Chem. Res. **49(21)** (2010) 10836–10844

[4] Gayubo, A. G., Alonso, A., Valle, B., Aguayo, A. T., Bilbao., J.: Deactivation Kinetics of a HZSM-5 zeolite catalyst treated with alkali for the transformation of bio-ethanol into hydrocarbons. American Inst. Chem. Eng. **58(2)** (2012) 526–537

[5] Gueguim Kana, E. B., Oloke, J. K., Lateef, A., Adesiyan, M. O.: Modeling and optimization of biogas production on saw dust and other co-substrates using artificial neural network and genetic algorithm. Renew. Energ. **46** (2012) 276–281

[6] Huber, G. W., Corma, A.: Synergies between bio- and oil refineries for the production of fuels from biomass. Angew. Chem. Int. Edit. **46(38)** (2007) 7184–7201

[7] Kashani, M. N., Shahhosseini, S.: A methodology for modeling batch reactors using generalized dynamic neural networks. Chem. Eng. J. **159(1–3)** (2010) 195–202

[8] Kordabadi, H., Jahanmiri, A.: Optimization of methanol synthesis reactor using genetic algorithms. Chem. Eng. J. **108(3)** (2005) 249–255

[9] Kordabadi, H., Jahanmiri, A.: A pseudo-dynamic optimization of a dual-stage methanol synthesis reactor in the face of catalyst deactivation. Chem. Eng. Process. **46(12)** (2007) 1299–1309

[10] Laguna, M., Martí, R.: Neural network prediction in a system for optimizing simulations. IIE Trans. **34** (2002) 273–282

[11] Molga, E. J.: Neural network approach to support modelling of chemical reactors: problems, resolutions, criteria of application. Chem. Eng. Process. **42(8–9)** (2003) 675–695

[12] Mujtaba, I. M., Aziz, N., Hussain, M. A.: Neural network based modelling and control in batch reactor. Chem. Eng. Res. Des. **84(8)** (2006) 635–644

[13] Nascimento, C. A. O., Giudici, R., Guardani, R.: Neural network based approach for optimization of industrial chemical processes. Comput. Chem. Eng. **24** (2000) 2303–2314

[14] Ong, Y. S., Nair, P. B., Keane, A. J., Wong, K. W.: Surrogate-assisted evolutionary optimization frameworks for high-fidelity engineering design problems. In: Knowledge Incorporation in Evolutionary Computation, (2004), 307–331, Springer Berlin Heidelberg

[15] Rahimpour, M. R., Shayanmehr, M., Nazari, M.: Modeling and simulation of an industrial ethylene oxide (EO) reactor using artificial neural networks (ANN). Ind. Eng. Chem. Res. **50(10)** (2011) 6044–6052

[16] Rajesh, J. K., Gupta, S. K., Rangaiah, G. P., Ray, A. K.: Multiobjective optimization of industrial hydrogen plants. Chem. Eng. Sci. **56(3)** (2001) 999–1010

[17] Rodemerck, U., Baerns, M., Holena, M., Wolf., D.: Application of a genetic algorithm and a neural network for the discovery and optimization of new solid catalytic materials. Appl. Surf. Sci. **223(1–3)** (2004) 168–174

[18] Yee, A. K. Y., Ray, A. K., Rangaiah, G. P.: Multiobjective optimization of an industrial styrene reactor. Comput. Chem. Eng. (2003) 111–130

[19] Yüzgeç, U., T urker, M., Hocalar, A.: On-line evolutionary optimization of an industrial fed-batch yeast fermentation process. ISA Trans. **48(1)** (2009) 79–92

J. Yaghob (Ed.): ITAT 2015 pp. 121–126
Charles University in Prague, Prague, 2015

Planning Fitness Training Sessions Using the Bat Algorithm

Iztok Fister Jr.[1], Samo Rauter[2], Karin Ljubič Fister[3], Dušan Fister[1], and Iztok Fister[1]

[1] University of Maribor, Faculty of Electrical Engineering and Computer Science,
Smetanova 17, 2000 Maribor, Slovenia,
iztok.fister1@um.si
[2] University of Ljubljana, Faculty of Sport
Gortanova 22, 1000 Ljubljana
[3] University of Maribor, Faculty of Medicine,
Taborska 8, 2000 Maribor, Slovenia

Abstract: Over fairly recent years the concept of an artificial sport trainer has been proposed in literature. This concept is based on computational intelligence algorithms. In this paper, we try to extend the artificial sports trainer by planning fitness training sessions that are suitable for athletes, especially during idle seasons when no competition takes place (e.g., winter). The bat algorithm was used for planning fitness training sessions and results showed promise for the proposed solution. Future directions for development are also outlined in the paper.

1 Introduction

Sport becomes highly addictive for many people in the world. A few decades ago, people around the world spend their free time doing different activities, like: short walks through the park, visiting the cinema or galleries, fishing, visiting a thermal spa and also meet friends. A lot of leisure studies also proved this. However, over recent decades, lifestyles have been substantially changed especially because of globalization that has transformed the whole earth into a global village. Due to a lack of time as well as personal willingness, people do not want to live like they used to. For this reason, different kind of new activities have emerged over recent past years. One of the bigger revitalizations has been sport which has became extremely in popular because of the emergence of different mass sports events. For instance, the main mass sport events are:

- Triathlon - This discipline consists of three sports: swimming, cycling and running. Additionally, there are different distances which vary from short via medium to long distances. One of the more famous distances is the Ironman triathlon [1, 2, 3], which is also known as the hardest one day sport event. Ironman consists of 3.8 km of swimming, 180 km of cycling and 42.2 km of running.

- Road marathons - A road marathon [4] consists of 42.2 km pure running and is a challenge for myriad of people. Big city marathons especially are the most popular and attract large numbers of runners. Some marathons can accommodate more than 40,000 runners [5, 6].

- Recreational cycling marathons - This kind of mass [7, 8] sports events was very popular approximately 10 years ago but still represents a challenge for numerous participants. The current world economic problems increased the prices of cycles and consequently less participants could participate on cycling marathons.

Participants of the mentioned events participate mostly because of two goals. The first goal is to enjoy (in other words: to have a nice time) and the second is to finish the trial. Usually, finishers are awarded with medals which are big stimulants for participants. In this case, every year many more participants have also began to take these competitions more seriously, i.e., semi professional. In line with this, they invest much more time in preparations for competitions. Unfortunately, there is a long way for good preparations for such kinds of competitions. This good preparation consists of proper sports training, good eating and also good resting. To maintain all these factors as high as possible is very hard for numerous athletes, since they do not have enough experience. Newbie athletes especially suffer from the unwanted effects of irregular training called over-training syndrome [9, 10, 11] which is reflected in reduced form. One of the possible solutions for avoiding this is to hire a personal trainer or join diverse training groups. However, these cost a lot of money and therefore many of them can not afford them.

In order to break this barrier, we began the development of an artificial sport trainer. An artificial sports trainer was presented recently in [12] and is based on computational intelligence [13] algorithms that are able for plan the sports training over both short-term and long-term. This trainer is also able to discover the different habits of athletes, avoid over-training, etc. Data for the artificial sports trainer are obtained from sports trackers [14] and sports watches like Garmin.

This paper, extends the artificial sport trainer with planning fitness training sessions. These kinds of training sessions are very important for athletes especially during idle seasons. In Europe, the idle season is usually during winter months when the athletes prepare their form for the whole season. Planning fitness sessions were performed using a bat algorithm [15, 16] which is a member of the computational intelligence family. The planning of the fitness sessions was defined as a constraint satisfaction problem, where the bat algorithm searches for feasible solutions arising when the number of constraint violations achieved the value of zero.

Organization of the remainder of this workshop paper is as follows: in section 2 we discuss about characterists of fitness training, while section 3 presents swarm intelligence algorithms and bat algorithm. Experiments and results are presented in section 4, while section 5 concludes the paper.

2   Characteristics of Fitness Training

This study, focused on fitness training in regard to cycling. Incorporation of strength training in cyclists preparatory periods has received more attention over the last two decades. Most of the serious and competitive cyclists also include strength training in their training programs. It is also evident in some previous research that adding strength training to an endurance training program can increase endurance performance [17, 18].

A combination of endurance and strength training (concurrent training) might therefore be a potential training strategy for promoting muscle oxidative capacity. It might be related to an improved cycling economy, as observed after adding strength training to the ongoing endurance training, namely because a stronger muscles at a certain intensity operates longer with a lower percentage of maximum capacity. It is well-known that adding strength training to endurance training can increase the maximal strengths and rate of force developments in cyclists. In theory, this may improve pedaling characteristics by increasing peak torque in the pedal stroke, reducing time to peak torque and reducing the pedaling torque relative to maximal strength, which in turn may allow for higher power output and/or increased blood flow.

The most important thing for developing a cycling strength program is to know, which muscle groups are the most active during the pedal stroke. Some previous studies have detected a strong correlation between cycling performance and some strength exercises, like leg presses, squats, and deadlifts [17, 18, 19, 20]. Some of the more useful exercises for fitness training are presented in Figs. 1 to 3.

Figure 1: Deadlift exercise

Figure 2: Squats exercise

Figure 3: Lunge exercise

For the smart planning of sports training, quantifications, regulating the intensity of a workout is the key for success as indicated as basic knowledge in sports training literature. This fact also holds for fitness training. As an estimate of the intensity of a fitness workout, two main measures are employed like a:

- the number of repetitions per set of exercises (NR),

- the maximum amount of weight that can be generated in one maximum contraction (1RM).

The logic behind the first measure is as follows. The heavier the weight, the higher the intensity and the fewer repetitions (also reps) an athlete will be able to lift it for. On the other hand, the 1RM determines the desired load for an exercise (typically as a percentage of the 1RM). Let us notice that a coach determines the measure of 1RM for a definite athlete using tests at the beginning of the fitness training and then calculates the number of repeats (NR) in regard to this characteristic value.

3   Swarm Intelligence Based Algorithms

Swarm intelligence (SI) is a paradigm that belongs to computational intelligence (CI). According to the [21], SI concerns the collective, emerging behavior of multiple, interacting agents that are capable of performing simple actions. While each agent may be considered as unintelligent, the whole system of multiple agents shows some self-organizational behavior and thus can behave like some sort of collective intelligence. The basic pseudo-code of SI-based algorithms is presented in Algorithm 1. Nowadays, the bat algorithm is one of the promising members of the SI family. It is very easy to implement and shows efficient results especially when solving small dimensional problems.

Algorithm 1 Swarm Intelligence

1: initialize_population_with_random_candidate_particles;
2: eval = evaluate_each_particle;
3: **while** termination_condition_not_meet **do**
4:     move_particles_towards_the_best_individual;
5:     eval += evaluate_each_particle;
6:     select_the_best_individuals_for_the_next_generation;
7: **end while**

Next subsection describes the mentioned algorithm in detail.

3.1 Bat Algorithm

The bat algorithm was developed by Yang in 2010. The main purposes of this algorithm were to be: simple, efficient and applicable to varios problem domains. The inspiration for the bat algorithm came from the phenomenon of the echolocation characteristics of some types of microbats. Developer used a three simplified rules describing the bat behavior, as follows [22]:

- All bats use echolocation to sense distance to target objects.

- Bats fly randomly with the velocity v_i at position x_i, the frequency $Q_i \in [Q_{min}, Q_{max}]$ (also the wavelength λ_i), the rate of pulse emission $r_i \in [0,1]$, and the loudness $A_i \in [A_0, A_{min}]$. The frequency (and wavelength) can be adjusted depending on the proximities of their targets.

- The loudness varies from a large (positive) A_0 to a minimum constant value A_{min}.

The algorithm's pseudo-code is presented in Algorithm 2. The main bat algorithm components [23] are summarized as follows:

- *initialization* (lines 1-3): initializing the algorithm parameters, generating the initial population, evaluating this, and finally, determining the best solution $\mathbf{x}_{best}$ in the population,

- *generate_the_new_solution* (line 6): moving the virtual bats in the search space according to the physical rules of bat echolocation,

- *local_search_step* (lines 7-9): improving the best solution using random walk direct exploitation (RWDE) heuristic,

---

Algorithm 2 Bat algorithm

Input: Bat population $\mathbf{x_i} = (x_{i1}, \ldots, x_{iD})^T$ for $i = 1 \ldots Np$, *MAX_FE*.
Output: The best solution $\mathbf{x}_{best}$ and its corresponding value $f_{min} = \min(f(\mathbf{x}))$.
1: init_bat();
2: *eval* = evaluate_the_new_population;
3: f_{min} = find_the_best_solution($\mathbf{x}_{best}$); {initialization}
4: **while** termination_condition_not_meet **do**
5:     **for** $i = 1$ **to** Np **do**
6:         y = generate_new_solution($\mathbf{x}_i$);
7:         **if** rand$(0,1) > r_i$ **then**
8:             y = improve_the_best_solution($\mathbf{x}_{best}$)
9:         **end if**{ local search step }
10:        f_{new} = evaluate_the_new_solution(y);
11:        $eval = eval + 1$;
12:        **if** $f_{new} \leq f_i$ **and** N$(0,1) < A_i$ **then**
13:            $\mathbf{x}_i = \mathbf{y}$; $f_i = f_{new}$;
14:        **end if**{ save the best solution conditionally }
15:        f_{min}=find_the_best_solution($\mathbf{x}_{best}$);
16:    **end for**
17: **end while**

- *evaluate_the_new_solution* (line 10): evaluating the new solution,

- *save_the_best_solution_conditionaly* (lines 12-14): saving the new best solution under some probability A_i,

- *find_the_best_solution* (line 15): finding the current best solution.

Generating the new solution is governed by the following equation:

$$
\begin{aligned}
Q_i^{(t)} &= Q_{min} + (Q_{max} - Q_{min})N(0,1), \\
\mathbf{v}_i^{(t+1)} &= \mathbf{v}_i^t + (\mathbf{x}_i^t - \mathbf{best})Q_i^{(t)}, \\
\mathbf{x}_i^{(t+1)} &= \mathbf{x}_i^{(t)} + \mathbf{v}_i^{(t+1)},
\end{aligned} \tag{1}
$$

where $N(0,1)$ is a random number drawn from a Gaussian distribution with zero mean and a standard deviation of one. A RWDE heuristic implemented in the function *improve_the_best_solution* modifies the current best solution according to the equation:

$$
\mathbf{x}^{(t)} = \mathbf{best} + \varepsilon A_i^{(t)} N(0,1), \tag{2}
$$

where $N(0,1)$ denotes the random number drawn from a Gaussian distribution with zero mean and a standard deviation of one, ε being the scaling factor, and $A_i^{(t)}$ the loudness.

Contemporary work on bat algorithms captures many variants and application domains. Some recent works are presented in papers [24, 25, 26, 27, 28]

3.2 Bat Algorithm for Planning Fitness Sessions

Based on the original bat algorithm, we have developed a modified bat algorithm for planning fitness sessions. Development of this algorithm demanded the following four steps:

- determining the fitness exercises,

- defining constraints,

- modifying the original bat algorithm,

- representing the results and their visualizations.

In the remainder of this paper, all these steps are described in detail.

Selecting Fitness Exercises. We need to determine specific exercises for different muscle groups before the fitness training can start. Although sports medicine recognizes more than 15 muscle groups that must be included within fitness training, we focus on four groups (i.e., legs, core, arms and back) in this preliminary study. Furthermore, some of these groups can be repeated during the training. Each muscle group is associated with three prescribed exercises as presented in Table 1.

Muscle groups	Exercise
LEGS	LEG PRESS, SQUATS, LUNGE
CORE	LEG SCISSORS, PLANK, LEG LIFTS
ARMS	PULLDOWN, PUSH UPS, UNDERARM ISOMETRIC EXERCISE
LEGS	LEG PRESS, SQUATS, LUNGE
BACK	BACK EXTENSION, DEADLIFT, BAR ROWS
CORE	LEG SCISSORS, PLANK, LEG LIFTS
LEGS	LEG PRESS, SQUATS, LUNGE
ARMS	PULLDOWN, PUSH UPS, UNDERARM ISOMETRIC EXERCISE

Table 1: Muscle groups and associated exercises

On the other hand, the intensities of the exercises must be determined in a fitness training plan. This intensity is associated with a measure 1RM measured for a specific athlete. Here, three levels of intensity are supported in our study, where each level is mapped according to the 1RM, as can be seen in Table 2.

Intensity	1RM measure
HIGH	1RM > 80%
MEDIUM	60% < 1RM ≤ 80%
LOW	1RM ≤ 60%

Table 2: Intensity mapping

Note that the data in Tables 1 and 2 were specified according to the suggestions of fitness trainers.

Defining constraints The purpose of the fitness training plan is to prescribe sufficient numbers of exercises for each of the prescribed muscle groups, their number of repeats (NR) and the proper intensities (%1RM) such that an athlete simultaneously develops all the muscle groups needed for building the cyclist's basic form. Therefore, trainers determine the proper amount of a specific exercise in the plan in regarding to the others. In order to regulate the relations between exercises in the fitness training plan, the following constraints are defined:

- at least four exercises must have the number of repeats over 25 times (i.e., NR>25),

- each training plan should have at least two exercises of high intensity,

- each muscle group repeating in Table 1 more than once does not have the same exercise,

- if the last exercise in the fitness training plan was of higher intensity, the next exercise should be of medium or high intensity.

In the remainder of this paper, these constraints were captured within the algorithmic structure of the original bat algorithm for planning the fitness training plan.

Modifying the original bat algorithm Each solution in the modified bat (MBA) algorithm consists of 24 floating-point elements representing the fitness training plans for some athlete. The elements of the solution are divided into three groups of elements. In other words, the solution is expressed as

$$\mathbf{x_i} = (x_{i1}, \dots, x_{i8}, x_{i9}, \dots, x_{i16}, x_{i17}, \dots, x_{i24})^T, \quad (3)$$

where elements $x_{i1}, \dots, x_{i8}$ denote exercises from Table 1, $x_{i9}, \dots, x_{i16}$ are the number of repeats NR and $x_{i17}, \dots, x_{i24}$ the corresponding intensity, respectively. This means, each fitness training plan consists of eight exercises with an assigned number of repeats and corresponding intensities. While the number of repeats is selected from interval NR $\in [1, 40]$, parameters exercises and intensities are drawn from the interval $[0, 1]$, and their proper values are encoded as indices into a discrete set of features according to the following equations

$$ex(x_{i,j}) = \lceil 3.0 \cdot x_{i,j} \rceil, \quad \text{for } j = 1, \dots, 8, \quad (4)$$

$$int(x_{i,j}) = \lceil 3.0 \cdot x_{i,j} \rceil, \quad \text{for } j = 17, \dots, 24, \quad (5)$$

where $ex(x_{i,j})$ and $int(x_{i,j})$ determine the element in the feature sets as represented in Tables 1 and 2. For instance, the function intensity can obtain the following values from the feature set

$$int(x_{i,j}) = \begin{cases} \text{HIGH,} & \text{if } 0 \le x_{i,j} < \frac{1}{3}, \\ \text{MEDIUM,} & \text{if } \frac{1}{3} \le x_{i,j} < \frac{2}{3}, \\ \text{LOW,} & \text{if } \frac{2}{3} \le x_{i,j} < 1, \end{cases}$$

respectively. The planning of the fitness training sessions is defined as a constraint satisfaction problem that is formally defined as

$$\text{Minimize} \quad f(\mathbf{x}_i) = \sum_{k=1}^{k<4} \chi_k(\mathbf{x}_i),$$

$$\text{subject to} \sum_{j=8}^{15} y_j \geq 4,$$

$$\sum_{j=16}^{24} z_j \geq 2,$$

$$x_{i,1} \neq x_{i,4} \neq x_{i,7} \wedge x_{i,2} \neq x_{i,6} \wedge x_{i,3} \neq x_{i,8},$$
$$int(x_{i,j}) \equiv \text{HIGH} \Rightarrow int(x_{i,j+1}) \neq \text{HIGH},$$

where

$$y_j = \begin{cases} +1 & \text{if } x_{i,j} \geq 25, \\ +0 & \text{otherwise,} \end{cases}$$

and

$$z_j = \begin{cases} +1 & \text{if } int(x_{i,j}) \equiv \text{HIGH}, \\ +0 & \text{otherwise,} \end{cases}$$

The proper solution to the problem is found, when the $f(\mathbf{x}) = 0$.

Representation of Results. Although the results could be visualized, the numerical results in the tables are presented only in this preliminary version of the modified bat algorithm.

4   Experiments and Results

The results of our experiments are illustrated in Tables 3 to Table 5, where the tables represent the three sets of exercises. An athlete has some free time for resting after finishing each set. We run algorithm 25 times and after the run we selected three generated training sessions which were successfully found by bat algorithm.

Exercise	Reps[NR]	Intensity[1RM]
LUNGE	26	HIGH
LEG SCISSORS	23	LOW
PULLDOWN	18	MEDIUM
SQUATS	27	HIGH
BAR ROWS	40	MEDIUM
LEG LIFTS	22	HIGH
LEG PRESS	35	MEDIUM
PUSH UPS	39	LOW

Table 3: First set of fitness workouts

The obtained results were evaluated by human trainer who evaluated and approved it. The obtained results confirm that the idea of automatic fitness training sessions was worth investigation and the promising results also show the potentials of the solution when used in practice. On

Exercise	Repeats	Intensity
SQUATS	34	LOW
LEG SCISSORS	22	MEDIUM
UNDERARM ISOMETRIC	40	HIGH
LEG PRESS	15	LOW
BAR ROWS	36	HIGH
LEG LIFTS	27	LOW
LUNGE	15	MEDIUM
PULLDOWN	22	LOW

Table 4: Second set of fitness workouts

Exercise	Repeats	Intensity
LEG PRESS	21	HIGH
LEG LIFTS	29	MEDIUM
UNDERARM ISOMETRIC	28	MEDIUM
LUNGE	39	MEDIUM
BACK EXTENSION	15	LOW
PLANK	35	MEDIUM
SQUATS	25	HIGH
PUSH UPS	38	MEDIUM

Table 5: Third set of fitness workouts

the other hand, we would also like to present some problems and bottlenecks which we encountered during development. Firstly, it seems that it will be good to test our idea with evolutionary algorithms in the future. Experiments showed that the success of the bat algorithm in satisfying all constraints was about 25% of runs only. The problem is that the bat algorithm is highly dependent on the best solution. From this reason, our algorithm went into local optimum a lot of times. We believe that advanced mechanisms e.g. arithmetic crossover would behave much better. Moreover, using adaptive and self-adaptive bat variants could also be suggested since we spent a lot of time tuning parameters. On the other hand, many more constraints should be defined in order to have very precise solutions which should be very similar to those solutions created by the human sport trainer.

5   Conclusions

In this workshop paper, we presented a simple, yet efficient solution for planning fitness training sessions automatically. The bat algorithm was employed in order to tackle this problem. This algorithm successfully generated training sessions which were evaluated and confirmed by a human trainer who had more than 20 years of experience. In the future, there are many tasks to do in this direction like for example testing with other nature-inspired algorithms, employing arithmetic crossover and taking more constraints and exercices into account.

Acknowledgement

The research reported in this paper has been partially supported by the Czech Science Foundation grant 13-17187S.

References

[1] Petschnig, S.: 10 Jahre Ironman Triathlon Austria. Meyer & Meyer, 2007

[2] Knechtle, B., Wirth, A., Baumann, B., Knechtle, P., Rosemann, T.: Personal best time, percent body fat, and training are differently associated with race time for male and female ironman triathletes. Research quarterly for exercise and sport **81(1)** (2010) 62–68

[3] McCarville, R.: From a fall in the mall to a run in the sun: One journey to ironman triathlon. Leisure Sciences **29(2)** (2014), 159–173

[4] Lehto, N.: Effects of age onmarathon finishing timeamong male amateur runners in stockholm marathon 1979–2014. Journal of Sport and Health Science, 2015

[5] Rauter, S.: Mass sport events as a way of life: differencess between participants in a cycling and running event. Kinesiologica Slovenica **20(1)** (2014), 5–15

[6] Shipway, R., Jones, I.: The great suburban everest: An "insiders" perspective on experiences at the 2007 Flora London marathon. Journal of Sport and Tourism **13(1)** (2008), 61–77

[7] Rauter, S., Topic, M. D.: Differences in travel behaviors of small and large cycling events participants. APSTRACT: Applied Studies in Agribusiness and Commerce **7(1)**, (2013)

[8] Rauter, S.: Socialni profil športnih turistov - udeležencev množičnih športnih prireditev v Sloveniji. PhD thesis, University of Ljubljana, Slovenia, 2012

[9] Budgett, R.: Fatigue and underperformance in athletes: the overtraining syndrome. British Journal of Sports Medicine **32(2)** (1998), 107–110

[10] Cardoos, N.: Overtraining syndrome. Current sports medicine reports **14(3)** (2015), 157–158

[11] Lehmann, M., Foster, C., Keul, J.: Overtraining in endurance athletes: a brief review. Medicine & Science in Sports & Exercise **25** (1993), 854–862

[12] Fister Jr., I., Ljubič, K., Suganthan, P. N., Perc, M., Fister, I.: Computational intelligence in sports: challenges and opportunities within a new research domain. Applied Mathematics and Computation **262** (2015), 178–186

[13] Engelbrecht, A. P.: Computational intelligence: an introduction. John Wiley & Sons, 2007

[14] Fister, I., Rauter, S., Yang, X -S., Ljubič, K., Fister Jr., I.: Planning the sports training sessions with the bat algorithm. Neurocomputing **149** (2015), 993–1002

[15] Yang, X. -S.: A new metaheuristic bat-inspired algorithm. In: Nature Inspired Cooperative Strategies for Optimization (NICSO 2010), 65–74, Springer, 2010

[16] Fister Jr, I., Fister, D., Yang, X. -S.: A hybrid bat algorithm. Elektrotehniški vestnik **80(1-2)** (2013), 1–7

[17] Aagaard, P., Simonsen, E. B., Andersen, J. L., Magnusson, P., Dyhre-Poulsen, P.: Increased rate of force development and neural drive of human skeletal muscle following resistance training. Journal of Applied Physiology **93(4)** (2002), 1318–1326

[18] Rønnestad, B. R., Hansen, E. A., Raastad, T.: Strength training improves 5-min all-out performance following 185 min of cycling. Scandinavian Journal of Medicine & Science in Sports **21(2)** (2011), 250–259

[19] Hausswirth, C., Bigard, A. X., Berthelot, M., Thomaidis, M., Guezennec, C. Y.: Variability in energy cost of running at the end of a triathlon and a marathon. International Journal of Sports Medicine **17(8)** (1996), 572–579

[20] Psilander, N., Frank, P., Flockhart, M., Sahlin, K.: Adding strength to endurance training does not enhance aerobic capacity in cyclists. Scandinavian Journal of Medicine & Science in Sports, 2015

[21] Fister Jr, I., Yang, X.-S., Fister, I., Brest, J., Fister, D.: A brief review of nature-inspired algorithms for optimization. Elektrotehniški vestnik **80(3)** (2013), 116–122

[22] Fister Jr., I., Fister, I., Yang, X.-S., Fong, S., Zhuang, Y.: Bat algorithm: recent advances. In: Computational Intelligence and Informatics (CINTI), 2014 IEEE 15th International Symposium on, 163–167, IEEE, 2014

[23] Fister Jr., I.: A comprehensive review of bat algorithms and their hybridization. Master's Thesis, University of Maribor, Slovenia, 2013

[24] Hassan, E. A., Hafez, A. I., Hassanien, A. E., Fahmy, A. A.: A discrete bat algorithm for the community detection problem. In: Hybrid Artificial Intelligent Systems, 188–199, Springer, 2015

[25] Soto, R., Crawford, B., Olivares, R., Johnson, F., Paredes, F.: Online control of enumeration strategies via bat-inspired optimization. In: Bioinspired Computation in Artificial Systems, 1–10, Springer, 2015

[26] Cai, X., Wang, L., Kang, Q., Wu, Q.: Adaptive bat algorithm for coverage of wireless sensor network. International Journal of Wireless and Mobile Computing **8(3)** (2015), 271–276

[27] Zhao, D., He, Y.: Chaotic binary bat algorithm for analog test point selection. Analog Integrated Circuits and Signal Processing, 1–14, 2015

[28] Premkumar, K., Manikandan, B. V.: Speed control of brushless dc motor using bat algorithm optimized adaptive neuro-fuzzy inference system. Applied Soft Computing **32** (2015), 403–419

J. Yaghob (Ed.): ITAT 2015 pp. 127–134
Charles University in Prague, Prague, 2015

Platform for Rapid Prototyping of AI Architectures

Peter Hroššo, Jan Knopp, Jaroslav Vítků, and Dušan Fedorčák

GoodAI, Czech Republic
contact author: peter.hrosso@keenswh.com

Abstract: Researching artificial intelligence (AI) is a big endeavour. It calls for agile collaboration among research teams, fast sharing of work between developers, and the easy testing of new hypotheses. Our primary contribution is a novel simulation platform for the prototyping of new algorithms with a variety of tools for visualization and debugging. The advantages of this platform are presented within the scope of three AI research problems: (1) motion execution in a complex 3D world; (2) learning how to play a computer game based on reward and punishment; and (3) learning hierarchies of goals. Although there are no theoretical novelties in (1,2,3), our goal is to show with these experiments that the proposed platform is not just another ANN simulator. This framework instead aims to provide the ability to test proactive and heterogeneous modular systems in a closed-loop with the environment. Furthermore, it enables the rapid prototyping, testing, and sharing of new AI architectures, or their parts.

1 Introduction and Related Work

The recent boom in the field of artificial intelligence (AI) was brought on by advances in so-called narrow AI, represented by highly specialized and optimized algorithms designed for solving specific tasks. Such programs can even sometimes surpass human performance when solving the single problem for which they were created. But these narrow AI programs lack one feature which has been so far widely omitted, partly due to its overwhelming difficulty: generality.

In order to compensate for this deficiency, the field of artificial general intelligence (AGI) is bringing the focus back to broadening the range of solvable tasks. The ultimate goal of AGI is therefore the creation of an agent which can perform well (at human level or better) at any task solvable by a human. For a more detailed description of AI/AGI, see e.g. [23].

Pursuing such a goal is a hard task. According to the scientific method – the only guideline we have – we need to come up with new theories, design experiments for testing them, and evaluate their results. Such a cycle needs to be repeated often, because it can be expected to reach more dead ends than breakthroughs. We don't know how to increase the rate of coming up with new ideas, but what can be improved is the efficiency of research. What we need is better tools which will simplify the implementation of new theories, speed up experiments, and help us understand the results better by visualizing obtained data.

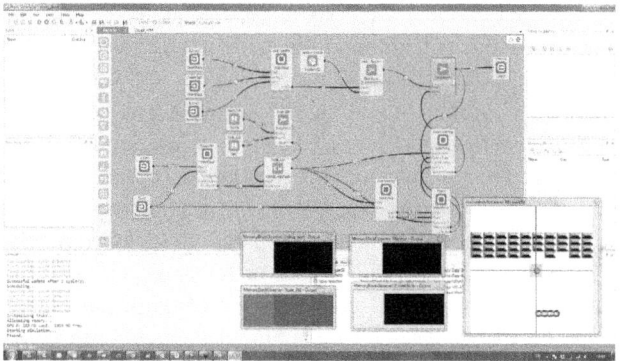

Figure 1: Print-screen of our simulation platform.

In this article, we would like to present our attempt to create such a tool. We here introduce a platform which allows:

- Easy prototyping of new models and fast sharing of existing ones (Sec. 4.1)

- Control of an agent in an environment on top of classic data processing (Sec. 3.3)

- Modular approach – seamless connecting of models inside a greater architecture (Sec. 2.1)

- Various tools for the visualization of data (Fig. 2)

- Simplified debugging (Sec. 2.2)

- User Friendly GPU programming (Sec. 2.1)

- Scalable due to GPU parallel computation

- Support of several scenarios such as an agent in an environment, classification, tic-tac-toe *etc.*

Our platform also includes a variety of visualization tools and enables easy access to diverse data sets not only for tasks such as image classification or recognition, but also scenarios where an agent interacts with its environment. Last but not least, it is open source and freely available under a non-commercial license[1]. The primary goal of this tool is easy collaboration among both specialists and laymen for developing novel algorithms, especially in the field of AI.

There are tools, languages, and libraries that are good in particular areas. In the research community, widely

---

[1]The platform is available as Brain Simulator at http://www.goodai.com/brainsimulator

used are Matlab [9] and Python [17] for prototyping by code, platforms that aim at high level graphical modeling (Simulink [20], Software Architect [8]), data analysis (Azure [10], Rapid Miner [18]), advanced visualization (ParaView [15]), rich graphical user interface (Blender [2], Maya [1]), modular computation (ROS [19]), or specific libraries for sharing [7] and parallel computation [3]. Each of these instruments is important in their specific domain, but there is none which would cover under one roof the most prominent features of all of those mentioned. Our platform is an attempt to fill this niche, and offers both high-level graphical coding and possibly, but not necessarily, also low-level (i.e. CUDA [3]) programming.

There are several tools for simulating neural networks (NN). Nengo [11] or OpenNN [14] focus on experimenting with all possible modifications of NN. Unfortunately, they either lack in visualization or focus on over-specific design approaches. Moreover, usage of these tools often requires extensive programing knowledge and installation of extension packages [4, 13]. In contrast, other tools that provide rich visualization focus purely on the functions of our brain. For example, PSICS [16] uses 3D synapse visualization to show data flow in parts of the brain, Digi-Cortex [6] nicely visualizes spike activations of the whole brain in time, and Cx3D [5] simulates growth of the cortex in 3D. Extensive comparison of various neural network simulators, including our platform (Brain Simulator), can be found in [12].

It is worth noting that the proposed platform is not limited to the design of neural networks only. Any algorithm useful for AI, machine learning, or control can be incorporated (various mathematical transformations, filters, PID controller, image segmentation, hashing functions, dictionary, *etc.*, are already included). The heterogeneous character of the platform is its main advantage.

Throughout this work, we describe our platform in the following Sec. 2. In Sec. 3, tasks where we show advantages of the platform are introduced. In Sec. 4, we discuss the experience with our tool and its advantages and weakness in the testing scenarios. The paper is concluded in Sec. 5.

2   Simulation Platform

Our modus operandi reflects our goals - we are aiming for a modular cognitive architecture, so we needed an environment which would efficiently support the whole life-cycle of experiments, starting with the testing of already existing algorithms, going through the design of a new algorithm, and ending with a results evaluation. We developed a platform where various algorithms from machine learning and narrow AI are available. It is easy to pick some, connect them, and start experimenting effortlessly. Agile development requires frequent testing of new hypotheses, which is facilitated by an easy way of prototyping new modules for the platform as well as their fast training and evaluation (accelerated on GPU) on data. After the experiment is

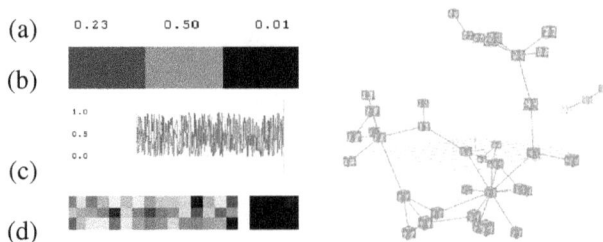

Figure 2: Visualization of data. Left: memory block of a 1×3 matrix can be visualized as: (a) value, (b) color-map, (c) value of each element in time, or (d) value as a color-map in time. Right: more complicated task-specific visualizations can be implemented too, i.e. growing neural gas [26].

running, it often happens that its outcome is not what was expected. Such a simulation platform could not be imagined without a tool for **runtime analysis of algorithms**. For this purpose various data observers can be displayed so the experimenter can visualize the computed data, evaluate performance of the model, and **change its parameters during runtime if needed**.

Our platform is tailored to suit two different points of view of the architecture development process:

- A user who desires quick architecture modeling and needs fast access to already existing state-of-the-art modules (such as PCA, NN, image pre-processing *etc.*) to experiment with. This perspective **requires no coding** and it's done through graphical modeling. Furthermore, it is often crucial to have good insight into the running model, and thus a large set of visualization tools is available (Fig. 2).

- On the other hand, a researcher/developer often requires the creation of a new module or the import of an already existing library. Our API provides an easy way for such a **module to be created and added** to the inner shared repository. Moreover, the API offers an opportunity to hook the code to the GUI and bring needed interactivity. Finally, the API defines a rigid interface, ensuring that the new module will be compatible with other modules.

The platform was designed to meet both needs. It is important to distinguish between them as a user can be a person interested in machine learning, but less experienced in programming. Our platform can be a good starting point, and the learning curve should therefore be smooth enough to bring the person in effortlessly.

From the experienced researcher/developer point of view, the platform should provide a convenient set of tools that can help with the development of novel algorithms and/or be able to envelope existing work into module that can be easily shared among a team.

Finally, the community-driven approach renders itself very powerful and we believe that it can speed up the research vastly. For this reason, we are **planning to in-build** a **"module market"** to allow for the sharing of state-of-the-art research results between many co-working teams.

2.1 Platform Meta Model

There are three basic concepts defined in the meta model: *a node, a task and a memory block*. The node encapsulates a functional block or algorithm that can "live" on its own (e.g. matrix operations, data transformations, various machine learning models, *etc.*). A node needs a memory for its function. The memory is organized into a set of memory blocks that are aggregated inside the node. Some of these memory blocks can be designated as output blocks and others as input blocks. The connection between input and output memory blocks is provided by the user.

From the functional point of view, the node behaviour can usually be divided into a set of tasks where each task is a part of the realized algorithm. Both nodes and tasks can define a set of parameters. Usually, node parameters describe structural properties (i.e. size of memory blocks) whereas task parameters affect behavior. At present, the memory model is constant during the simulation, and therefore structural properties are editable only in design time. On the other hand, it is useful to **change task parameters during simulation and observe changes** in behavior of the algorithm/node.

Memory blocks are located at GPU (device memory) and every task can be seen as a collection of kernel calls (methods executed on GPU). If two nodes are connected in the GUI, it means that they have a pointer to the same memory block (input in one node, output in the other).

If one requires dynamically allocated memory, the user can either define a memory block that is large enough, or implement the node only for the CPU (which is more flexible than GPU) using all data structures supported by C#. The only mandatory requirement is usage of input/output memory blocks.

All concepts described above can be easily implemented through rich API that is provided. The actual implementation relies heavily on annotated code describing various aspects of the model (UI interactivity, constraints, persistence, *etc.*). It allows the user to be extremely efficient in creating model prototypes. Sometimes, this can lead to unreadable, over-annotated code which is hard to maintain but this can be eliminated by applying standard software design patterns like MVC when needed.

2.2 Computation

As described above, the prototyped model forms an oriented graph with nodes and data connection edges. As the connections between nodes can be any of $M \rightarrow N$ and recurrent connections are also possible, the resulting graph can be very complex. Moreover, the usual model is connected to the world node from which "perception" inputs are taken, and control outputs are passed, forming the main loop of the simulation.

Before running the simulation, the order of nodes execution needs to be evaluated. There are other aspects that level the problem up (e.g. inner cycles, clustering and balancing of the model in HPC environment) but it usually boils down to various forms of dependency ordering, cycle detection, or the job shop problem [34]. Solving these tasks is automated and user/developer assistance is usually discouraged, but there are use cases where user aid is necessary or can simplify the problem substantially.

There is also another view of the problem of execution order when faced in the area of machine learning. It turns out that many of ML methods are surprisingly noise resistant (i.e.neural nets). Therefore, *if approached with caution*, one can run the model asynchronously and let inner parts of the model deal with sometimes temporally inconsistent data. We made some experiments and the preliminary results show that relatively complex models can be run completely without synchronization.

Another aspect of the model execution is GPU enhanced computation which can speed up the simulation substantially. The main purpose of our simulation platform is fast prototyping and testing of hypotheses. With increasing generality the efficiency usually decreases, so one should not expect top execution speed from our simulation platform. The devised practice is to design, test, and analyze new architectures, and once the final model is tested and working, it can be replaced by a specialized, highly optimized implementation still within the platform environment. Finally, it can be argued that the overall time necessary to get from an idea to the final product is much shorter compared to the classic approach of writing a specific program from scratch for each new experiment.

An important part of the development process is the easy visualization of what is happening at each part of the designed system. This is especially important for debugging as the most frequent problem is due to the difference between what the programmer thinks the program should do and what it does in reality. In addition to the variety of observers that have already been discussed (Fig. 2), the platform contains its own debugger, where one can walk through the execution of all components used in the model.

3 Testing Scenarios

Whether the ultimate goal of AGI (a general autonomous machine) is achievable or not [35], researchers focus on its sub-goals such as learning how to play games [37], *etc.* One of the prominent building blocks for these sub-goals are neural networks in the form of deep learning and CNN, and which have recently made big progress in speech recognition [40], computer vision [33], medical analyses [43], or language translation [28]. Le and colleagues [36] used a deep network to learn in unsupervised manner what an ordinary "cat" looks like only by watching youtube videos. While NN can also learn how to play simple games [32, 37], they usually fail in structured problems which demand learning hierarchies or chains of goals. From this perspective, it seems promising to focus on machines which can control another machine, such as NN that learn how to control a Turing Machine [27]. They designed a neural network which learns a procedure to control a Turing Machine to sort numbers.

As the goal of this paper is to provide a tool that shortcuts the research path to an autonomous machine, we will show how it performs on three selected AI tasks solved by our team: learning motion control, game playing of the Atari game Breakout, and learning hierarchies of goals. Our solutions are highly inspired by current machine learning literature with a stress on the usage of neural networks, which are one of the basic building blocks for bigger architectures.

The first experiment (Sec. 3.1) will demonstrate how our platform can be connected to an external source of input data and how various modules of narrow AI can be combined together to form a functioning system which can drive a robot in a virtual world with simulated physics.

The second experiment (Sec. 3.2) will be situated in much simpler simulation environment – an Atari [38] game called Breakout. In this experiment a more advanced adaptive system will be showcased. The system works directly on raw image input. It takes advantage of the semantic pointer architecture [25] for representing its perceptions and for converting them into a long term memories such as goals. This knowledge is then used for learning necessary actions for playing the game.

In the third experiment (Sec. 3.3), we move a bit higher in the level of abstraction. The presented problem consists of an agent in a simple 2D environment which needs to satisfy a chain of preconditions before reaching a reward, such as if the agent wants to turn on a light, it needs to press a switch, but to get to the switch he also needs to overcome an obstacle (a door controlled by another switch). The task is solved by hierarchical reinforcement learning [30].

To clarify why we selected these experiments, one could imagine the three systems as parts of a future higher-order cognitive architecture where they will work together. The system from the first experiment could be thought of as a basic motoric and sensory system driven by reflexes and higher-level commands. These would come from the system used in the second experiment, which would allow the agent to learn how to reach a specific goal. And finally, the third system should discover the hierarchy of goals and preconditions, and thus could resemble a simplified version of the agent's central executive. Such connection of the systems remains for our future work.

3.1 SE Robot

We took advantage of a sandbox game called Space Engineers [21], which provides a physically realistic 3D environment where various structures can be built. We built a six-legged robot within the game and connected it bi-directionally with our simulation platform. In one direction the game sends visual data from the robot's view and a description of the state of the robot's body. In the other direction motoric commands from our control module inside the simulation platform are sent to the robot, which executes them in the game.

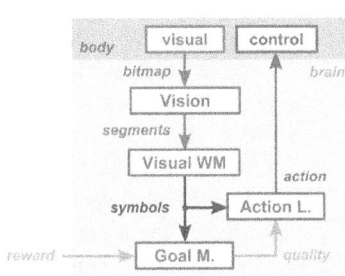

Figure 3: Overall architecture. Raw visual signals are processed into symbols, which are then added to the working memory. States corresponding to reward and punishment are accumulated and later used as teaching signals for training the action selection network.

The control module was trained to associate visual input with motor commands in a supervised way. The associative memory was implemented with a Self-Organizing Map [31], which found the most similar representative of the received input in the visual memory and returned the associated high-level motoric command (turn left/right, move forward/backward). These high-level commands were then unrolled into sequences of body states consisting of joint angles of all of the robot's limbs using a recurrent neural network (RNN) [39]. These body states were afterwards used as waypoints for a control RNN which was trained to act as an inverse dynamics model of the robot's body. In order to reach a specified waypoint, the control network generated full motoric inputs to the robot - the desired angular velocities of joints.

The training phase consisted of a mentor leading the robot from various starting locations towards a goal location in the 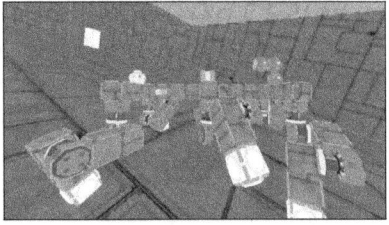 environment, which was identified by an easily distinguishable 3D symbol located on that position. The mentor was implemented by a hard-coded navigation system. In this way the hexapod was trained to look for the goal symbol and when it appeared in the robot's field of view, to navigate successfully towards the destination through the environment.

3.2 Atari Game

The Breakout game was chosen as our second testing scenario. The game consists of a ball, a paddle, and

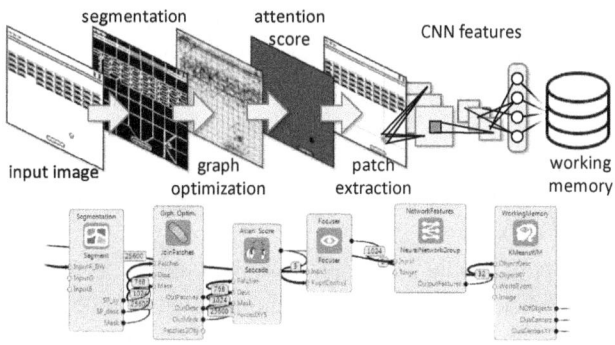

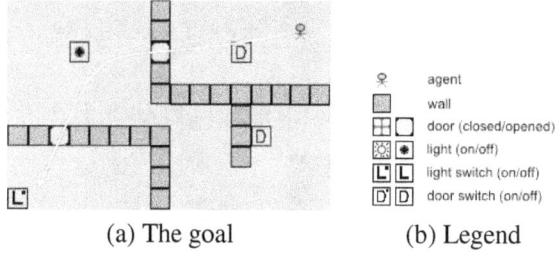

(a) The goal   (b) Legend

Figure 5: Multiple goals. (a) The agent's current goal is to reach the light switch and turn on the lights. (b) objects of the environment

Figure 4: Top: Image processing. First, an input image is segmented into super-pixels (SP) using SLIC [22]. Second, each SP is connected with its neighbors and nearby SP are assigned into the same object id. Third, the attention score (s_A) is estimated for each object. Fourth, features are estimated for the object with the highest s_A by a hierarchy of convolutions and fully-connected layers. Fifth, the object features are clustered into Visual Words [41] to constitute a "Working Memory". Bottom: Corresponding implementation in our platform.

Details of the vision system and semantic point architecture are described in Appendix II.

3.3 2D World with Hierarchical Goals

In previous testing scenarios we wanted to test if our system was able to coordinate complex motoric commands in a 3D environment, learn simple goals, and act towards maximizing the received reward. Our goal for the third scenario is to increase the generality of the designed system to enable identification and satisfaction of chained preconditions before the final goal can be reached.

bricks. The ball bounces from walls, can destroy bricks, and can fall to the ground, for which the player is penalized by losing a life. After losing 4 lives, the game is over. When all bricks are destroyed, the player successfully finishes the level and enters the next one consisting of a different arrangement of bricks. The player has three actions available which accelerate the paddle to the right, to the left, or decelerate. Even though our modular approach uses pure unstructured data input

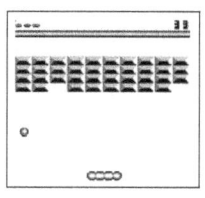

(raw image as in [37]), it later extracts the structure, so we can understand the inner workings of the model as opposed to the cited work. The architecture of the system consists of four main parts: image processing, working memory, accumulators of reward and penalty, and an action selection network (Fig. 3).

Relevant information about the objects is extracted from the raw bitmap in the Vision System (Fig. 4).

Working memory (WM) is the agent's internal representation of the environment. It contains all of the objects detected by vision. WM is kept up to date by adding new objects which haven't been seen yet, and by updating those already seen. The identity of objects is detected through a comparison of visual features. Contents of the working memory are transformed into a symbolic representation and passed to the goals memory and action selection network.

The goals memory is trained by accumulating states associated with reward and punishment in their respective semantic pointers, goal$^+$ and goal$^-$. These are then used for evaluating the quality of game-states, which is necessary for training the action selection network.

We present a task, consisting of a simple 2D world, where a single source of reward is located – a light bulb, which starts in the "off" position and should be turned on by the agent. This can be achieved by pressing a switch, but the switch is hidden behind a locked door. The door can be unlocked through a switch, but this switch is hidden behind another locked door. It would be possible to chain the preconditions further in this manner, but without the loss of generality we use only two locked doors with two matching switches. The setup can be seen in Fig. 5.

We approached this problem by employing HARM (Hierarchical Action Reinforcement Motivation system) [30]. It is an approach based on a combination of a hierarchical Q-learning algorithm [42] and a motivation model. The system is able to learn and compose different strategies in order to create a more complex goal.

Q-learning is able to "spread information about the reward" received in a specific state (e.g. the agent reaching a position on the map) to the surrounding space, so the brain can take proper action by climbing the steepest gradient of the Q function later. However, if the goal state is far away from the current state, it might take a long time to build a strategy that will lead to that goal state. Also, a high number of variables in the environment can lead to extremely long routes through the state space, rendering the problem almost unsolvable.

There are several ideas that can improve the overall performance of the algorithm. First, this agent *rewards itself for any successful change to the environment*. The motivation value can be assigned to each variable change so the agent is constantly motivated to change its surroundings.

Figure 6: Learnt strategy. Visualization of the agent's knowledge for a particular task, which changes the state of the lights. It tells the agent what to do in order to change the state of the lights in all known states of the world. The heat map corresponds to the expected utility ("usefulness") of the best action learned in a given state. A graphical representation of the best action is shown at each position on the map.

Second, for each variable that the agent is able to change, it *creates a Q-learning module* assigned to the variable (e.g. changing the state of a door). Therefore, it can learn an underlying strategy defining how this change can be made again. In such a system, a whole network of Q-learning modules can be created, where each module learns a different strategy.

Third, in order to lower the complexity of each sub-problem (strategy), the brain can analyze its "experience buffer" from the past and eventually drop variables that are not affected by its actions or are not necessary for the current goal (i.e. strategy to fulfill the goal).

A mixture of these improvements creates a hierarchical decision model that is built online (first, the agent is left to (semi-)randomly explore the environment). After a sufficient amount of knowledge is gathered, we can "order" the agent to fulfill a goal by manually raising the motivation that corresponds to a variable that we want to change. The agent then will execute the learned abstract action (strategy) by traversing the network of Q-learning modules and unrolling it into a chain of primitive actions that lie at the bottom.

4 Discussion

Throughout the work on the testing scenarios (Sec. 3) we have observed several advantages and weakness of the platform. In this section, our experience with the usage of the platform is discussed. The discussion is focused especially on the end-user experience, i.e. experience of a person that did not develop the platform but wants to use it for solving her problem.

First, a list of identified features is presented, and then further experience is described.

+ Fast and easy online observation and interaction with the simulation.

+ Created modules can be easily understood and shared with collaborators due to the same interface. Moreover,

the persistence capabilities allow easy sharing of whole models (projects) and merging of them together.

+ It is easy to replace an existing module with its improved version, as the architecture is separated from the implementation. As the backward compatibility becomes crucial at this point, the inner versioning system was implemented.

+ Provided interface drives users to follow design patterns when developing low-level optimized modules.

- Current version runs on MS Windows only, but a port to MacOS and Linux is planned for the future.

- The user has an option to either develop optimized modules in code (CUDA [3] or C#) or use the graphical interface for connecting existing modules into bigger architectures. There is no middle layer which would support scripting.

4.1 Experience of Newcomers

The expertise of people that have started to use our platform varies from C++ experts to Matlab users only. We found that users with a very short training can connect existing modules into simple architectures (like neural network MNIST image recognizer) as the graphical modeling is somehow natural and easy to understand. The linear learning curve of newcomers is supported by a video tutorial as well as several examples of how to implement simple and more advanced tasks[2].

For development of new modules, it is necessary to understand a programming language (C#, C++, CUDA) at the basic level at least. Once the definitions of inputs, outputs, tasks, and kernels (four lines of code each) are understood, developers soon start creating their own nodes. Their learning curve then equals learning how to use a new library.

4.2 Our Observations on the Testing Scenarios

In the first scenario (Sec. 3.1), we have shown that our platform successfully connects with the open source game Space Engineers [21]. Modules created in the platform controlled the hexapod in the world of the game. It was the understanding of the game's communication module that took the most time in this case. Otherwise the development of the controller did not raise any challenges for the platform.

The second scenario (Sec. 3.2) consisted of several modules that were developed independently. We found it extremely useful that each module communicates with others using only the pre-defined interface (memory blocks) that correspond to a sketched diagram (i.e. Fig. 4). Modules were merged into one big architecture right before the deadline without any complications. As the final model was quite large and performance-demanding,

---

[2]Documentation available at http://docs.goodai.com/brainsimulator/

we were forced to profile, find bottlenecks, and optimize in the process. It was extremely useful to visualize data flowing between (and inside) the modules.

In the third scenario (Sec. 3.3), HARM constituted a single module with complex insides. Therefore this scenario presented an ideal example for designing a number of task-specific visualization tools (for example, the agent's knowledge in Fig. 6).

5 Conclusion

We have presented a platform for prototyping AI architectures. The platform is tailored both for users with no mathematical/programming background but with a high desire to experiment with AI modules, and for researchers/developers who want to improve and experiment with their existing state-of-the-art techniques.

To show the usage of our platform we have presented three development scenarios: linkage with a 3D game world and controlling an agent there; playing an Atari game using the raw bitmap input processed by computer vision techniques, attention model, and semantic pointer architecture; and learning a complex hierarchy of goals.

The proposed platform opens up possibilities to share ideas not only within the community but also with non-experts who can boost the research via rapid testing, or utilize fresh, out-of-the-box solutions. There is also the prospect of support from the open source community - if not directly in the development, then at least in assessing missing features, so we can incorporate them and thus provide a tool that can be used at many levels of expertise.

We believe that by providing an open platform for AI and ML experiments along with a smooth learning curve, we can bring together many enthusiasts across different fields of interest, potentially leading to unexpected advancements in research.

Acknowledgement. This material is based upon work supported by GoodAI and Keen Software House.

References

[1] Autodesk maya. Available at http://www.autodesk.com/products/maya/overview.

[2] Blender. Available at https://www.blender.org/.

[3] CUDA. Available at https://developer.nvidia.com/cuda-zone.

[4] CVX: Software for Disciplined Convex Programming. Available at http://cvxr.com/.

[5] Cx3D: Cortex simulation in 3D. Available at http://www.ini.uzh.ch/~amw/seco/cx3d/.

[6] DigiCortex: Biological neural network simulator. Available at http://www.dimkovic.com/node/1.

[7] GitHub. Available at https://github.com/.

[8] IBM Rational Software Architect. Available at http://www.ibm.com/developerworks/downloads/r/architect/index.html.

[9] MATLAB: The language of technical computing. Available at http://www.mathworks.com/products/matlab/.

[10] Microsoft Azure. Available at http://azure.microsoft.com/en-us/.

[11] The nengo neural simulator. Available at http://nengo.ca/.

[12] Neural networks simulators. Available at https://goo.gl/hRf4KA.

[13] OpenCV: Open source computer vision. Available at http://opencv.org/.

[14] OpenNN: Open neural networks library. Available at http://www.intelnics.com/opennn/.

[15] ParaView. Available at http://www.paraview.org/.

[16] PSICS: The parallel stochastic ion channel simulator. Available at http://www.psics.org/.

[17] Python. Available at https://www.python.org/.

[18] Rapid miner. Available at https://rapidminer.com/.

[19] ROS. Available at http://www.ros.org/.

[20] Simulink: Simulation and model-based design. Available at http://www.mathworks.com/products/simulink/.

[21] Space Engineers, open source code. Available at https://github.com/KeenSoftwareHouse/SpaceEngineers.

[22] Achanta, R., Shaji, A., Smith, K., Lucchi, A., Fua, P., Susstrunk, S.: Slic superpixels compared to state-of-the-art superpixel methods. PAMI (2012), 2274–2282

[23] Ben, G.: Artificial general intelligence: concept, state of the art, and future prospects. Journal of Artificial General Intelligence **5** (2014), 1–48

[24] Bishop, C. M.: Pattern recognition and machine learning. Springer, 2006

[25] Eliasmith, C.: How to build a brain: a neural architecture for biological cognition (Oxford Series on Cognitive Models and Architectures). Oxford University Press, 2013

[26] Fritzke, B.: A growing neural gas network learns topologies. In: NIPS, 1995

[27] Graves, A., Wayne, G., Danihelka, I.: Neural turing machines. CoRR, 2014

[28] He, X., Gao, J., Deng, L.: Deep learning for natural language processing and related applications (tutorial at ICASSP). ICASSP, 2014

[29] Huang, F. J., Boureau, Y. L., Lecun, Y.: Unsupervised learning of invariant feature hierarchies with applications to object recognition. In: CVPR, 2007

[30] Kadlecek, D., Nahodil, P.: Adopting animal concepts in hierarchical reinforcement learning and control of intelligent agents. In: Proc. 2nd IEEE RAS & EMBS BioRob, 2008

[31] Kohonen, T., Schroeder, M. R., Huang, T. S.: Self-organizing maps. 3rd edition, 2001

[32] Koutník, J., Cuccu, G., Schmidhuber, J., Gomez, F.: Evolving large-scale neural networks for vision-based reinforcement learning. In: GECCO, 2013

[33] Krizhevsky, A., Sutskever, I., Hinton, G. E.: Imagenet classification with deep convolutional neural networks. In: NIPS, Curran Associates, Inc., 2012

[34] Tsaia, M. -J., Chang, H.-Y., Huang, K. -C., Huanga, T -C., Tung, Y. -H.: Moldable job scheduling for hpc as a service with application speedup model and execution time information. Journal of Convergence, 2013

[35] Kurzweil, R.: The singularity is near: when humans transcend biology, 2006

[36] Le, Q., Ranzato, M. 'A., Monga, R., Devin, M., Chen, K., Corrado, G., Dean, J., Ng, A.: Building high-level features using large scale unsupervised learning. In: ICML, 2012

[37] Volodymyr et al. Mnih: Human-level control through deep reinforcement learning. Nature **518** (2015), 529–533

[38] Naddaf, Y.: Game-independent AI agents for playing Atari 2600 console games. Masters, University of Alberta, 2010

[39] Rojas, R.: Neural networks: a systematic introduction. Springer-Verlag New York, Inc., New York, NY, USA, 1996

[40] Sak, H., Vinyals, O., Heigold, G., Senior, A., McDermott, E., Monga, R., Mao, M.: Sequence discriminative distributed training of long short-term memory recurrent neural networks. In: Interspeech, 2014

[41] Sivic, J., Zisserman, A.: Video Google: A text retrieval approach to object matching in videos. In: ICCV **2** (2003), 1470–1477

[42] Sutton, R. S., Barto, A. G.: Introduction to reinforcement learning. MIT Press, Cambridge, MA, USA, 1st edition, 1998

[43] Xu, Y., Mo, T., Feng, Q., Zhong, P., Lai, M., Chang, E. I. -C.: Deep learning of feature representation with multiple instance learning for medical image analysis. ICASSP, 2014

Appendix I: Details of Image Processing

Unlike the major stream of Computer Vision, our approach has to be *unsupervised* without any training data. Thus, we have *no* prior knowledge and the system has to learn everything on-the-fly. Our system is a pipeline visualized and implemented in Fig. 4. It consists of the following parts:

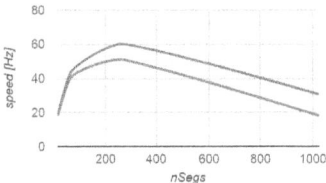

The input image is first segmented into a set of super-pixels [22] (SP). Then, each SP is connected to its vicinity constituting a graph where nodes are SPs and edges connect neighboring SPs. SPs with similar color are merged into connected components. Note that

Figure 7: Performance of SLIC (blue) and SLIC+graph optimization (red) w.r.t the number of segments.

while more SP speeds up the segmentation, it slows down the graph optimization algorithm, see Fig. 7. Once we have object proposals, we estimate an attention score (s_A) for each object, $s_A(\mathbf{o}_i) = \psi_{time}(\mathbf{o}_i) + \psi_{move}(\mathbf{o}_i)$, where $\psi_{time}(\mathbf{o}_i)$ is time since we have focused on the object $\mathbf{o}_i$, $\psi_{move}(\mathbf{o}_i)$ is the object's movement. The object with the highest s_A is selected[3] and its position together with its size define an image patch. The image patch is processed into a CNN features [29], and this feature representation

---

[3]Once the object is selected, its ψ_{time} is decreased and then it won't be selected in the next time step.

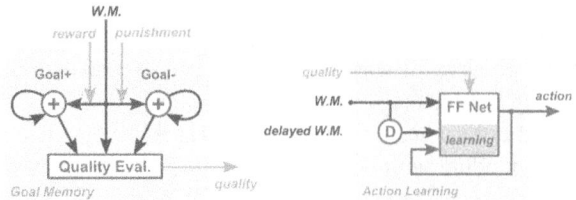

Figure 8: Left: **Goal+ and goal-.** States associated with received reward and punishment are accumulated in semantic pointers goal+ and goal-, respectively. Right: **Action learning.** Actions selection is learnt in a supervised way with fitness computed from goal+ and goal- as a teaching signal.

is then clustered into a "Working Memory" (WM). The WM stores feature id together with the object position for the last 10 seen objects.

For CNN features, we used two convolutions layers of 8 and 5 neurons and patch sizes 5×5 followed by a fully-connected layer of 16 neurons. Learning converged in 6K iterations. We observed no performance improvement with bigger networks. WM was implemented as a simple K-means [24].

Appendix II: Semantic Pointer Architecture

As was already mentioned in section 1, one of the main features of our method is the semantic pointer architecture (SPA), which merges the symbolic and connectionist approach. Artificial neural networks are very powerful adaptive tools, but their usage usually comes at the expense of losing detailed insight into how exactly the task is solved. Such a drawback can be mitigated by using the SPA and its variable binding. It introduces composite symbols of the form X bind x, where X is the name of the variable and x is its value. It is then possible to train a network to perform complicated transforms such as:

$$V \otimes (X \otimes x + Y \otimes y) \to C \otimes (Y \otimes x) \tag{1}$$

which could be interpreted as an action selection rule for the pong game:

$$\begin{aligned} Visual \otimes (Ball \otimes x_1 + Paddle \otimes x_2) \to \\ \to Move \otimes (Paddle \otimes x_1) \end{aligned} \tag{2}$$

If ball is seen at position x_1 and paddle at x_2, execute command 'move paddle to position x_1'. Without SPA it would be much harder to maintain understanding of the transformed symbols, if not entirely impossible.

Goals Memory. The accumulated states g^+ and g^- are used for calculation of quality q of the state x, using dot product '·', $q = g^+ \cdot x - g^- \cdot x$, see Fig. 8 left.

Action Learning. Training of action selection (Fig. 8 right) is delayed by one simulation step to allow the system to observe results of its actions. Actions leading to higher quality states are labeled as correct, otherwise incorrect.

J. Yaghob (Ed.): ITAT 2015 pp. 135–142
Charles University in Prague, Prague, 2015

Comparing Non-Linear Regression Methods
on Black-Box Optimization Benchmarks

Vojtěch Kopal[1] Martin Holeňa[2]

[1] Charles University in Prague, Faculty of Mathematics and Physics,
Malostranské nám. 25, 118 00 Praha 1, Czech Republic
vojtech.kopal@gmail.com,
[2] Institute of Computer Science, Academy of Sciences of the Czech Republic,
Pod Vodárenskou věží 2, 182 07 Praha, Czech Republic
martin@cs.cas.cz

Abstract: The paper compares several non-linear regression methods on synthetic data sets generated using standard benchmarks for continuous black-box optimization. For that comparison, we have chosen regression methods that have been used as surrogate models in such optimization: radial basis function networks, Gaussian processes, and random forests. Because the purpose of black-box optimization is frequently some kind of design of experiments, and because a role similar to surrogate models is in the traditional design of experiments played by response surface models, we also include standard response surface models, i.e., polynomial regression. The methods are evaluated based on their mean-squared error and on the Kendall's rank correlation coefficient between the ordering of function values according to the model and according to the function used to generate the data.

1 Introduction

In this paper, we compare non-linear regression methods that could be used as surrogate models for optimization tasks. The methods are compared on synthetic data sets generated using standard benchmarks for continuous black-box optimization, for which we used implementations based on definitions from *Real-Parameter Black-Box Optimization Benchmarking 2009* [8].

A continuous black-box optimization is a task where we try to minimize a continuous objective function $f : X \subseteq \mathbb{R}^n \to \mathbb{R}$ for which we do not have an analytical expression. Such problems arise, for example, if the values of the objective function are results of experimental measurements.

For that comparison, we have chosen regression methods that have been used as surrogate models in such optimization: *radial basis function networks* [3] [18], *Gaussian processes* [6] [11], and *random forests* [4].

We measure the accuracy of each methods based on mean square error and Kendall's rank coefficient and based on the results we suggest which methods work better as surrogate models. We are interested in properties of each method to be used as a surrogate model, though our experiments do not replace a direct evaluation in optimization or

in evolutionary algorithms. This is a subject of two other papers included in this proceedings.

Other comparisons of non-linear models have been presented. A numerical comparison of neural networks and polynomial regression has been performed in [2] and in [16], in the latter one also classification and regression tree (CART) model has been compared. An evaluation of Gaussian processes with other non-linear methods has been done in [15] and in [10]. These studies compared accuracy of each model for prediction and have not paid attention to surrogate models for optimization. Example of such works can be found in [7], where they have compared quadratic polynomial regression with other methods based on prediction accuracy and mean-squared error, and in [13], where is polynomial regression compared with radial basis function networks based on accuracy and also on optimization results. In this paper, we compare the methods by means of mean-squared error and also Kendall's rank coefficient.

We briefly describe the theoretical background for each of these methods: how the corresponding models are being induced and how they are used to predict new values. For the synthetic data, we added an overview of how the functions look like in a 3-dimensional space (Figure 1).

The paper is organised as follows. In Section 2, we recall the theoretical fundamentals of the employed regression methods. In Section 3, we describe the setup of our experiments and summarise the results, before the paper concludes in Section 4.

2 Regression Methods in Data Mining

With a continuously increasing amount of gathered data, data mining techniques allow us to search for patterns in the data sets and model the underlying reality. Various models have been introduced in the past, starting from a linear regression to complex nonlinear methods such as neural networks, or Gaussian processes. These models are used to approximate a function that describes the relationship between target and input values.

We now introduce the methods compared in this paper. Each of these methods has its strengths and weaknesses

	Parameters	Hyper-parameters	Strengths	Weaknesses	Complexity
Polynomial regression	order of the polynom	×	fast and simple	too simple	$\Theta(M^2N)$
Gaussian processes	covariance function, mean function (usually constantly zero)	depends on cov. fun. length-scale (l) noise-level (SN)	robust, generalising well	time complexity, black box	$\Theta(N^3)$
Random forests	# of trees (NT), min. data in leaves (ML), # of randomly selected variables	×	interpretability, allows parallel computations	slower to compute predictions	$\Theta(MK\tilde{N}\log^2\tilde{N})$ [12]
Radial basis functions network	spread constant (CS) maximum of neurons (MAX) error goal (EG)	×	robust	black box	polynomial time [17]

Table 1: Summary of regression methods. $N = \#$ of samples, $\tilde{N} = 0.632N$, $M = \#$ of variables, $K = \#$ of variables randomly drawn at each node (in random forests)

which we point out in Table 2 and later we will discuss them in the context of results of our experiments.

We assume to have a pair $((X),Y)$, where $\mathbf{X}$ is p-dimensional data set with n points, i.e. $\mathbf{X}$ is a matrix $p \times n$, or it is a vector $\mathbf{X} = (\mathbf{x_1},\mathbf{x_2},...,\mathbf{x_n})$, where $\mathbf{x_i}$ is a column vector of size p, i.e. $\mathbf{x_i} = (x_{i1},x_{i2},...,x_{ip})$, and $Y = (Y_1,Y_2,...,Y_n)$ is a vector of size n of target values to corresponding rows in matrix $\mathbf{X}$. We use $\|\mathbf{x}\|$ as the Euclidean norm of vector $\mathbf{x}$. In the paper, we use following notation:

- X,Y,β are vectors with elements X_i,Y_i,β_i, respectively, also $\beta_{j,k}$ is a scalar denoting a parameter in polynomial regression for interaction x_jx_k,

- f is a function and $f(x)$ is an output of the functions corresponding to input x, for multivariate function f, we have either matrix notation $Y = f(\mathbf{X})$, or vector notation $Y_i = f(\mathbf{x}_i)$,

- $\bar{f}(\mathbf{X})$ is an average output over $f(\mathbf{x}_i), \forall i \in \{1,...,N\}$.

2.1 Polynomial Regression

The most simple form of *polynomial regression* (PR) is linear regression in which the model is described by $p+1$ parameters $\beta_0,\beta_1,...,\beta_p$,

$$f(X_i) = \beta_0 + \sum_{j=1}^{p} x_j\beta_j$$

which can be computed by [9]

$$\beta = (\beta_0,...,\beta_p) = (\mathbf{X}^T\mathbf{X})^{-1}\mathbf{X}^T\mathbf{y} \qquad (1)$$

Polynomial regression is still part of the linear regression family, because the dependence on the model parameters is linear. However, we consider also higher powers of input variables. For example, in the quadratic case we

add both x_i^2 form for $i \in \{1,...,p\}$, and also as an interaction x_ix_j for $i,j \in \{1,...,p\}, i \neq j$. Consequently we have $(p^2+p)/2$ new variables.

For our experiments, we will restrict attention to quadratic regression,

$$f(X_i) = \beta_0 + \sum_{j=1}^{p} x_j\beta_j + \sum_{j=1}^{p} x_j^2\beta_{p+j} + \sum_{j=1}^{p}\sum_{\substack{k=1 \\ k<j}}^{p} x_jx_k\beta_{i,k}$$

This is also the standard restriction in response surface modeling [14].

2.2 Random Forests

Random Forests (RF) is a model proposed by Breiman [5], and it is based on ensembles of decision trees. Due to our interest in surrogate models for continuous black-box optimization, we are interested in ensembles of regression trees.

A regression tree is a function defined by means of a binary tree with inner nodes representing predicates, and edges from a node to its children representing whether the predicate is or is not fulfilled. The leaf nodes give the predicted target value. The tree is built recursively starting with a root node and searching for an optimal binary predicate over the input variables. Regression trees can be applied to data sets with both categorical/discrete variables, and real-valued variables. Since we focus on surrogate models for continuous black-box optimization, we only consider real-valued predicates. For a real-valued variable, the data set is split into two parts through minimizing following formula

$$\sum_{x_i \in R_1(j,s)} (y_i - c_1)^2 + \sum_{x_i \in R_2(j,s)} (y_i - c_2)^2$$

where R_1,R_2 are the two linearly bounded regions with axes-perpendicular borders into which the data set is split using a j-th variable x_j and its splitting point s, and c_1,c_2

are the averages of function values of points belonging to R_1, R_2, respectively. After finding the optimal splitting point we recursively apply this process to both regions R_1 and R_2, and for each of them, only the data points in the region are considered. This process continues until a stopping criterion is met. This can be either the *minimum number of data points in leaves* or inner nodes, or the depth of the tree.

If the regression tree finally splits the input space into the regions $R_1, \ldots, R_m$, we can compute the prediction for a new data point using the following formula:

$$f(x) = \sum_{m=1}^{M} c_m I(x \in R_m)$$

where c_m is an average target value of data points in region R_m.

An ensemble of regression trees averages the predictions when presented with a new data point.

There are several options how to induce a number of trees over the same data set that will lead to low correlation. In traditional bagging, independent subsets of the original data used for individual trees are obtained by sampling from the data set uniformly and with replacement. In addition, random subsets of input variables can be used. In Matlab implementation of random forests, a square root of number of input variables are selected by default, which is also a setting we have used for our experiments.

The model parameters are *number of trees* (NT) which are added to the ensemble and the minimum number of data in leaves (ML).

2.3 Gaussian Processes

A *Gaussian process* (GP) is a random process such that its restriction to any finite number of points has a Gaussian probability distribution. A Gaussian process $\mathbb{GP}(\mu(\mathbf{x}), \kappa(\mathbf{x}, \mathbf{x}'))$ is defined by its mean function $\mu(\mathbf{x})$ and a covariance function $\kappa(\mathbf{x}, \mathbf{x}')$.

$$f(\mathbf{x}) \sim \mathbb{GP}(\mu(\mathbf{x}), \kappa(\mathbf{x}, \mathbf{x}')) \qquad (2)$$

These functions determine the mean and covariance of the process because

$$\mathbb{E}[f(\mathbf{x})] = \mu(\mathbf{x}),$$
$$Cov[f(\mathbf{x}), f(\mathbf{x}')] = \mathbb{E}[(f(\mathbf{x}) - \mu(\mathbf{x}))(f(\mathbf{x}') - \mu(\mathbf{x}'))] \quad (3)$$
$$= \kappa(\mathbf{x}, \mathbf{x}')$$

The important part of modelling functions with Gaussian processes is choosing the covariance function. An important feature of covariance functions is that they can be combined together using addition and multiplication, i.e. for κ, κ' covariance functions, $\kappa \times \kappa'$ and $\kappa + \kappa'$ are again covariance functions. Frequently used covariance functions are: linear, periodic, squared-exponential, and rational quadratic.

- **Linear**:
$$\kappa_{lin}(\mathbf{x}, \mathbf{x}') = \mathbf{x}\mathbf{x}'$$

- **Periodic**:
$$\kappa_{per}(r) = exp(-\frac{2}{l^2} sin^2(\pi \frac{r}{p}))$$

- **Squared-exponential**:
$$\kappa_{SE}(r) = exp(-\frac{r^2}{2l^2})$$

- **Rational Quadratic**:
$$\kappa_{RQ}(r) = (1 + \frac{r^2}{2\alpha l^2})^{-\alpha}$$

where $r = |\mathbf{x} - \mathbf{x}'|$ and c, l, p, α are parameters of the covariance function (because the covariance function itself is a parameter of the Gaussian process, they are called hyper-parameters of the process). l is a length-scale, p defines period, and α changes the smoothness of rational quadratic function. An additional parameter in the model is the *noise level* (SN) which is an additive Gaussian noise in the model.

When working with multivariate data sets, the covariance functions which have the *length-scale* as a parameter, can either apply the same length-scale l to all dimensions, or i-th dimension has its length-scale l_i. In the first case, the covariance functions have isotropic distance measure, the latter case uses *automatic relevance determination (ARD)*.

2.4 Radial Basis Network Functions

Radial basis network functions (RBF) is a feed-forward neural network with one hidden layer in which the nodes have radial transfer function ρ. The output of the network is given by

$$\varphi(\mathbf{x}) = \sum_{i=1}^{N} a_i \rho(||\mathbf{x} - \mathbf{c_i}||) \qquad (4)$$

or its normalized version:

$$\varphi(\mathbf{x}) = \frac{\sum_{i=1}^{N} a_i \rho(||\mathbf{x} - \mathbf{c_i}||)}{\sum_{i=1}^{N} \rho(||\mathbf{x} - \mathbf{c_i}||)} \qquad (5)$$

where $\rho(||\mathbf{x} - \mathbf{c_i}||)$ is usually in form of *gaussian*:

$$\rho(||\mathbf{x} - \mathbf{c_i}||) = \exp\left(\frac{||\mathbf{x} - \mathbf{c_i}||^2}{2\sigma_i^2}\right),$$

$\mathbf{c_i}$ is a center vector of the respective neuron, a_i is a weight of the neuron, and $||\mathbf{x} - \mathbf{c_i}||$ is a norm, typically the Euclidean norm. The model parameters are the *spread constant* σ_i^2 (SC), the *maximum of neurons* (MAX) that can be added to network during iterative learning process, and the *error goal* (EG) which is a mean-squared error on training set. The maximum neurons or the error goal are stopping criteria for the network induction.

2.5   Model Selection and Evaluation

The parameters for regression models were selected by 10-fold cross validation based on the *mean-squared error (MSE)*

$$\overline{err} = MSE = \frac{1}{N} \sum_{i=1}^{N} (Y_i - \bar{f}(\mathbf{X}))^2$$

The cross validation is suited for limited data samples, but it is also justified method for synthetic data.

3   Experiments with Synthetic Data

As we are interested primarily in the suitability of the considered regression methods for surrogate models in black-box optimization, we compared them on synthetic data generated using standard benchmarks for continuous black-box optimization [8].

All performed experiments were implemented in Matlab. For each function, we have sampled 5000 p-dimensional data points where $p \in \{5, 10, 20, 40\}$ and used it for a 10-fold cross-validation to compare the considered models. The result of cross validation is MSE for training set, MSE for testing set and the Kendall's rank correlation coefficient. The significance of the difference between results obtained for two models m, m' was tested using independent sample t-test

$$t = \frac{\overline{res}_m - \overline{res}_{m'}}{\sqrt{\frac{1}{k}(\sigma_m^2 + \sigma_{m'}^2)}} \tag{6}$$

which we compare for a significance level $\alpha \in (0, 1)$ against the $(1 - \frac{\alpha}{2})$-quantile of the Student distribution with $2(k\text{-}1)$ degrees of freedom, where k is the number of cross-validation folds, degrees of freedom, and $\overline{res}_m, \sigma_m$ are computed as follows:

$$\overline{res}_m = \frac{1}{k} \sum_{i=1}^{k} res_{m,i}$$

$$\sigma_m = \sqrt{\frac{1}{(k-1)} \sum_{i=1}^{k} (res_i - \overline{res})^2}$$

For a comparison of two models, it would have been better to use paired t-test, which provide better estimates, but since we have decided to use unpaired t-test at the beginning of our experiment, we haven't had necessary subresults to perform it.

We have used MSE together with *Kendall's rank correlation coefficient* [1] between the ordering of function values according to the model $(y_1, \ldots, y_n)$ and according to the function used to generate the data $(t_1, \ldots, t_n)$

$$\tau_m = \frac{(\# \text{ of concordant pairs}) - (\# \text{ of discordant pairs})}{\frac{1}{2}n(n-1)} \tag{7}$$

where for (t_j, y_j) and (t_k, y_k) different pairs of target value t and predicted value y, (t_j, y_j) and (t_k, y_k) are concordant if $t_j < t_k$ and $y_j < y_k$, or $t_j > t_k$ and $y_j > y_k$, and discordant otherwise.

3.1   Selection of Model Parameters

For each dataset, we have searched for optimal model parameters (in the case of a Gaussian process, these are its hyper-parameters) minimizing the MSE. With regression trees, we have considered different settings for the number of trees and minimum number of data points in leaves. With Gaussian processes, we have tried rational quadratic and squared exponential in their isomorphic form, and also the ARD version of squared exponential. With radial basis function networks, as a radial function, we have considered different settings of the parameters: spread constant, MSE goal, maximum of neurons. As to polynomial regression, we have used quadratic regression. See Table 3 for overview of selected parameters for each model.

3.2   Results

We will now present the results of our experiments. First, we have included a detailed Table 3 with measured values of the MSE and the Kendall's coefficient for each dataset and each model. We can see how optimal combinations of values of parameters for each model change with higher dimensions. Random forests have lower number of trees (NT) and higher minimum number of data in leaves (ML). The comparison of the performance of each method across different dimensions of the data sets follows.

Table 4 shows the results of our experiments where we have compared four different models across 40 different data sets. For each model, we entered the number of times the model was better than the other model and we also added how many times the result was significantly better on the significance level 0.05.

A summary of the results can be seen in Table 2 and additional comments on the results follow. With 10 dimensions, the radial basis functions started performing better, although not significantly. With 20 dimensions, there are even less methods that outperforms polynomial regression significantly according to MSE and random forests were the weakest model from the triple of models RBF, GP and RT. With 40 dimensions, there is a surprising result since MSE values are much lower comparing to lower dimensions and we would expect the MSE to be growing with higher dimensions. This may be an artifact of the function definitions which suppress higher dimensions and that may lower the MSE values.

In summary, when comparing the MSE over all dimensions, the Gaussian processes were the best model for our data followed by radial basis functions and random forests before polynomial regression at the last position. With Kendall's coefficient, the results are not that clear. Even though the Gaussian processes have the most wins, they do

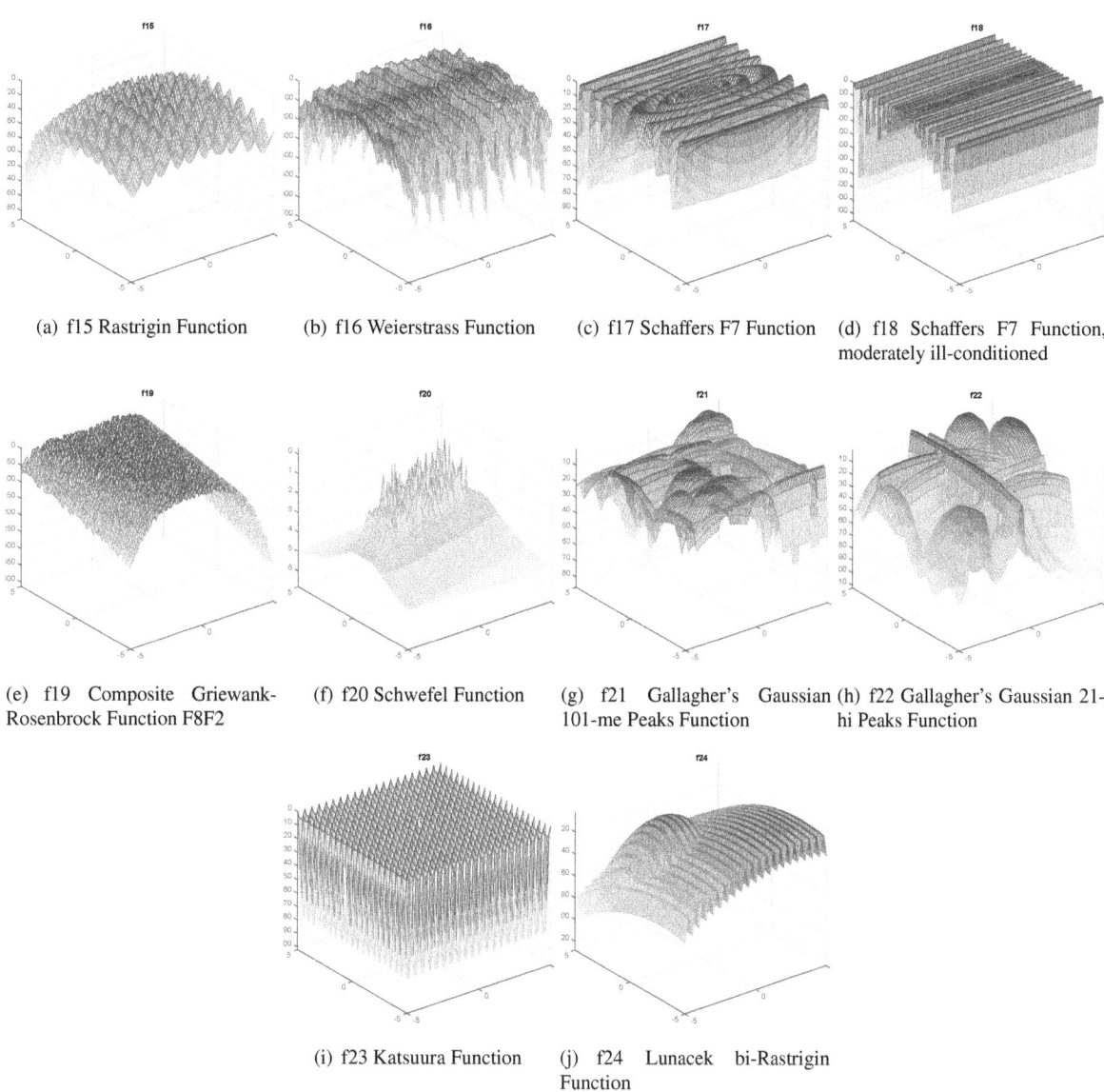

(a) f15 Rastrigin Function   (b) f16 Weierstrass Function   (c) f17 Schaffers F7 Function   (d) f18 Schaffers F7 Function, moderately ill-conditioned

(e) f19 Composite Griewank-Rosenbrock Function F8F2   (f) f20 Schwefel Function   (g) f21 Gallagher's Gaussian 101-me Peaks Function   (h) f22 Gallagher's Gaussian 21-hi Peaks Function

(i) f23 Katsuura Function   (j) f24 Lunacek bi-Rastrigin Function

Figure 1: Benchmark functions for continuous black-box optimization. The graphs has been created in accordance with [8]. Since we has been focused on multimodal functions, we chose functions 15 to 24.

Dimension	Mean squared error	Kendall's rank correlation coefficient
5	GP outperformed other models in most of the cases, all methods performs better most of the time comparing to polynomial regression	less conclusive results, the best performing model were RF
10	GP outperformed other models	less conclusive results, the best performing model were RF
20	GP outperformed other models	RBF was the weakest model outperformed even by PR
40	GP still outperformed others, but with smaller difference	GP outperformed others

Table 2: Results summary. For each dimension, we briefly describe the outcome.

Dataset	Random forests				Gaussian Processes				Polynomial		Radial basis functions				
	NT	ML	MSE	Kendall	CF	SN	MSE	Kendall	MSE	Kendall	SC	EG	MAX	MSE	Kendall
f15-05d	1000	1	**2.1±0.2e2**	**7.1±0.1e-1**	SEiso	1	2.6±0.1e2	6.6±0.2e-1	*3.1±0.2e2*	*6.4±0.2e-1*	8e1	1	500	2.5±0.1e2	6.7±0.2e-1
f15-10d	1600	1	*6.9±0.5e2*	*6.4±0.2e-1*	SEiso	1	5.8±0.5e2	**6.7±0.2e-1**	6.2±0.3e2	6.5±0.2e-1	5e1	1	250	**5.7±0.4e2**	6.6±0.1e-1
f15-20d	1200	1	*1.9±0.2e3*	*5.7±0.2e-1*	RQiso	5e-1	1.4±0.1e3	6.3±0.1e-1	1.31±0.07e3	**6.4±0.2e-1**	1e2	1	150	**1.28±0.06e3**	6.4±0.2e-1
f15-40d	800	2	*5.1±0.2e3*	*4.9±0.2e-1*	RQiso	5e-1	3±0.2e3	6.1±0.1e-1	3±0.2e3	6.1±0.2e-1	5e1	1	75	**2.7±0.2e3**	**6.3±0.1e-1**
f16-05d	1600	1	**1.1±0.1e3**	**5.3±0.1e-1**	SEiso	1	1.8±0.1e3	3.4±0.3e-1	*1.9±0.1e3*	*2.7±0.3e-1*	7.5e1	1	800	1.7±0.2e3	3.5±0.4e-1
f16-10d	800	1	**7.1±0.5e2**	**4.3±0.3e-1**	SEiso	1	*9.4±0.08e2*	*2±1e-1*	8.6±0.6e2	2.7±0.2e-1	1.8e1	1	40	9±1e2	2.5±0.3e-1
f16-20d	800	2	**4.1±0.4e2**	3.6±0.2e-1	SEiso	1e-2	*7.8±0.2e3*	*-4±4e-2*	4.6±0.3e2	2.3±0.2e-1	3.2e1	1	60	5±0.3e2	1.3±0.4e-1
f16-40d	400	5	**2.3±0.2e2**	2.9±0.4e-1	RQiso	5e-1	2.5±0.2e2	8±3e-2	*2.7±0.2e2*	*1.7±0.2e-1*	7.5	1	50	2.5±0.2e2	*7±2e-2*
f17-05d	1800	1	**2.3±0.3e1**	**4.6±0.2e-1**	SEiso	1.3	3.2±0.3e1	2.2±0.3e-1	3.3±0.2e1	*2.1±0.2e-1*	4.8	1	60	*3.3±0.2e1*	2.1±0.2e-1
f17-10d	800	1	**1.35±0.07e1**	**3.2±0.3e-1**	SEiso	1	*1.5±0.1e1*	*2.2±0.2e-1*	1.48±0.09e1	2.2±0.4e-1	4e1	1	50	1.4±0.1e1	2.4±0.3e-1
f17-20d	800	1	**6.8±0.4**	2.6±0.3e-1	SEiso	1e-2	6.5±0.2e1	*-1.3±0.3e-1*	6.9±0.5	2.2±0.3e-1	6.4e1	1	250	7.2±0.4	2±0.3e-1
f17-40d	400	1	3.4±0.2	2.2±0.3e-1	RQiso	5e-1	**3.2±0.2**	2±0.4e-1	*4.1±0.3*	*1.3±0.2e-1*	8e1	1e-3	40	3.3±0.2	**2.3±0.3e-1**
f18-05d	1200	1	**5.2±0.3e2**	**4.9±0.3e-1**	SEiso	2	8±1e2	2.4±0.2e-1	8.5±0.7e2	*2.3±0.3e-1*	5e1	1	500	*9.2±0.08e2*	2.3±0.2e-1
f18-10d	800	2	**2.7±0.2e2**	**3±0.2e-1**	SEiso	1	2.9±0.2e2	2.3±0.3e-1	*2.9±0.3e2*	*2.3±0.2e-1*	6e1	1	50	2.9±0.2e2	2.3±0.3e-1
f18-20d	1200	1	**1.26±0.06e2**	**2.4±0.4e-1**	SEiso	1e-2	*1.06±0.03e3*	*-1.3±0.3e-1*	1.41±0.05e2	2.2±0.2e-1	1e-3	1e-3	20	1.5±0.1e2	0±0
f18-40d	400	2	**6.4±0.5e1**	2.1±0.2e-1	RQiso	5e-1	6.5±0.4e1	1.9±0.2e-1	*7.6±0.6e1*	*1.5±0.3e-1*	8e1	1e-3	40	6.4±0.5e1	2±0.2e-1
f19-05d	800	50	5±0.3e2	5.2±0.2e-1	RQiso	5e-1	**4.4±0.3e2**	**5.4±0.2e-1**	4.8±0.3e2	5±0.2e-1	1e-2	1	500	*1.08±0.08e3*	*0±0*
f19-10d	2000	1	4.5±0.3e2	3.8±0.3e-1	RQiso	5e-1	**4.4±0.3e2**	3.9±0.3e-1	4.4±0.3e2	3.8±0.3e-1	1e-2	1	500	6.8±0.4e2	*0±0*
f19-20d	800	50	4.6±0.2e2	2.3±0.2e-1	RQiso	5e-1	**4.3±0.3e2**	2.7±0.2e-1	4.4±0.3e2	2.7±0.3e-1	1e-1	1	1000	5.9±0.3e2	*0±0*
f19-40d	800	50	4.3±0.3e2	1.6±0.3e-1	RQiso	5e-1	**4.2±0.1e2**	1.8±0.2e-1	4.5±0.2e2	1.3±0.4e-1	5	1	50	4.5±0.2e2	*-1.3±0.3e-1*
f20-05d	2000	1	1.2±0.8e-2	9.23±0.04e-1	SEiso	1	**1.1±0.8e-2**	**9.35±0.04e-1**	*7±3e-2*	8.5±0.06e-1	4.5	1e-3	1000	1.1±0.6e-2	9.13±0.06e-1
f20-10d	1800	1	8±1e-3	*8.3±0.1e-1*	SEiso	1e-1	**3.5±0.7e-3**	**9.18±0.03e-1**	*1.2±0.2e-2*	8.98±0.05e-1	8.6	1e-4	1150	3.8±0.5e-3	8.92±0.07e-1
f20-20d	1800	1	*8.6±0.6e-3*	*7±0.1e-1*	RQiso	5e-1	**1.5±0.2e-3**	9.01±0.04e-1	2.4±0.3e-3	**9.1±0.05e-1**	1.5e1	1e-4	800	1.8±0.2e-3	8.84±0.06e-1
f20-40d	800	5	*6.9±0.5e-3*	*5.99±0.07e-1*	RQiso	5e-1	**7.7±0.6e-4**	9.03±0.05e-1	8.3±0.7e-4	9.03±0.06e-1	8e1	1e-4	700	8.1±0.6e-4	9.02±0.06e-1
f21-05d	1600	1	1.4±0.2e2	5.7±0.3e-1	RQiso	5e-1	**6.7±0.6e1**	**6.9±0.2e-1**	2.5±0.3e2	*1.9±0.5e-1*	2	1	1800	9.6±0.09e1	6.1±0.2e-1
f21-10d	800	5	3±0.6e1	4.1±0.3e-1	SEiso	1	**2.4±0.4e1**	**5.1±0.2e-1**	*3.1±0.6e1*	*3.9±0.2e-1*	8	1	600	2.5±0.3e1	4.9±0.2e-1
f21-20d	800	10	3±0.4	3.5±0.3e-1	SEiso	1e-2	*7.187±0.009e3*	*-2.1±0.1e-1*	2.6±0.4	**4.6±0.3e-1**	8	1	800	**2.4±0.4**	4.4±0.3e-1
f21-40d	200	20	*2.5±0.9e-1*	2.9±0.3e-1	RQiso	5e-1	**1.8±0.7e-1**	5.3±0.3e-1	2±0.6e-1	4.8±0.3e-1	1.5e1	1e-3	400	2±0.7e-1	4.6±0.3e-1
f22-05d	1800	1	5.1±0.7e1	7.1±0.1e-1	SEiso	1	**3.5±0.5e1**	**7.4±0.2e-1**	1.3±0.2e2	*4.1±0.3e-1*	3	1	700	3.8±0.5e1	7±0.2e-1
f22-10d	800	2	9±0.3	5.3±0.3e-1	SEiso	1e-1	**8±3**	5.9±0.1e-1	*1.1±0.3e1*	*5.1±0.3e-1*	1e1	1	500	9±3	5.2±0.1e-1
f22-20d	800	10	7±1e-1	4.5±0.1e-1	SEiso	1e-2	*7.385±0.005e3*	*-2.1±0.2e-1*	6±1e1	**6±0.2e-1**	5.8	1e-2	1500	**6±1e-1**	5.9±0.1e-1
f22-40d	200	20	*5±1e-2*	3.5±0.2e-1	RQiso	5e-1	3±0.8e-2	6±0.2e-1	3.6±0.8e-2	5.5±0.2e-1	2e1	1e-3	400	3.7±0.9e-2	5.2±0.1e-1
f23-05d	1200	1	**6.7±0.3e1**	**3.7±0.2e-1**	SEiso	5e-1	8.5±0.5e1	0±0	8.4±0.4e1	*-1±3e-2*	1e-3	1e-3	200	8.4±0.5e1	0±0
f23-10d	1000	1	**4.1±0.2e1**	2±0.4e-1	SEiso	1.2	4.5±0.2e1	1±40e-2	4.4±0.1e1	1±30e-2	1.2	1e-3	100	4.4±0.3e1	*0±300*
f23-20d	200	2	**2.7±0.2e1**	1±0.3e-1	SEiso	1e-2	*2.63±0.07e2*	*-1±30e-2*	2.9±0.2e1	2±4e-2	1e-3	1e-3	5	2.7±0.2e1	0±0
f23-40d	800	10	1.49±0.09e1	6±2e-2	RQiso	5e-1	1.54±0.07e1	2±2e-2	*1.8±0.1e1*	3±3e-2	5	1	50	**1.5±0.1e1**	*1±30e-2*
f24-05d	1200	1	4.5±0.4e2	7.2±0.1e-1	SEiso	1.2	**2.7±0.2e2**	**7.67±0.1e-1**	4.1±0.4e2	7.2±0.1e-1	5.5e1	1e-3	400	2.8±0.2e2	7.6±0.1e-1
f24-10d	800	2	2.3±0.1e3	6.4±0.1e-1	SEiso	7.5e-1	9±5e2	7.8±0.4e-1	9±0.05e2	7.7±0.1e-1	1.2e1	1e-3	300	**7±0.5e2**	**7.94±0.09e-1**
f24-20d	800	2	8.9±0.7e3	5.9±0.2e-1	RQiso	5e-1	1.8±0.1e3	8±0.1e-1	1.8±0.1e3	8±0.1e-1	8e1	1	300	**1.8±0.2e3**	8±0.1e-1
f24-40d	400	5	*2.8±0.2e4*	5.2±0.3e-1	RQiso	1	4.6±0.3e3	7.9±0.1e-1	**4.5±0.4e3**	**7.9±0.1e-1**	8e1	1e-3	400	5.8±0.4e3	7.6±0.1e-1

Table 3: The values of MSE and Kendall's rank correlation coefficient for each dataset and that combination of model parameters used in each considered method that lead to the minimal test-data MSE for the performed crossvalidation of the respective dataset. The best results are shown in bold, the worst results in italics. *Random forests parameters: NT = number of trees, ML = minimum of data in leaves. Gaussian Processes parameters: CF = covariance function, SN = noise level (hyperparameter, additional hyperparameters for GP covariance functions were omitted). Radial basis functions parameters: SC = spread constant, EG = error goal, MAX = maximum neurons.*

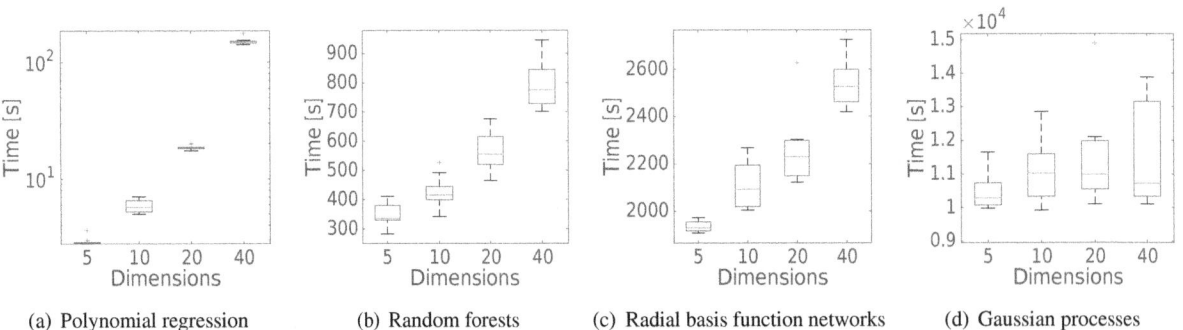

(a) Polynomial regression   (b) Random forests   (c) Radial basis function networks   (d) Gaussian processes

Figure 2: An overview how long does it take to compute 10-fold cross validation for each regression model and each of the considered dimensions of data. (a) Polynomial regression (quadratic) (b) Random forests (number of trees: 800, minimum number of data in leaves: 1) (c) Radial basis function networks (spread constant: 5.5, mean square error goal: 1, maximal number of neurons: 500) (d) Gaussian processes (covariance functions: rational quadratic (isomorphic), noise level: $sn = 0.5$, length-scale: $ell = log(1.1)$, signal variance $sf2 = log(1)$, shape parameter: $\alpha = log(1.1)$)

Dimension	Method	Mean squared error				Kendall's rank correlation coefficient			
		GP	RBF	RF	Polynom	GP	RBF	RF	Polynom
5	GP	×	9 (3)	5 (3)	10 (6)	×	8 (5)	4 (4)	10 (6)
	RBF	1 (0)	×	5 (3)	8 (6)	2 (0)	×	2 (2)	8 (6)
	RF	5 (5)	5 (5)	×	8 (8)	6 (5)	8 (6)	×	9 (8)
	Poly	0 (0)	2 (0)	2 (0)	×	0 (0)	2 (0)	1 (0)	×
10	GP	×	6 (2)	5 (4)	8 (4)	×	7 (3)	5 (5)	9 (6)
	RBF	4 (0)	×	6 (4)	9 (4)	3 (0)	×	5 (4)	10 (3)
	RF	5 (2)	4 (3)	×	8 (5)	5 (4)	5 (4)	×	7 (7)
	Poly	2 (0)	1 (0)	2 (2)	×	1 (0)	0 (0)	3 (2)	×
20	GP	×	6 (2)	6 (5)	6 (3)	×	8 (5)	5 (5)	5 (2)
	RBF	4 (1)	×	6 (4)	7 (1)	2 (1)	×	5 (5)	2 (0)
	RF	4 (3)	4 (3)	×	5 (3)	5 (3)	5 (4)	×	5 (4)
	Poly	4 (1)	3 (2)	5 (3)	×	5 (2)	8 (3)	5 (5)	×
40	GP	×	7 (1)	7 (5)	8 (6)	×	5 (3)	5 (5)	6 (4)
	RBF	3 (1)	×	7 (3)	8 (6)	5 (3)	×	6 (5)	4 (3)
	RF	3 (0)	3 (1)	×	5 (5)	5 (4)	4 (2)	×	5 (5)
	Poly	2 (0)	2 (1)	5 (3)	×	4 (1)	6 (2)	5 (5)	×
Summary	GP	×	28 (8)	23 (17)	32 (19)	×	28 (16)	19 (19)	30 (17)
	RBF	12 (2)	×	24 (14)	32 (17)	12 (4)	×	18 (16)	24 (12)
	RF	17 (10)	16 (12)	×	26 (21)	21 (16)	22 (16)	×	26 (24)
	Poly	8 (1)	8 (3)	14 (8)	×	10 (3)	16 (5)	14 (12)	×

Table 4: Results of experiments comparing 4 different models across 40 different data sets. For each model (in a row), we entered the number of times the model was better than the other model (in a column) and we also added how many times the result was significantly better on the significance level 0.05 (in the brackets).

not have the most significant wins. Based on the significant wins, the best performing model were random forests.

With higher dimensions, when comparing the models based on the MSE, we may notice that the results for Gaussian processes and random forests are less significant. Which is also the case with Kendall's coefficient, where the polynomial regression gets more wins with higher dimension.

Now we have a look at how long does it take to evaluate 10-fold cross validation for selected parameters settings for each model (see Figure 2). With higher dimensions, each method takes more time to evaluate. All the computations were performed on PC (x86-64) Intel Core i7 920 (4x 2.66 GHz + HyperThreading), 6 GB RAM.

4 Discussion and Conclusion

The figures and tables presented in Results compared four different regression methods over 40 synthetic data sets (10 functions × 4 different dimensions) generated using standard benchmark functions for continuous black-box optimization. We have shown how the performance of these methods changes with increasing dimensionality and how the time to cross-validate the models grows. We have compared the methods based on the MSE and on Kendall's coefficient. We will now comment on each of them.

Gaussian process is probably the most complex method. With its time complexity $O(N^3)$ it takes the longest time to compute, some of the cross-validations, i.e. 10 constructions of the model, took up to 24 hours. This model was

better then the others according to both MSE and Kendall's coefficient comparison.

Random forests ended up with poorer results for 40 dimensional data and overall they were slightly behind Gaussian Processes based on the MSE. According to Kendall's coefficient results, they were comparable with Gaussian processes and, according to the number of significant wins, they even outperformed GP. With some data sets (f19-10d, f20-05d), we have learnt 2000 trees out of 4500 samples. In these cases, we could have compare the results with nearest neighbor method.

Radial basis functions network has the clearly poorer results compared to Gaussian processes and random forests according to both the MSE and the Kendall's coefficient.

Even though the *polynomial regression* was included due to its importance as traditional response surface model, the method was not always worse then all other methods. For the dimensions 20 and 40, their MSE was comparable to that of random forest. Also with higher dimensions, the results based on the Kendall's coefficient are comparable to both GP and RBF and it even outperformed RBF.

In this paper, we have compared a selection of non-linear methods on synthetic data sets based on their mean-squared error and on the Kendall's rank correlation coefficient. We have chosen regression methods that have been used as surrogate models in such optimization: radial basis function networks, Gaussian processes, random forests, and polynomial regression. A better accuracy of the models suggests better applicability of the models as a surrogate model for optimization. From the results we have learnt that Gaussian processes had better results in most cases, thus, would be better surrogate model compared to the others, although random forests were only slightly behind.

Acknowledgements

This research was partially supported by SVV project number 260 224.

References

[1] Springer Encyclopedia of Mathematics

[2] Alessandri, A., Cassettari, L.,Mosca. R.: Nonparametric nonlinear regression using polynomial and neural approximators: a numerical comparison. Computational Management Science **6(1)** (2009), 5–24

[3] Bajer, L., Holeňa, M.: Surrogate model for continuous and discrete genetic optimization based on rbf networks. In: Colin Fyfe, Peter Tino, Darryl Charles, Cesar Garcia-Osorio, and Hujun Yin, (eds), Intelligent Data Engineering and Automated Learning – IDEAL 2010, volume 6283 of Lecture Notes in Computer Science, 251–258, Springer Berlin Heidelberg, 2010

[4] Bajer, L., Pitra, Z., Holeňa, M.: Benchmarking gaussian processes and random forests surrogate models on the bbob noiseless testbed. In: GECCO 2015, 2015

[5] Breiman, L.: Random forests. Mach. Learn. **45(1)** (October 2001), 5–32

[6] Buche, D., Schraudolph, N. N., Koumoutsakos, P.: Accelerating evolutionary algorithms with gaussian process fitness function models. Systems, Man, and Cybernetics, Part C: Applications and Reviews, IEEE Transactions on **35(2)** (May 2005), 183–194

[7] Gano, S. E., Kim, H., Brown, D. E.: Comparison of three surrogate modeling techniques: datascape, kriging and second order regression. In: 11th AIAA/ISSMO Multidisciplinary Analysis and Optimization Conference, 2006

[8] Hansen, N., Finck, S., Ros, R., Auger, A.: Real-parameter black-box optimization benchmarking 2009: noiseless functions definitions. Research Report RR-6829, 2009

[9] Hastie, T., Tibshirani, R., Friedman, J.: The elements of statistical learning. Springer Series in Statistics, Springer New York Inc., New York, NY, USA, 2001

[10] Hultquist, C., Chen, G., Zhao, K.: A comparison of gaussian process regression, random forests and support vector regression for burn severity assessment in diseased forests. Remote Sensing Letters **5(8)** (2014), 723–732

[11] Kleijnen, J. P. C., van Beers, W., van Nieuwenhuyse, I.: Constrained optimization in expensive simulation: novel approach. European Journal of Operational Research **202(1)** (2010), 164 – 174

[12] Louppe, G.: Understanding random forests: from theory to practice. PhD thesis, 2014

[13] Luo, J., Lu, W.: Comparison of surrogate models with different methods in groundwater remediation process. Journal of Earth System Science **123(7)** (2014), 1579–1589

[14] Myers, R. H., Montgomery, D. C., Anderson-Cook, C. M.: Response surface methodology: proces and product optimization using designed experiments, 2009

[15] Rasmussen, C. E.: Evaluation of gaussian processes and other methods for non-linear regression. Technical Report, 1996

[16] Razi, M. A., Athappilly, K.: A comparative predictive analysis of neural networks (nns), nonlinear regression and classification and regression tree (cart) models. Expert Systems with Applications **29(1)** (2005), 65–74

[17] Roy, A., Govil, S. Miranda, R.: A neural-network learning theory and a polynomial time rbf algorithm. Neural Networks, IEEE Transactions on **8(6)** (Nov. 1997), 1301–1313

[18] Zhou, Z., Ong, Y. S., Nair, P. B., Keane, A. J., Lum, K. Y.: Combining global and local surrogate models to accelerate evolutionary optimization. Systems, Man, and Cybernetics, Part C: Applications and Reviews, IEEE Transactions on **37(1)** (Jan. 2007), 66–76

J. Yaghob (Ed.): ITAT 2015 pp. 143–149
Charles University in Prague, Prague, 2015

Evaluation of Association Rules Extracted during Anomaly Explanation

Martin Kopp[1,2,3] Martin Holeňa[1,3]

[1] Faculty of Information Technology, Czech Technical University in Prague
Thákurova 9, 160 00 Prague
[2] Cisco Systems, Cognitive Research Team in Prague
[3] Institute of Computer Science, Academy of Sciences of the Czech Republic
Pod Vodárenskou věží 2, 182 07 Prague

Abstract: Discovering anomalies within data is nowadays very important, because it helps to uncover interesting events. Consequently, a considerable amount of anomaly detection algorithms was proposed in the last few years. Only a few papers about anomaly detection at least mentioned why some samples were labelled as anomalous. Therefore, we proposed a method allowing to extract rules explaining the anomaly from an ensemble of specifically trained decision trees, called sapling random forest.

Our method is able to interpret the output of an arbitrary anomaly detector. The explanation is given as conjunctions of atomic conditions, which can be viewed as antecedents of association rules. In this work we focus on selection, post processing and evaluation of those rules. The main goal is to present a small number of the most important rules. To achieve this, we use quality measures such as lift and confidence boost. The resulting sets of rules are experimentally and empirically evaluated on two artificial datasets and one real-world dataset.

Keywords: Anomaly detection, anomaly interpretation, association rules, confidence boost, random forest

1 Introduction

According to an IBM research [13] there were 2,7 zettabytes of data in the digital universe at April 2012 and this amount is doubling approximately every 40 months.

Not only it is almost impossible to process such huge amounts of data, we are actually not interested in the raw data, but rather in the salient knowledge and interesting patterns contained in them. This is the reason why anomaly detection, especially unsupervised anomaly detection, becomes more and more important [1, 23]. Despite it can be formalised as a binary classification, it entails different issues and challenges than those in supervised classification. For example, anomalous events often adapt to appear normally and even normal behaviour evolve over time. Furthermore, defining a normal regions is very difficult, especially when the boundary between normal and anomalous is not always precise.

For the purposes of this paper consider anomalies equal to outliers as defined by Hawkins [11]: "*An outlier is an observation which deviates so much from the other observations as to arouse suspicions that it was generated by a different mechanism.*"

The more formal definition would necessarily reduce the amount of plausible anomaly detectors and/or application domains. This is in conflict with our goal to provide a solution as general as possible.

Even though anomaly detection techniques are aimed at only a minority of samples, the importance and demand for them grows rapidly. The real world applications range from the network security [10], bioinformatics [24] or financial fraud detection [22] to the astronomy and space exploration [9].

The identification of anomalies is only a half of the whole task. The second and equally important half is the interpretation. In high dimensional domains, like the network security or bioinformatics, where hundreds or even thousands of features are common, the proper interpretation is crucial. Therefore, anomalies have to be interpreted clearly, as a feature subset that explains its deviation from ordinary data, or even better as a set of association rules.

In [21] we proposed method of anomaly explanation based upon specifically trained ensembles of decision trees called sapling random forest (SRF). The main idea behind it is to view the explanation as a feature selection and classification problem. Specifically, the goal is to find features in which the margin between anomalous sample and the normal samples is maximised. Therefore, SRF returns subset of features, respectively rules on these features describing why this sample has been identified as an anomaly.

The main drawback of the direct rule extraction from our sapling random forests is the big number of rules with some of them introduced by unfortunate training set selection. Partially, these issues can be solved by confidence and / or support thresholds. But for our ultimate goal to present the minimal number of rules containing the maximal amount of useful information, such a simple approach is insufficient. Therefore, in this paper we focus on proper selection, post processing and evaluation of rulesets extracted from sapling random forests during anomaly explanation. We tested association rules quality measures such as lift and confidence boost. This paper is work in progress and we would like to extend the number of tested quality measures by some subjective measures like novelty.

The rest of this paper is organised as follows. The next section briefly reviews related work. Section 3 describes the SRF principles and its training followed by the rule extraction process in Section 4. The selected quality measures of association rules are presented in Section 5. Experimental evaluation is described in Section 6 and Section 7 concludes the paper.

2   Related Work

For more information about the anomaly detection we refer to the recent book of Aggarwall [1]. This book provides an exhaustive listing of anomaly detection algorithms and their applications in different domains. Another source may be [5], which is briefer but very well written. To our best knowledge, there have been only few works addressing not only identification of anomalies, but also their explanation.

Knorr et al. [15] focused on what kind of knowledge should be extracted and provided to the user. Strong and weak outliers were defined and searched within data by distance-based algorithms described in detail in [14].

Dang et al. [7] presented an algorithm identifying and explaining anomalies. The algorithm starts by selecting a set of neighbouring samples based on quadratic entropy that are presented to a Fisher linear discriminant classifier to seek for an optimal half-space, in which a detected anomaly is well separated. The process of interpretation is entangled with the presented method of identification of anomalies. The difference to our work is that SRF can be used after an arbitrary anomaly detection algorithm to interpret its results.

The most similar to our approach and most recent is [20]. Their approach, as well as ours, can interpret output of an arbitrary anomaly detector as a subset of features. They use classification accuracy for outlier ranking. The main drawback of this approach is that it needs balanced training sets which are created by sampling artificial samples around the anomalous point. With respect to this work, our approach can handle unbalanced training sets easily and returns not only feature subsets but feature subsets with rules on them, providing even more information about the anomaly. Furthermore, we simplify the analysis by clustering, which enables to interpret similar anomalies at once [16].

On the other hand, there are many papers about association rules. This paper was inspired mainly by [3], which is about measuring redundancy and information quality of sets of association rules. The author presents a measure called confidence boost and an algorithm to produce a small set of association rules using this measure. A really extensive list of interestingness measures can be found in [12]. There is a lot of inspiration for our future work.

An alternative approach, well described in [6], may be so called subjective measures. A typical example is the novelty, sometimes called unexpectedness, of a rule with respect to user provided domain knowledge or against the another rule set. Because these terms are ambiguous there are multiple approaches of measuring them. An approach in [18] inspired us for our future work.

3   Sapling Random Forest

This section outlines principles of sapling random forests. SRF is a method able to explain an output of an arbitrary anomaly detector, proposed by us in [21]. It is a random forest of specifically trained decision trees. Because produced trees are small they are called saplings rather than trees. Produced explanations show features in which inspected samples differ the most from the rest of data. These features are used to produce association rules, which are more informative than only a set of features. An outline of the whole method is at Algorithm 1.

Algorithm 1 Algorithm summary
___
$y \leftarrow anomalyDetector(data)$
for all data($y ==$anomaly) **do**
   $T \leftarrow createTrainingSet(size, method)$
   $t \leftarrow trainTree(T)$
   $SRF \leftarrow t$
end for
$extractRules(SRF)$

3.1   Training Set Selection

Dataset $\mathcal{X} = \{x_1, x_2, \ldots, x_l\}$, where $x \in \mathbb{R}^d$, can be split into two disjoint sets $\mathcal{X}^a$, containing anomalous samples, and $\mathcal{X}^n$, containing normal samples. Then, a training set $\mathcal{T}$ contains the anomaly x^a as one class and a subset of $\mathcal{X}^n$ as the other. The first strategy of creating training sets is to select k nearest neighbours of x^a from $\mathcal{X}^n$. This strategy is sensible for algorithms detecting local anomalies, as according to [8] they are more general than algorithms detecting global anomalies. The drawback of this strategy is a computational complexity.

The second strategy is to select k samples randomly from $\mathcal{X}^n$ with uniform probability. The advantage of this approach is a possibility to generate more than one training set per anomaly by repeating the sampling process. More training sets lead to more saplings per anomaly and to more robust explanation, but at the expense of the more complicated aggregation of rules extracted from them (see Section 4). A comparison of both approaches can be found in [21].

3.2   Training a Sapling

For simplicity consider sapling a binary decision tree with typical height 1-3. In the SRF method, there are always two leaves at the maximal depth, one of which contains only an anomaly x^a and the other containing only normal samples. The saplings small height has two reasons. First, training sets are relatively small. Second, according to the anomaly isolation approach [19], if the analysed sample is an anomaly, it should be separated easily from the rest of data, resulting again into small trees. Therefore, if the height of a sapling is higher than expected it should be taken into consideration that the explained sample may not be an anomaly.

The standard procedure, to find the splitting function h for a new internal node, is maximising an information gain over the space of all possible splitting functions $\mathcal{H}$ as

$$\arg\max_{h\in\mathcal{H}} - \sum_{b\in\{L,R\}} \frac{|\mathcal{S}^b(h)|}{|\mathcal{S}|} H(\mathcal{S}^b), \qquad (1)$$

where $\mathcal{S}$ is the subset of the training set $\mathcal{T}$ reaching the leaf being split, $\mathcal{S}^L(h) = \{x\in\mathcal{S}|h(x) = +1\}$ and $\mathcal{S}^R(h) = \{x\in\mathcal{S}|h(x) = -1\}$ and $H(\mathcal{S})$ is an entropy of $\mathcal{S}$.

The second commonly used approach involves minimising the Gini impurity.

$$\arg\min_{h\in\mathcal{H}} \sum_{b\in\{L,R\}} 1 - \left(\frac{|x^a|}{(\mathcal{S}^b(h))}\right)^2 - \left(\frac{|x^n|}{(\mathcal{S}^b(h))}\right)^2 \qquad (2)$$

For experiments presented in this paper we used information gain.

4 Extraction and Evaluation of Rules

Once a sapling is grown, it is used to explain the anomaly x^a. Let $h_{j_1,\theta_1},\ldots,h_{j_d,\theta_d}$ be the set of splitting functions, with features $j_1,\ldots,j_d$ and threshold $\theta_1,\ldots,\theta_d$, used in the inner nodes on the path from the root to the leaf with the anomaly x^a. Then x^a is explained as a conjunction of atomic conditions :

$$c = (x_{j_1} > \theta_1) \wedge (x_{j_2} > \theta_2) \wedge \ldots \wedge (x_{j_d} > \theta_d), \qquad (3)$$

which is the output of the algorithm. This conjunction can be read as "the sample is anomalous because it is greater than threshold θ_1 in feature j_1 and greater than θ_2 in feature j_2 and ... than majority (or nearest neighbour) samples." Because resulting trees are very small, the explanation is compact.

The situation is more difficult, when more saplings per anomaly have been grown, as each sapling provides one conjunction of type (3). Using more than one sapling per anomaly improves robustness for training sets created by uniform sampling. The problem is that returning set of all conjunctions $\mathcal{C}$ is undesirable, as the primary objective — explanation of the anomaly to a human — would not be met. Hence, the algorithm needs to aggregate conjunctions in $\mathcal{C}$.

For simplicity of the following notation consider $2d$ items, in such a way that 2 items are assigned to each feature, one for "<" rules, the other for ">" rules. Denote this $2d$ set of items $\mathcal{F}$. Then we can group rules into the rule sets $\mathcal{R}_f$ according to the item set $f \subseteq \mathcal{F}$ they share.

Based on $|\mathcal{R}_f|$ the algorithm discards groups of low importance by sorting them in descend order, and then using only the first k groups such, that their cumulative frequency is greater than a threshold τ, which we recommend

to be 0.90 or 0.95. Using the adopted notation, k is determined as

$$k = \arg\min_k \frac{1}{\sum_{f\in\mathcal{F}}|\mathcal{R}_f|} \sum_{i=1}^{k} |\mathcal{R}_{f_i}| > \tau, \qquad (4)$$

where it is assumed, that $\mathcal{R}_f$ are sorted according to their size to simplify the notation. We have also investigated the complementary approach, where groups are selected, if they were used with a frequency higher than a specified threshold. But the presented strategy based on the cumulative frequency showed more consistent results in our experiments.

Once the set of groups with decision rules is selected, we create one rule $\bar{r}_f$ for every rule set $\mathcal{R}_f$. Thresholds for each item f_j are calculated as an average of all thresholds within the rule set $\mathcal{R}_f$.

$$\bar{\theta}_j = \frac{1}{|\mathcal{R}_f|} \sum_{i=1}^{|\mathcal{R}_{f_i}|} \theta_{i,j} \qquad (5)$$

By this approach we obtain one representative rule for each feature set f as:

$$\bar{c}_f = x_{j_1} > \bar{\theta}_1 \wedge x_{j_2} > \bar{\theta}_2 \wedge \ldots \wedge x_{j_t} > \bar{\theta}_t. \qquad (6)$$

5 Measuring Quality of Rules

This section reviews selected quality measures of association rules. Typical association rules are in the form $\mathcal{A} \to \mathcal{Y}$, where $\mathcal{A}, \mathcal{Y}$ are item sets. In rules extracted from SRF, items are atomic conditions and $\mathcal{Y}$ always means: "is anomalous". Therefore, our rules are in the form:

$$r_f = c_f \to y, \qquad (7)$$

where c is a conjunction of atomic conditions like (3), $y = x \in \mathcal{X}^a$ and $f \subseteq \mathcal{F}$. The r_f in its full form then look as:

$$r_f(x) = x_{f_1} > \theta_{f_1} \wedge x_{f_2} > \theta_{f_2} \wedge \ldots \wedge x_{f_n} > \theta_{f_n} \to x \in \mathcal{X}^a, \qquad (8)$$

where n is a maximal index in the itemset f.

For this kind of rules support [2] is calculated as:

$$\text{supp}(c_f) = \frac{|\{c_f(x)|x \in \mathcal{X}\}|}{|\mathcal{X}|}, \qquad (9)$$

$$\text{supp}(y) = \frac{|\mathcal{X}^a|}{|\mathcal{X}|}. \qquad (10)$$

and gives the proportion of data points which satisfy the antecedent c, respectively the consequent y. It is used to measure the importance of a rule or as a frequency constrain. The disadvantage of support is that infrequent rules are often discarded. This is much bigger problem than it could seem because we are generating rules for anomalies, which are rare by definition.

Another frequently used measure is confidence [2]:

$$\text{conf}(c_f \to y) = \frac{\text{supp}(c_f \to y)}{\text{supp}(c_f)}. \quad (11)$$

It estimates the conditional probability of the consequent being true on condition that the antecendent is true. The trouble with confidence is caused by its sensitivity to the frequency of y. Because all rules extracted from SRF have the same consequent the rule ranking produced by lift a confidence would be the same.

The third measure we used is lift [4]:

$$\text{lift}(c_f \to y) = \frac{\text{conf}(c_f \to y)}{\text{supp}(y)}, \quad (12)$$

which measures how many times more often the antecedent c and consequent y occur together than expected if they were statistically independent. Lift does not suffer from the rare items problem. Because in our experiments the consequent will always the same frequency there is no need to measure both

Finally, confidence boost introduced by Balcázar [3] is calculated as:

$$\beta(r_f) = \frac{\text{conf}(r_f)}{\max\{\text{conf}(r'_{f'})|\text{supp}(r'_{f'}) > \sigma, r_f \not\equiv r'_{f'}, f' \subseteq f\}}, \quad (13)$$

where σ is support threshold and $r_f \not\equiv r'_{f'}$ denotes the inequivalence of rules r_f and $r'_{f'}$, which for our simple case where all consequents are the same, means that $f \neq f'$. From (13) is evident that $f' \subset f$.

If the set of confidences in denominator is empty, the confidence boost is by convention set to infinity.

6 Experiments

For the experimental evaluation we used the synthetical three layer donut, the well known Fisher's iris and the Letter recognition data set from the UCI repository [17].

6.1 Three Layer Donut

The three layer donut dataset contains 1000 normal samples forming the two dimensional toroid (donut). There are 200 anomalies, one half inside the toroid and the second half out of it. For this dataset we created 10 rules per anomaly, using SRF, resulting in 2000 rules. After the simple aggregation, described in Section 4, only 8 rules left. All of them printed in Table 1, sorted by their respective support. All rules before aggregation were used to calculate confidence boost. Otherwise, many rules would have confidence boost equal to infinity because there are no rules defined on any subset of their item sets f.

The fact is that rules $r_5 - r_8$ have quite small support but, according to the other measures and our intuition, they are very important. The small supports is due to the small

rule	supp	lift	β
$r_1 = x_1 > -0.33 \wedge x_2 > -0.39$	0.38	1.64	0.27
$r_2 = x_1 > -0.33 \wedge x_2 < 0.3$	0.37	1.60	0.27
$r_3 = x_1 < 0.4 \wedge x_2 < 0.34$	0.37	1.49	0.25
$r_4 = x_1 < 0.37 \wedge x_2 > -0.37$	0.37	1.57	0.26
$r_5 = x_2 > 2.2$	0.02	6.00	1.00
$r_6 = x_1 > 2.3$	0.02	6.00	1.00
$r_7 = x_2 < -2.4$	0.01	6.00	1.00
$r_8 = x_1 < -2.4$	0.01	6.00	1.00

Table 1: Aggregated rules with their quality measures for the three layer donut dataset, sorted by their respective supports.

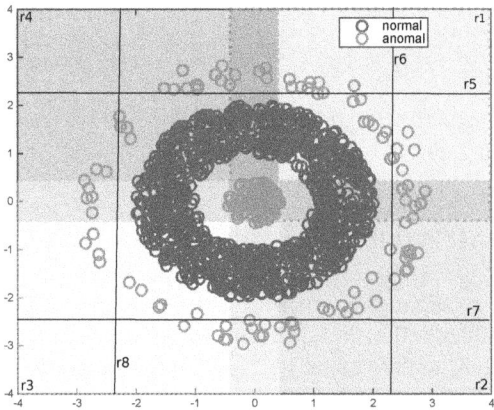

Figure 1: Three layer donut dataset with plotted rules $r_1 - r_8$. Rules $r_1 - r_4$ are plotted as filled squares and rules $r_5 - r_8$ as half-planes delimited by solid lines.

number of data points explained, lets recall that anomalies are only one sixth of all data points in this dataset. Both, lift and confidence boost, mostly reflects our subjective expectations.

All rules are depicted at Figure 1. Its evident that presented rules cannot separate anomalies from normal samples perfectly. Especially difficult are anomalies inside the donut. To separate those inner anomalies perfectly it would be necessary to combine more rules together, for example r_1 and r_3 or r_2 and r_4.

6.2 Iris

The virginica species were selected as anomalous class for the iris data set. Five rules per anomaly were produced using SRF, resulting in 250 rules. After aggregation we have got 6 rules. They are written in Table 2 with their respective quality measures. Confidence boost was calculated using all 250 rules.

The main problem with all those rules is that almost every one of them can sufficiently separate anomalies from

rule	supp	lift	β
$r_1 = x_1 > 6 \wedge x_4 > 1.7$	0.25	3.00	1.00
$r_2 = x4 > 2$	0.19	3.00	1.00
$r_3 = x_3 > 5.5$	0.17	3.00	1.00
$r_4 = x_2 < 2.8 \wedge x_4 > 1.6$	0.06	3.00	1.00
$r_5 = x_1 > 7.3$	0.05	3.00	1.00
$r_6 = x_2 < 2.2$	0.03	0.75	0.25

Table 2: Aggregated rules with their quality measures for the iris dataset , sorted by their respective supports.

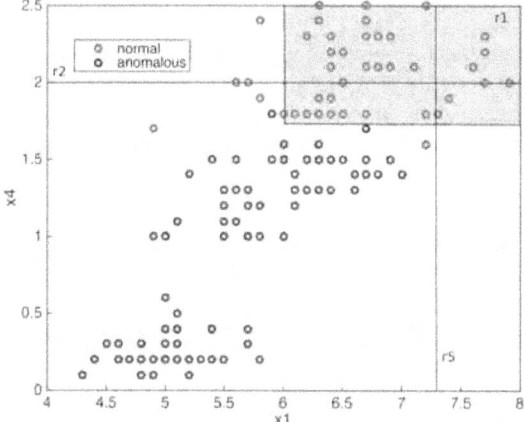

Figure 2: The iris dataset with r_1 plotted as a filled rectangle, and rules r_2 and r_5 as half-planes delimited by solid lines.

normal samples. No one of presented measures could help in selecting the most informative, yet small as possible, set of rules. Because presented rules are seen informationally equivalent by all quality measures. This doesn't say much about difference between the quality measures but it justifies the rule extraction process, because all generated rules have high score.

Figure 2 shows the iris dataset with rules r_1, r_2 and r_5.

6.3 Letter Recognition

This dataset was created as a classification problem with 26 classes, one class for each letter in the English alphabet. The charactersÒ were obtained from 20 different fonts and randomly distorted to produce 20,000 unique samples presented as 16 dimensional numerical vectors. Letter X was selected as the anomaly class. SRF produced more than 15,000 rules, which were reduced by aggregation to 1955. Aggregated rules with support higher than 0.10 are presented in Table 3. The ranking of those rules is plotted at Figure 3. Its evident that the ranking given by lift and confidence boost differs substantially.

It is nearly impossible to evaluate all rules. Therefore, we have selected only those with confidence boost higher than one (202 rules) and those with lift higher than one

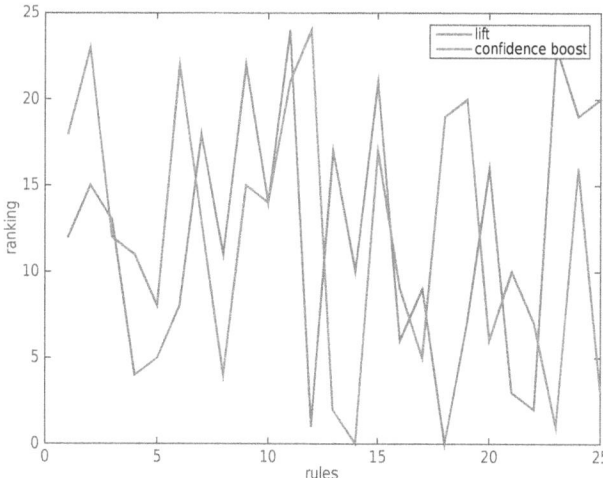

Figure 3: Ranking (higher number means better ranking) of rules from Table 3 by its lift and confidence boost.

rule	supp	lift	β
$x_2 > 14$	0.29	1.44	0.29
$x_{11} < 0.62$	0.24	0.27	1.82
$x_9 > 8.8 \wedge x_1 0 < 5.2$	0.21	1.67	0.96
$x_3 > 10$	0.18	0.89	0.35
$x_1 < 0.38$	0.17	0.19	0.14
$x_{13} > 6.2 \wedge x_{15} > 8.6$	0.16	0.21	0.17
$x_6 > 9.8 \wedge x_8 < 1.2$	0.15	2.39	1.91
$x_1 > 9.5 \wedge x_2 > 14$	0.15	1.10	0.22
$x_2 > 13 \wedge x_{10} > 12$	0.15	0.44	0.04
$x_4 > 7.8 \wedge x_8 < 1.2 \wedge x_9 > 7.2$	0.15	5.75	0.71
$x_8 < 1.2 \wedge x_{16} < 5.2$	0.14	1.62	0.39
$x_2 > 11 \wedge x_4 > 7.8 \wedge x_9 > 6.4$	0.13	5.32	0.88
$x_8 < 1.2 \wedge x_9 > 8.8$	0.13	8.06	1.80
$x_9 > 8 \wedge x_{10} < 5.2 \wedge x_{15} > 6.8$	0.13	2.20	1.31
$x_1 > 9.5 \wedge x_3 > 9$	0.12	1.04	0.06
$x_9 > 8.8 \wedge x_{12} < 5.5$	0.12	0.78	0.22
$x_3 > 6.1 \wedge x_{14} < 6.2 \wedge x_{15} > 7.2$	0.12	2.91	1.09
$x_1 > 4.2 \wedge x_8 < 1.2 \wedge x_{12} < 6.8$	0.12	0.55	0.05
$x_1 > 8.5 \wedge x_{10} > 12$	0.12	0.84	0.10
$x_1 > 5.5 \wedge x_7 > 7.8 \wedge x_8 < 1.2$	0.11	3.39	0.22
$x_2 < 1.5 \wedge x_9 > 7.5 \wedge x_{15} > 5.5$	0.11	3.43	0.84
$x_1 > 7.8 \wedge x_6 > 8.8 \wedge x_7 > 6$	0.11	0.29	0.02
$x_6 > 8.2 \wedge x_{15} > 7.9 \wedge x_{16} < 6.2$	0.11	0.29	0.24
$x_2 > 11 \wedge x_3 > 7.6 \wedge x_{10} < 5.2$	0.11	1.21	0.05
$x_3 > 10 \wedge x_5 > 6.2$	0.11	0.61	0.24

Table 3: Rules extracted from Letter recognition by SRF with support higher than 0.10 with their quality measures sorted by their respective supports.

(446 rules). The confidence boost selected the smaller rule set where almost all rules lookcd plausible. On the other hand, they missed some really interesting ones most highly rated by lift. The confidence boost tend to choose shorter more similar rules, whereas lift prefer richer and more heterogenous rules. Therefore, from our point of view the

top 10 lift rules	top 10 β rules
$x_8 < 1.8 \wedge x_{15} > 7.8$	$x_4 < 1.2 \wedge x_9 > 8$
$x_2 < 1.8 \wedge x_3 > 4.5 \wedge x_9 > 8.8$	$x_2 < 1.5 \wedge x_9 > 8.8$
$x_5 > 6.8 \wedge x_8 < 1.8 \wedge x_{15} > 7$	$x_8 < 1.2 \wedge x_9 > 8.8$
$x_5 > 5 \wedge x_9 > 8.2 \wedge x_{10} < 5.2$	$x_9 > 8 \wedge x_{14} < 5.2$
$x_4 < 2.2 \wedge x_6 > 8.2 \wedge x_9 > 7.8$	$x_{11} > 12 \wedge x_{16} < 4.5$
$x_2 < 5.6 \wedge x_3 > 7.2 \wedge x_8 < 1.2$	$x_6 > 9.8 \wedge x_8 < 1.2$
$x_1 > 4.5 \wedge x_8 < 1.8 \wedge x_{15} > 7.8$	$x_4 > 8.8 \wedge x_{14} < 5.2$
$x_7 < 5.8 \wedge x_8 < 2.5 \wedge x_{15} > 7.8$	$x_5 > 8.5 \wedge x_{14} < 5.2$
$x_2 < 4.8 \wedge x_9 > 8 \wedge x_{11} > 9.8$	$x_8 < 1.2 \wedge x_{16} > 9.9$
$x_{11} > 9 \wedge x_{14} > 13$	$x_{12} > 10 \wedge x_{16} < 5.2$

Table 4: Comparison of top 10 rules extracted from the Letter recognition dataset by SRF selected by lift and confidence boost.

best selection strategy is choosing top k rules according to the lift ranking. The top 10 rules chosen from the whole set by lift and confidence boost, regardless their support, are in Table 4.

Still there are too much rules to make some conclusions, in our future work we are going to investigate more measures of interestingness and novelty, which will hopefully help us to reduce the amount of extracted rules even more.

7 Conclusion

In this paper, we presented a novel approach for the explanation of an output of an arbitrary anomaly detector using sapling random forests. The explanation is given as conjunctions of atomic conditions, which can be viewed as antecedents of association rules. Due to an extraction method, the individual rules are short and comprehensible. The main drawback was that the rule sets for the bigger dataset were large and redundant. Therefore, we applied multiple quality measures to evaluate them and select those rules with desired properties. Performed experiments showed that no one of presented measures reflect our expectation. From the considered measures the lift looks the most promising. But this paper is just a work in progress and we don't view this observation as a final conclusion.

For our future work we would like to have a measure that will rate the novelty of a rule with respect to the set of previously selected rules. The first idea is to chose those rules that describe anomalies not covered by the already selected rules. The second idea is to select rules which may describe already covered anomalies but using completely different set of features. The last thing we would like to work on is finding a way of concatenating mined rules to make smaller yet precise rule sets.

Acknowledgement

The research reported in this paper has been supported by the Czech Science Foundation (GA ČR) grant 13-17187S.

References

[1] Aggarwal, C. C.: Outlier analysis. Springer, 2013

[2] Agrawal, R., Imieliński, T., Swami, A.: Mining association rules between sets of items in large databases. In: ACM SIGMOD Record, volume 22, 207–216, ACM, 1993

[3] Balcázar, J. L.: Formal and computational properties of the confidence boost of association rules. ACM Transactions on Knowledge Discovery from Data (TKDD) 7(4) (2013), 19

[4] Brin, S., Motwani, R., Ullman, J. D., Tsur, S.: Dynamic itemset counting and implication rules for market basket data. In: SIGMOD 1997, Proceedings ACM SIGMOD International Conference on Management of Data, 255–264, Tucson, Arizona, USA, May 1997

[5] Chandola, V., Banerjee, A., Kumar, V.: Anomaly detection: a survey. ACM Computing Surveys (CSUR) 41(3) (2009), 15

[6] Christen, P., Goiser, K.: Quality and complexity measures for data linkage and deduplication. In: Quality Measures in Data Mining, 127–151, Springer, 2007

[7] Dang, X.-H., Micenková, B., Assent, I., Ng, R. T.: Local outlier detection with interpretation. In: European Conference on Machine Learning and Principles and Practice of Knowledge Discovery in Databases (ECML/PKDD 2013), 2013

[8] de Vries, T., Chawla, S., Houle, M. E.: Finding local anomalies in very high dimensional space. In: IEEE 10th International Conference on Data Mining (ICDM 2010), 2010

[9] Fujimaki, R., Yairi, T., Machida, K.: An approach to space-craft anomaly detection problem using kernel feature space. In: Proceedings of the Eleventh ACM SIGKDD International Conference on Knowledge Discovery in Data Mining, 2005

[10] Garcia-Teodoro, P., Diaz-Verdejo, J., Maciá-Fernández, G., Vázquez, E.: Anomaly-based network intrusion detection: techniques, systems and challenges. Computers & Security, 2009

[11] Hawkins, D. M.: Identification of outliers, volume 11. Springer, 1980

[12] Jalali-Heravi, M., Zaïane, O. R.: A study on interestingness measures for associative classifiers. In: Proceedings of the 2010 ACM Symposium on Applied Computing, 1039–1046, ACM, 2010

[13] Karr, D.: Big data brings marketing big numbers. [https://www.marketingtechblog.com/ibm-big-data-marketing/], 2012 [Online; accessed 19-June-2015].

[14] Knorr, E. M., Ng, R. T.: Algorithms for mining distance-based outliers in large datasets. In: Proceedings of the International Conference on Very Large Data Bases, 1998

[15] Knorr, E. M., Ng, R. T.: Finding intensional knowledge of distance-based outliers. In: VLDB, 1999

[16] Kopp, M., Pevný, T., Holeňa, M.: Interpreting and clustering outliers with sapling random forests. In: Information Technologies – Applications and Theory Workshops, Posters, and Tutorials (ITAT 2014), 2014

[17] Lichman, M.: UCI machine learning repository. [http://archive.ics.uci.edu/ml/]. University of California, Irvine, School of Information and Computer Sciences, 2013

[18] Liu, B., Hsu, W., Chen, S.: Using general impressions to analyze discovered classification rules. In: KDD, 31–36, 1997

[19] Liu, F. T., Ting, K. M., Zhou, Z. -H.: Isolation forest. In: Eighth IEEE International Conference on Data Mining (ICDM 2008), 2008

[20] Micenková, B., Ng, R. T., Dang, X. -H., Assent, I.: Explaining outliers by subspace separability. In: IEEE 13th International Conference on Data Mining (ICDM 2013), 2013.

[21] Pevný, T., Kopp, M.: Explaining anomalies with sapling random forests. In: Information Technologies – Applications and Theory Workshops, Posters, and Tutorials (ITAT 2014), 2014

[22] Pradnya, K., Khanuja, H. K.: Article: A survey on outlier detection in financial transactions. International Journal of Computer Applications **108(17)** (December 2014), 23–25

[23] Rousseeuw, P. J., Leroy, A. M.: Robust regression and outlier detection. John Wiley & Sons, 2005

[24] Tibshirani, R., Hastie, T.: Outlier sums for differential gene expression analysis. Biostatistics, 2007

J. Yaghob (Ed.): ITAT 2015 pp. 150–158
Charles University in Prague, Prague, 2015

Search for Structure in Audiovisual Recordings of Lectures and Conferences

Michal Kopp[1], Petr Pulc[1], and Martin Holeňa[2]

[1] Faculty of Information Technology, Czech Technical University in Prague
Thákurova 9, 160 00 Prague
koppmic2@fit.cvut.cz, petrpulc@gmail.com
[2] Institute of Computer Science, Czech Academy of Sciences
Pod Vodárenskou věží 2, 182 07 Prague
martin@cs.cas.cz

Abstract: With the quickly rising popularity of multimedia, especially of the audiovisual data, the need to understand the inner structure of such data is increasing. In this case study, we propose a method for structure discovery in recorded lectures. The method consists in integrating a self-organizing map (SOM) and hierarchical clustering to find a suitable cluster structure of the lectures. The output of every SOM is evaluated by various levels of hierarchical clustering with different number of clusters mapped to the SOM. Within these mapped levels we search for the one with the lowest average within-cluster distance, which we consider the most appropriate number of clusters for the map. In experiments, we applied the proposed approach, with SOMs of four different sizes, to nearly 16 000 slides extracted from the recorded lectures.

1 Introduction

In the past few decades, the amount of audiovisual data has grown rapidly. Every day, modern technology is accessible to more and more people. With these modern devices, such as smartphones or cameras, people are able to create large amounts of audiovisual data. Growing popularity of audiovisual data is the main reason why many new sites containing such data are created. From the consumer's point of view, the main purpose of such a site is to find a video concerning the topic they are interested in.

Usually, any video file that is being uploaded to some video sharing site is annotated with some keywords by the uploader. Beside that, it also contains some metadata such as resolution, frame rate or length of the video. Consequently, any search in the site is restricted to the keywords provided by the uploader and/or some technical data about the video. It does not rely on any relationship between the contents of the videos themselves. This would need to first discover some structure in the data available at the site.

We investigated the use of a self-organizing map (SOM) [1], which is a kind of artificial neural network, for clustering the videos with respect to their content. First, some features are extracted from the videos and then the map is built as a two-dimensional grid of neurons, so it can be easily visualized. According to the training principle of SOMs, the videos which are similar in their features are placed close to each other in the map. Consequently, the closer the videos represented by their respective vectors in the map are, the more similar they should be.

The most simple clustering of the original data would be based on individual neurons, i.e., all feature vectors mapped to the same neuron would form one cluster. However, there is more information involved in the SOM than could be represented by such a simple clustering. To have a clue what SOM-based clustering would be most suitable, we complement SOM with hierarchical clustering.

2 Related Work

Studies for utilizing SOM as a tool for clustering multimedia data were proposed in [3, 4]. In the PocketSOM [3], the SOM is utilized for a creation of the map of songs. The songs are characterized by their acoustic attributes using Rhythm Patterns [5]. These attributes are used for the creation of a high dimensional vector of features which represents the original song. The PocketSOM utilization of the song's acoustic attributes leads to the extraction of many of these features. In [3], the extraction of the acoustic attributes generated more than thousand features. This is quite similar to our approach to the multimedia clustering proposed in this paper. However, we included even more features describing the content of multimedia files because we were able to use more than one kind of features due to the multimodality of our audiovisual data.

There are also some other differences, in particular between the main purpose of the PocketSOM and our clustering method. The main purpose of the PocketSOM is to provide the user with an ability to build his or her own playlists. The user can choose the songs by virtually drawing a path in the map. The songs which are mapped to the nodes (neurons) on the path, are selected to be included in the playlist and the user has to decide which of them will be selected for the final playlist creation. It is quite obvious that the user is involved a lot in the generation of the playlists. However, our structure discovery method is user-independent.

Till the SOM has been trained, PocketSOM and our application of SOM proceed in a similar way. However, after the map has been trained, PocketSOM starts involving the user, whereas our application integrates the SOM with hierarchical clustering.

Our method has also some similarities with the Music Map [4], which also utilizes SOM for the music playlist generation. Both, the Music Map and our application utilize SOM to create a two dimensional map of similarities in the input data. However, the purpose of each of both applications is rather different. The Music Map is used for playlist generation from a collection of songs. The user's preferences for the songs that should be included in the playlist are transformed to an optimization criterion for the path planing. Then the algorithm that traverses the map and optimizes the criterion is executed, and the resulting optimal path determines the generated playlist. Although the user is not involved in the Music Map playlist generation as much as in the PocketSOM playlist generation, the main idea remains the same – to provide the user a possibility to create a playlist that corresponds to his or her demands. In the Music Map, the trained SOM is used as a background layer which is able to discover and preserve some structure in the data, although the main purpose is to traverse the map and generate the final product – the playlist. Differently to the Music Map and Pocket-SOM, our application does not primarily focus on any final product generated from the map. It rather tries to discover structure in the employed data.

There is also another difference between the Music Map and our method – the dimensionality of the feature vector. In the Music Map, every song is described by nine features. However, in our application we use more than five thousands features. This also means that the training of the map takes much more time.

Studies utilizing SOM as a clustering tool were also proposed in [6, 7]. In [6], a SOM is used as a one of the clustering tools for the analysis of embryonic stem cells gene expression data. The important difference compared to our approach is that each neuron represents one cluster with a strict boundaries, so there are as many clusters as neurons in the SOM. Then it is easy to measure the within cluster distance and between cluster distance to provide an information about the overall clusters' quality.

The study[7] deals with cluster quality evaluation. Similarly as in [6], a SOM is used as a clustering tool where every neuron corresponds to a one cluster and it is treated this way when quality of a clustering solution is being measured. Thus, the overall number of clusters is determined by the size of the 2-D map.

Our approach, on the other hand, relies on combining the information from SOM with information from hierarchical clustering.

3   Case Study Audiovisual Data

The data in which we would like to search structure has been created over the period of several years during the Weeks of Science and Technology, a two-week science festivals organized by the Czech Academy of Sciences. Altogether, the data includes more than 100 hours of multimedia content. For the purposes of further preprocessing

and data consistency, only lectures and their related content in the Czech language have been chosen.

All lectures were recorded with Mediasite recorder, a platform maintained by Sonic Foundry, Inc. The idea behind the Mediasite platform is to enable streaming of lectures as easy as possible. The main element of the system is Mediasite server which stores the lectures and streams them to the internet. On the side of the lecturer, only audio and video sources are connected to the device.

The data consist of three main modalities – synchronized video with sound and slides from lecturer's presentation. The slides are captured directly from the lecturer's presentation through the VGA/DVI card of Mediasite, which is supplied with the same signal as the projector.

The signal input for the slides is continuously monitored and any change to it above given threshold is considered as a new slide. Mediasite stores the image alongside with the timestamps of slide transition used for the synchronization during playback.

Audio is recorded through Mediasite just as lecturer speaks. It is synchronized with the video and slides by the use of timestamps of slide transitions.

The Mediasite platform only provides a video signal in the resolution up to 240 rows and the bitrate of 350Kbps. However, for purpose of the lectures the high-definition video is not necessary, because we have the slides in a good resolution synchronized with the videos. Also, during the lecture, there are usually not many changes happening on the scene, since only the lecturer is supposed to move significantly.

4   Proposed Structuring Approach

4.1   Data Preprocessing

The multimedia data described in the previous section is saved by the Mediasite system as audio, video and slides files in their respective formats. However, for the purposes of knowledge discovery in that data, including the discovery of its inner structure, we need all the data to be preprocessed to a suitable form. Because we have three modalities in the data, we treated each one separately, so we can apply appropriate feature extraction tools.

First, we used Optical Character Recognition (OCR) system for the slides. The slides usually contain some text related to the topic or the subtopic of the lecture, so an OCR can give us a good insight into the content of the slides. The downside of this approach are slides, which contains, for example, just an image. For these slides OCR often returns nothing. However, we still have the other modalities we can rely on.

To this end, we used the OCR system Tesseract, which is an open-source software made publicly available by Google. The problem of the OCR output is that it sometimes contains a lot of unrecognized characters and misspelled words, so a basic text processing was applied. All punctuation is removed. To reduce the dimensionality of the data before further processing, only the words found in the spellchecking dictionary are considered. Also, a stemmer is applied.

Second, a speech recognition tool was used for the audio files. Because we consider only lectures in the Czech language, Google's API for speech recognition, which supports this language, was used. The result of the API call is a recognized speech converted to a text. To reduce misspelled words and the dimensionality of the data, the same methods as for OCR are applied, resulting in a text document for every slide, aggregated from the OCR and the speech recognition.

The text data returned from the speech and slide recognition have to be transformed into term-by-document matrices. For term weighting on cleaned text data, a tf-idf (term frequency – inverse document frequency) scheme has been used [9]:

$$W_{t,d} = (n_{t,d}/max_t(n_{t,d})) \cdot log_2(n_D/n_{D,t}) \qquad (1)$$

where $n_{t,d}$ is the count of term t in document d, n_D denotes the number of documents in the employed corpus and $n_{D,t}$ is count of documents from the corpus that contain at least one occurrence of term t. Logarithmic weighting in idf part is used to not penalize relatively frequent terms as much.

As the resulting matrices had tens of thousands of columns and we would like to minimize the impact of the curse of dimensionality, the Latent Semantic Analysis [8] has been used. To this end, the k largest eigenvalues of the term-by-document matrix of weights (1), corresponding to the most significant concepts, and their associated eigenvectors are found first. Let U_k, Σ_k and V_k be the matrices resulting from the singular value decomposition of the term-by-document matrix, and k denotes the number of the largest eigenvalues.

To obtain dense matrices of significantly lower dimensions, a concept-by-document matrix [10] is then computed as:

$$C_k = \Sigma_k V_k^\top \qquad (2)$$

Third, the video is used by two different extraction methods – Speeded Up Robust Features (SURF) and color histogram. Both methods work with an image, so just one frame from the center of the video's time span related to a slide is taken. SURF is used to find visual descriptors in the image. We have almost 16 000 slides, therefore the same number of images taken from the videos, and each of these images produced nearly a hundred SURF descriptors – summing up to over a million descriptors in total. To reduce the dimensionality, a dictionary of SURF descriptors concepts was created by k-means clustering of the original SURF descriptors with k empirically set to 400. Because the matrix of these 400 descriptor concepts was still very sparse, we decided to cluster it even further, by hi-

erarchical clustering. This resulted in 32 final descriptor concepts.

The histogram describes each image (taken from the video) by count of pixels with certain color value in each color channel. We use the RGB coding with 1 byte depth. Each color count is relatively scaled to the size of the image.

All three kinds of features are combined into feature vectors, their resulting dimension is 5800 features.

4.2 Hierarchical Clustering

General purpose of clustering is grouping objects that are similar into same group. There are 2 main kinds of clustering – the number of clusters is a priori known or it is unknown, in which case the clusters are formed hierarchically. Hierarchical clustering can also capture all levels of similarities, with respect to how the similarity is defined.

The result of hierarchical clustering is a multilevel hierarchy of clusters, where two clusters at one level are joined as a cluster at the next level. At the bottom level every single cluster corresponds to a single observation.

Decision, which clusters will be joined on the higher level, is based on a similarity between these clusters, which is measured in respect of a chosen distance metric between observations / clusters at the previous level of the cluster hierarchy and a linkage criterion. The distance metric determines how similarity between two observation is calculated, while the linkage criterion determines how the distance metric is employed to calculate the similarity between two clusters. The Euclidean distance and the Ward's criterion were chosen as the metric and the linkage criterion in all our experiments. The Euclidean distance was chosen because it the most common distance metric being used. The Ward's criterion (also called the minimum variance criterion or the inner squared distance) was chosen because it minimizes the total within-cluster variance after a potential merge of the two clusters. This variance is calculated as a weighted Euclidean distance between clusters' centers. The weight w is calculated as:

$$w_{r,s} = \sqrt{\frac{2n_r n_s}{(n_r + n_s)}} \qquad (3)$$

where n_r and n_s are the number of elements in clusters r and s.

4.3 SOM Clustering

Self-organizing map (or Self-organizing feature map)[1] is a kind of an artificial neural network. The main idea behind SOM is that it can preserve topological relations of the input space, which is typically high-dimensional, in a low dimensional map (typically 2-D). That means the data which is similar in their original high-dimensional input space is also similar in the map. Due to this characteristics, the high-dimensional data whose relationships are often very hard to visualize in the original high-dimensional space can be relative easily visualized in the low-dimensional map. SOM does not only learn the distribution of the input data, but it also learns a topology.

Since in all our experiments we used a 2-D map, which is also the most common type of SOM, the following description will be restricted just to that specific case.

SOM consists of one layer of neurons which are organized in some regular topology according to a topology function, which determines how the neurons in the map are organized. Most common are the grid, hexagonal and random topology. Also, a distance function, which measures some kind of distance between neurons needs to be selected. Common distance functions are Euclidean distance, Manhattan distance or the length of the shortest path between neurons.

Each neuron in the map is assigned a weight vector which is of the same dimension as the input vectors. The weight vectors are first initialized, most easily with samples from the input data or with small random values.

Training the SOM is an iterative process. In each iteration, one sample vector X is selected and it is presented to the map. A similarity between the selected feature vector and all weight vectors in the map is calculated, usually as Euclidean distance. The neuron that has the most similar weight vector is selected as a winning neuron c.

When the winning neuron is found, the weight vectors of the winning neuron and its neighbors are updated. They are moved closer to the presented input vector, with decreasing magnitude of changes according to the growing distance from the winning neuron. The update rule for a neuron i with weight vector $W_i(t)$ is:

$$W_i(t+1) = W_i(t) + \alpha(t)\theta(c,i,t)(X(t) - W_i(t)) \quad (4)$$

where $\alpha(t)$ is learning rate, which is from the interval $(0,1)$ and it is decreasing with increasing iterations, and $\theta(c,i,t)$ is the neighborhood function which depends on the distance between the winning neuron c and the neuron i, and may depend also on the iteration t.

Another possibilty for SOM training is the batch algorithm[2]. In each epoch, this algorithm presents the whole training data set to the network at once. The winning neurons for all the input data are selected and each weight vector is modified according to the position of all the input vectors for which it is a winner or for which it is in the neighborhood of a winner.

In our experiments with SOM, we used the MATLAB's Neural network toolbox implementation of SOM, which uses the batch algorithm by default.

4.4 Integrating Hierarchical Clustering with SOM

During each epoch of the batch algorithm that is used in the training process of the SOM, each vector of values of input features is assigned to the neuron that is the winner for that input feature vector. Thus, after the training

process is completed, each neuron in the map could be interpreted as a cluster which is formed by a subset of the input feature vectors, those for which it is a winner.

In many scenarios, for example in studies [6] and [7], the output of SOM is interpreted in that way. However, this means that every neuron in the map is interpreted as a single cluster.

In this case study, we use a different approach – an integration of SOM and hierarchical clustering. First, we train SOM with all the input feature vectors. The resulting map serves as a reference for all clustering solutions obtained by cutting the cluster hierarchy. These solutions are mapped to that reference SOM and then evaluated. Algorithm 1 and the following paragraphs describe how the average withing cluster distance of a clustering solution is calculated.

Algorithm 1 An average within-cluster distance
___

$map \leftarrow$ trained SOM
$S \leftarrow$ clustering solution obtained by cutting cluster hierarchy
for all $c \in S$ **do**
    $WCD[c] \leftarrow$ WITHINCLUSTERDISTANCE(map,c)
end for
$SolutionAvgWCD \leftarrow avg(WCD)$ //over all clusters c

function WITHINCLUSTERDISTANCE(map,c)
    for each pair p of input vectors $u,v \in c$ **do**
        $N_u \leftarrow$ FINDNEURON(u,map)
        $N_v \leftarrow$ FINDNEURON(v,map)
        $Dist[p] \leftarrow dist(N_u,N_v)$
    end for
    return $avg(Dist)$ between all pairs
end function

function FINDNEURON(v, map)
    return Neuron N in the map map to which the input vector v is mapped.
end function

After the reference SOM has been created, hierarchical clustering is performed. The cluster hierarchy is cut at many different levels and the clusters at the respective level are used as a starting point to calculate the distance between their respective vectors positioned in the map. When evaluating a solution, all feature vectors that form a cluster by hierarchical clustering are taken, the positions of their respective neurons in the map are found and the distance between each pair of them is calculated according to the distance function used in the map. From these distances between all pairs, we calculate the average distance, which is interpreted as a within-cluster distance.

From those within-cluster distances, an average distance of the whole solution is calculated. That average distance is used as a measure of quality of each clustering solution obtained by cutting the cluster hierarchy. We also interpret it as a measure of similarity between the map output and the hierarchical clustering output – the smaller the within-cluster distance is, the more similar the outputs of SOM and hierarchical clustering are. We also consider that the more similar they are (the smaller the within-cluster distance is), the better the number of clusters produced from cutting the tree suits the map.

5 Experimental Evaluation

5.1 Evaluation Methodology

For our experimental evaluation, we need to train several differently sized SOMs first. Data from all available lectures has been used for the training to make the SOMs as precise as possible. However, with 5800 features, the data dimensionality is quite high and so is the number of feature vectors – 15953. Due to the time complexity of SOM training and also with respect to the size of our data, only relatively small maps were used. Namely, we considered only the squared maps with sizes of 8×8, 12×12, 16×16 and 20×20 neurons. The number of training epochs was set empirically, taking into account the time taken by the training process, to 200 for the three smaller maps and 150 for the largest one.

Even though we need outputs from many levels of the hierarchical clustering, the cluster hierarchy was constructed only once. It is important to realize that the hierarchy must be constructed from the same set of feature vectors as the SOMs to make the comparison possible. In all considered clustering solutions, the same cluster hierarchy was considered, cut at an appropriate level to obtain a solution with the desired number of clusters.

A different number of clustering solutions for different sized SOMs was used. We started comparison with hierarchical clustering output containing just a few clusters and ended with an output containing three or for times more clusters than the number of neurons present in the respective map. We also introduced different steps for increasing the number of clusters, depending on the map size.

In Table 1, the empirically chosen values for of our experiments are shown. The number of clusters range is an interval which determines, together with the step size, how many clustering solutions produced by cutting the tree were evaluated.

SOM size	# of clusters range	step	# of solutions
8×8	16 – 256	4	61
12×12	16 – 500	4	122
16×16	32 – 800	8	97
20×20	32 – 1000	8	122

Table 1: Settings used in the performed experiments

5.2 Evaluation Results

In this section we are going to use three types of figures to illustrate the measured results. Because these figure types are used repeatedly, their interpretation will be first shortly described.

The first type shows the map topology (which is hexagonal in all our experiments) with a relative distribution of the input feature vectors. Each neuron is represented as a white hexagon with a blue patch. The bigger the blue patch at each neuron is, the more feature vectors has been mapped to that neuron. Also in smaller maps, the numbers representing the total number of feature vectors mapped to the neurons are shown in each hexagon.

The second type is a neighbor distances map. It shows the distances between weight vectors of the neurons. The small regular hexagons are the neurons and the lines connecting them represents their direct neighbor relations. The colored patches between the neurons show how close each neuron's weight vector is to the weight vectors of its neighbors. Color range varies from yellow to black, where the darker color means the greater distance between the weight vectors.

The third type is a figure showing an average within-cluster distance depending on the number of clusters produced from the hierarchical clustering, which were mapped to the SOM as described in Section 4.4.

We started the evaluation with the smallest, 8×8 map. A relative distribution of the input feature vectors onto the map is shown in Figure 1.

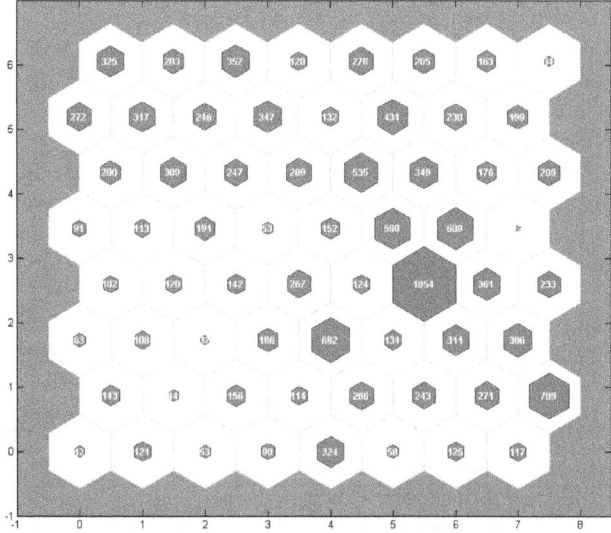

Figure 1: Relative distribution of the input feature vectors alongside with the total numbers of vectors mapped to each neuron in the 8×8 map.

From the distribution of feature vectors shown in Figure 1 we can see that from the total number of 15 953 feature vectors, there are 1854 feature vectors mapped to a single neuron. That indicates that there are many quite similar input vectors in the data. This is even more evident when we look at the neurons in its neighborhood, which are also quite a lot populated with the input vectors in contrast with the lower-left part of the map.

The neuron with the most mapped feature vectors has them mapped more than twice as many as any other neuron in this map. When we look at the input data, we find that many different slides are mapped to this neuron.

Another reason for the large number of feature vectors mapped to this neuron is that these feature vectors provide just one modality of the available input data. Namely, nearly a half of input data mapped to this neuron seems to be slides created from video playback or they are just images without text. So, even though these slides look different, the OCR produces little to no text, which can result in great similarity between them, depending on other modalities.

These feature vectors also have in common that there is little of recognized speech, thus they are very similar in the speech-to-text modality. The only really important difference in feature vectors is due to the SURFs and histograms produced from one frame taken from the video, but this difference can also be just a small one. In the case of lectures, the majority of lecture halls seem very similar – white walls and brown desks accompanied by projection screen.

However, there are also some inputs mapped to this neuron that seem to be typical examples of a lecture – a slide with some text, video from lecture hall and an average speech-recognized audio.

Quite interesting is that usually not the whole sequence of similar input data from one lecture is mapped to this neuron. However, the input data from the sequences, which are not present in this neuron, are in the most cases mapped to its direct neighbours. This behaviour supports the idea of large clusters spread over more neighbouring neurons.

In Figure 2, which shows the distances between neighbors, we can see that the distances are the shortest between the neurons with the highest number of vectors. Conversely, neurons in the lower-left corner are the most distant from each other. When we look at the input data, we find that the segments of video-recordings and associated slides projected during that segment indeed substantially differ from the remaining segments of all recorded lectures. First, the camcorder during the time span of respective segment was set to shoot only the projected slide, instead of the whole lecture hall. Second, the projected slides in most cases were examples of web pages. Moreover, in 35 out of 42 slides mapped to the neuron in the bottom-left corner, the examples of web pages were taken from the Wikipedia, so they contain a lot of text and hyperlinks. The image taken from the video, containing just a slide instead of the whole lecture hall, caused the visual features and histogram to be very different. A lot of text on a Wikipedia page instead of only few lines of text on a typical slide caused the main difference in OCR.

Figure 3 shows that in the map of size 8×8 neurons,

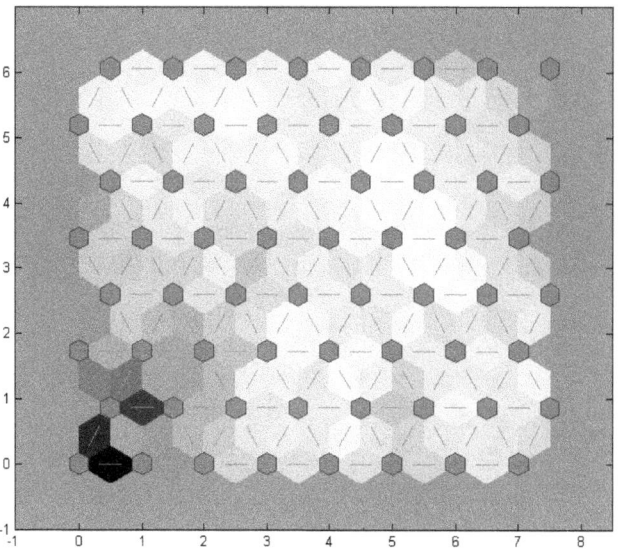

Figure 2: The neighbor distances in the 8×8 map.

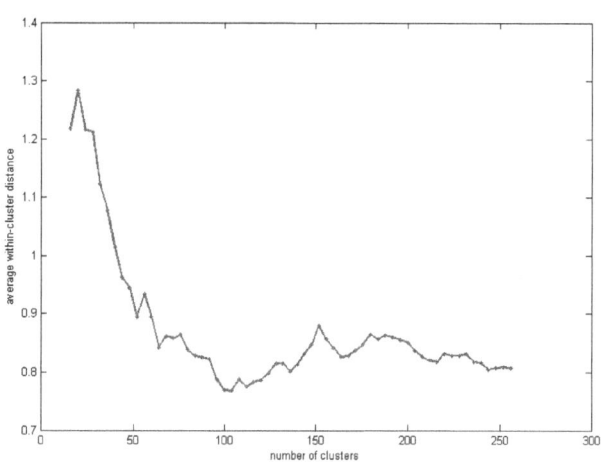

Figure 3: A within-cluster distance, measured as the length of the shortest path between neurons, dependent on the number of clusters produced from the hierarchical clustering, mapped to the 8×8 map.

an average within-cluster distance drops down to its local minimal value when the cluster hierarchy is cut at an appropriate level to form about a hundred clusters. It means that the hierarchical clustering should produce more clusters than the number of neurons in the map, which is only 64.

Let us now pass to the 12×12 and 16×16 maps. In Figures 4 and 7 we can see similar, one highly populated neuron, like in the 8×8 map. In bigger maps, the total number of vectors mapped to that neuron is lower, but in comparison to other neurons, the relative count is the same or even higher.

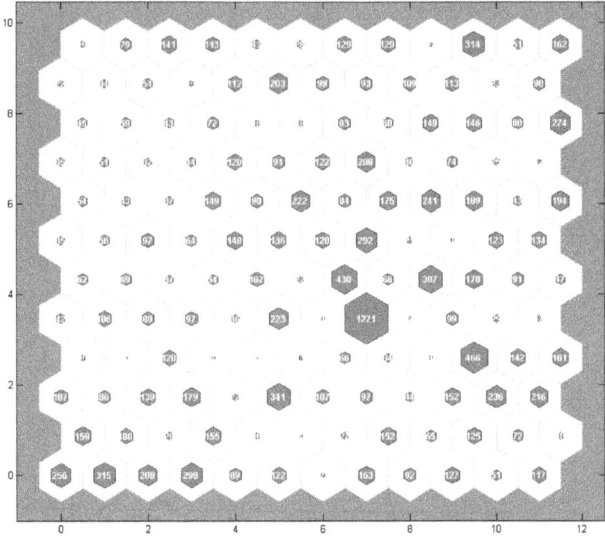

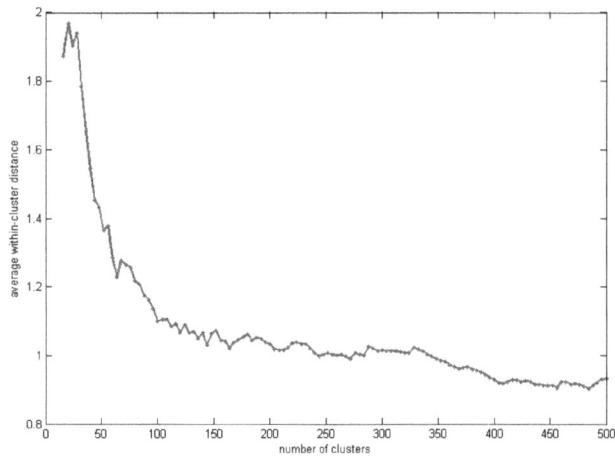

Figure 6: A within-cluster distance, measured as the length of the shortest path between neurons, dependent on the number of clusters produced from the hierarchical clustering, mapped to the 12×12 map.

Figure 4: Relative distribution of the input feature vectors alongside with the total numbers of vectors mapped to each neuron in the 12×12 map.

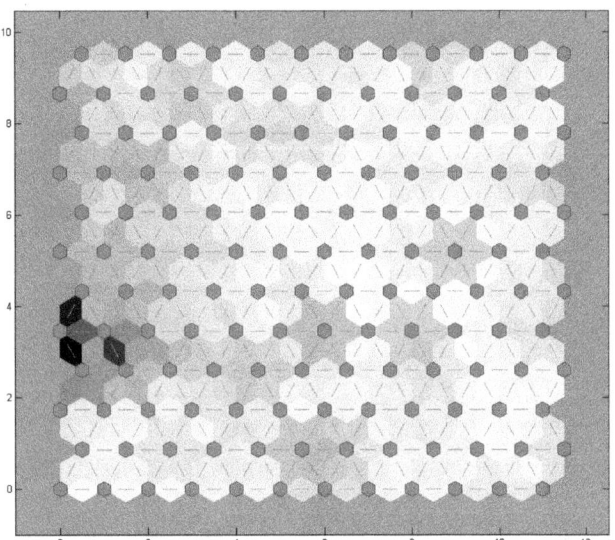

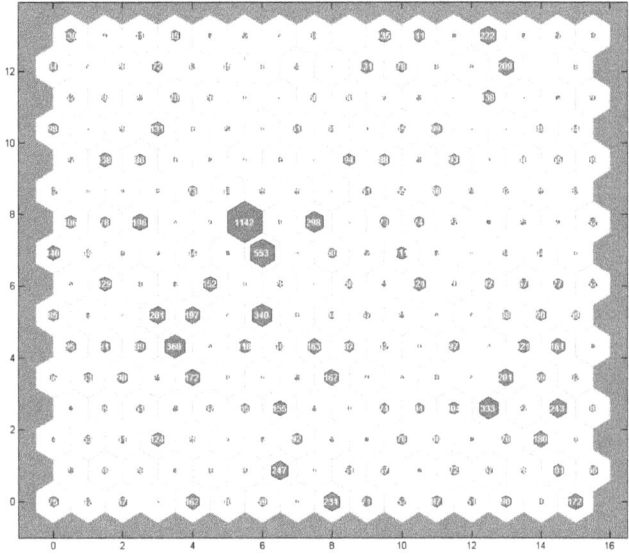

Figure 5: The neighbor distances in the 12×12 map.

Figure 7: Relative distribution of the input feature vectors alongside with the total numbers of vectors mapped to each neuron in the 16×16 map.

In Figures 5 and 8, we can see that there are some neurons, mostly on the left side of the map 5 and in the top-center, lower-left and top-right corners in 8, that have a large distance to their respective neighbors. The situation is quite similar to that of the smaller map in Figure 2. The darkest areas just moved around to another physical location in each of those two maps, but the relationships seems to be similar and those neurons tend to stay together.

It is also interesting, that in those larger maps a few neurons are more distant from their respective neighbors than others in their neighborhoods, which indicates that there is some input data which is quite different from the rest in

the most features. This can be seen in the map as a darker "star" in a lighter neighborhood.

In Figures 6 and 9, which show the within-cluster distance dependence on the number of clusters formed by cutting the cluster hierarchy, we can not find a clear local minimum. The value of distance drops down quickly with the increasing number of clusters, forming an elbow in the figure, which is at just about a few less clusters than is the number of neurons in the map.

Even though the most suitable number of clusters in the map could not be define precisely this way when there is

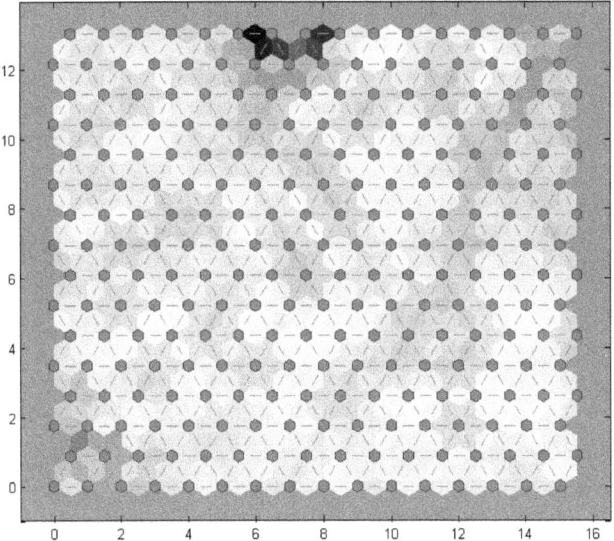

Figure 8: The neighbor distances in the 16×16 map.

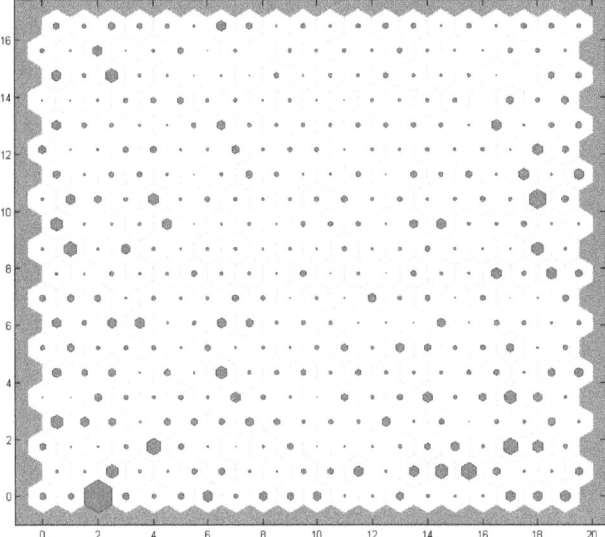

Figure 10: Relative distribution of the input feature vectors alongside with the total numbers of vectors mapped to each neuron in the 20×20 map.

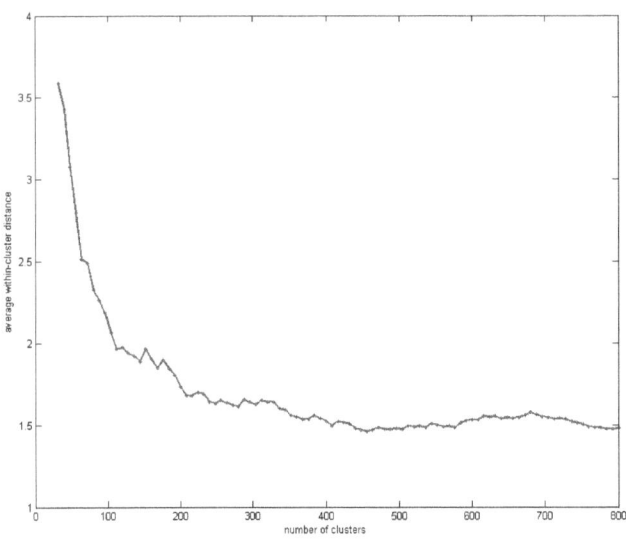

Figure 9: A within-cluster distance, measured as the length of the shortest path between neurons, dependent on the number of clusters produced from the hierarchical clustering, mapped to the 16×16 map.

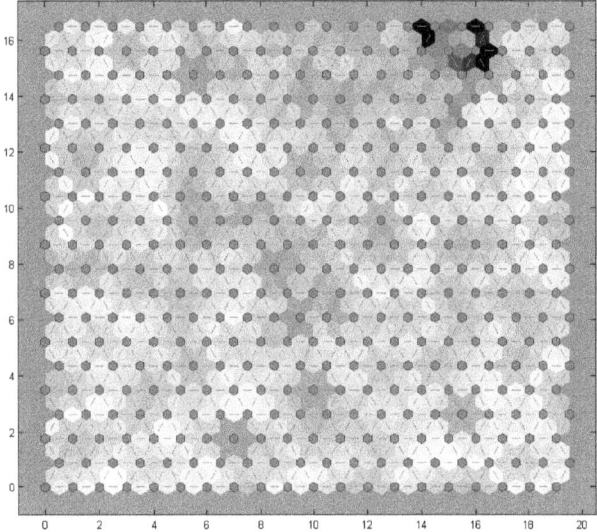

Figure 11: The neighbor distances in the 20×20 map.

no distinct local minimum, locating an elbow on the curve gives us a great possible candidate when we are searching for a solution with a good average within-cluster distance and the least number of clusters.

Finally, let us pass to the largest, 20×20 map. Figure 10 shows that there is still one neuron with significantly more input vectors mapped to it than to any other neuron. A great part of the map is low populated with feature vectors, even though there are some neurons with small neighborhoods which are populated a little more.

In Figure 11, we can see that the small area of neurons

distant from each other is still present, situated at the top of the map, almost in the right corner. Alongside with Figures 1, 4 and 7, we can also see that as the map grows, there are more distant neurons and the whole map becomes darker. This is due to the higher number of neurons which allows more distinct input feature vectors to form their own clusters.

The 20×20 map also has other interesting characteristics. In Figure 12, there is once again a local minimum, though in this case it can be found within the interval of 400 and 500 clusters. It seems that clustering solutions produced by hierarchical clustering and clusters formed in

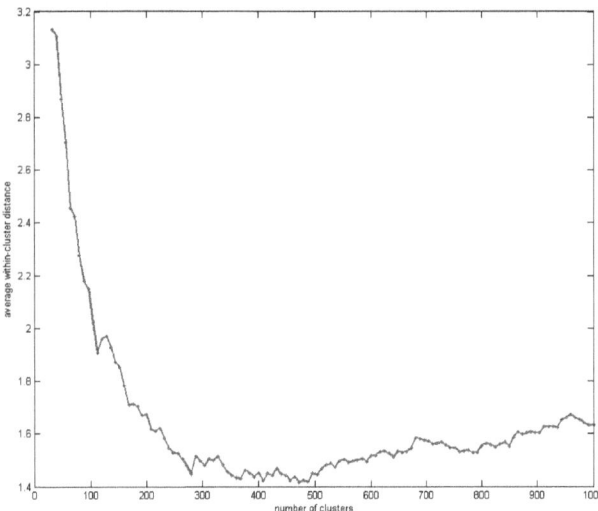

Figure 12: A within-cluster distance, measured as the length of the shortest path between neurons, dependent on the number of clusters produced from the hierarchical clustering, mapped to the 20×20 map.

the map are very similar in that interval, because there is exactly 400 neurons in this map.

6   Conclusion

The objective of this paper was to provide a proof of concept for the possibility to improve structure discovery in complex multimedia data through the integration of two unsupervised approaches – hierarchical clustering and self-organizing maps. For that proof of concept, we have used a case study of more then 100 hours of audiovisual recordings of lectures from several years of a Czech science festival called Week of Science and Technology. The approach we propose relies on the one hand on the flexibility and universality of hierarchical clustering, on the other hand on the suitability of SOM for multimedia data, convincingly documented by Pitoyo Hartono. Our approach has been much inspired by his paper [4]. However, the need to integrate SOM with hierarchical clustering is a consequence of the fact that our data are much more complex, both from the point of view of involved modalities, and from the point of view of the number of attributes.

We have tested the proposed approach with 4 differently sized self-organizing maps. Although space limitations allowed us to present only two examples of the interpretaion of the obtained results in the context of the original audio-visual data, even these examples indicate that SOMs indeed help to recognize structure in that data. In the first provided example, the SOM organized the most common data into a large cluster spread over few neighbouring neurons and in the second example, the SOM discovered a specific group among the nearly 16000 considered

segments of video-recordings and their associated slides, which substantially differs from the others.

In the future, we want to further elaborate our approach, taking into account the results obtained in the case study presented here. We also intend to apply it to further complex multimedia data sets, in particular to data nowadays being collected within the project Open Narra at the Film and TV School of the Academy of Performing Arts in Prague.

Acknowledgement

The research reported in this paper has been supported by the Czech Science Foundation (GAČR) grant 13-17187S. Access to computing and storage facilities owned by parties and projects contributing to the National Grid Infrastructure MetaCentrum, provided under the programme "Projects of Large Infrastructure for Research, Development, and Innovations" (LM2010005), is greatly appreciated.

References

[1] Kohonen, T.: Self-organized formation of topologically correct feature maps. Biological Cybernetics **43** (1982), 59–69

[2] Kohonen, T.: Self-organizing maps. Springer Series in Information Sciences, 1995, 2nd ed. 1997

[3] Neumayer, R., Dittenbach, M., Rauber, A.: PlaySOM and PocketSOM player: alternative interfaces to large music collections. In: Proceedings of the 6th International Conference on Music Information Retrieval, 618–623, 2005

[4] Hartono, P., Yoshitake, R.: Automatic playlist generation from self-organizing music map. Journal of Signal Processing **17 (1)** (2013), 11–19

[5] Rauber, A., Pampalk, E., Merkl, D.: Using psycho-acoustic models and self-organizing maps to create a hierarchical structuring of music by musical styles. In: Proceedings of the 3rd International Symposium on Music Information Retrieval, 2002, 71–80,

[6] Chen, G., Jaradat, S. A., Banerjee, N., et al.: Evaluation and comparison of clustering algorithms in analyzing ES cell gene expression data. Statistica Sinica **12.1** (2002), 241–262

[7] Lamirel, J. C., Cuxac, P., Mall, R., et al.: A new efficient and unbiased approach for clustering quality evaluation.

[8] Deerwester, S. C., Dumais, S. T., Landauer, T. K., et al.: Indexing by latent semantic analysis. JASIS **41 (6)** (1990), 391–407

[9] Ramos, J.: Using tf-idf to determine word relevance in document queries. In: Proceedings of the First Instructional Conference on Machine Learning, 2003

[10] Skopal, T., Moravec, P.: Modified LSI model for efficient search by metric access methods. In: Advances in Information Retrieval, 2005, 245–259

J. Yaghob (Ed.): ITAT 2015 pp. 159–166
Charles University in Prague, Prague, 2015

ITAT

Investigation of Gaussian Processes in the Context
of Black-Box Evolutionary Optimization

Andrej Kudinov[1], Lukáš Bajer[2], Zbyněk Pitra[3], and Martin Holeňa[4]

[1] Faculty of Information Technology, Czech Technical University, Prague, kudinand@fit.cvut.cz,
[2] Faculty of Mathematics and Physics, Charles University, Prague, bajeluk@gmail.com,
[3] Faculty of Nuclear Sciences and Physical Engineering, Czech Technical University, Prague, z.pitra@gmail.com,
[4] Czech Academy of Sciences, Prague, martin@cs.cas.cz

Abstract: Minimizing the number of function evaluations became a very challenging problem in the field of black-box optimization, when one evaluation of the objective function may be very expensive or time-consuming. Gaussian processes (GPs) are one of the approaches suggested to this end, already nearly 20 years ago, in the area of general global optimization. So far, however, they received only little attention in the area of evolutionary black-box optimization.

This work investigates the performance of GPs in the context of black-box continuous optimization, using multimodal functions from the CEC 2013 competition. It shows the performance of two methods based on GPs, Model Guided Sampling Optimization (MGSO) and GPs as a surrogate model for CMA-ES. The paper compares the speed-up of both methods with respect to the number of function evaluations using different settings to CMA-ES with no surrogate model.

1 Introduction

Evolutionary computation became very successful during the past few decades in continuous black-box optimization. In such optimization, we have no mathematical definition of the optimized function available, thus we can calculate analytically neither that function itself, nor its derivatives. In such cases, there is no option but to empirically evaluate the objective function through measurements, tests or simulations.

In various real-world optimization problems, the evaluation of the objective function is very expensive or time-consuming, e.g., protein's folding stability optimization [6], computer-assisted design [1] and job allocations in a computational grid [13]. In such cases, we need to keep the number of function evaluations as low as possible, without impairing the quality of expected results.

In this paper, we employ two approaches addressing that task. The first, called Model Guided Sampling Optimization (MGSO) [2], is one of the recent implementations of GPs. The second employed approach is surrogate modeling, recalled in Subsection 2.2, which we will use in conjunction with CMA-ES.

This work investigates the performance of both methods on the set of niching functions from the CEC 2013 competition [10], characterized by a high number of local optima,

which makes evolutionary search for the global optimum difficult because evolutionary algorithms (EAs) tend to get trapped in one of the local optima.

The following section describes the theoretical fundamentals of GPs and introduces the MGSO method. It also explains surrogate modeling and using GPs as a surrogate model for CMA-ES. Section 3 presents results of experimental evaluation of the considered methods. Section 5 summarizes the results and concludes the paper.

2 Gaussian Processes in Optimization

GP is a random process such that any finite sequence $X_1, \ldots, X_k$ of the involved random variables has a multivariate Gaussian distribution. GP is defined by its mean value and covariance matrix described by a function with relatively small number of hyper-parameters, which are usually fitted by the maximum likelihood method. Firstly, GP is trained with N data points from the input space $\mathbb{X}$,

$$\mathbf{X}_N = \{\mathbf{x}_i | \mathbf{x}_i \in \mathbb{R}^D\}_{i=1}^N$$

with known input-output values $(\mathbf{x}_N, \mathbf{y}_N)$, then it is used for predicting the $(N+1)$-st point. The conditional density of the extended vector $\mathbf{y}_{N+1} = (y_1, \ldots, y_N, y_{N+1})$, conditioned on $\mathbf{X}_{N+1} = \mathbf{X}_N \cup \{\mathbf{x}_{N+1}\}$ is

$$p(\mathbf{y}_{N+1}|\mathbf{X}_{N+1}) = \frac{\exp(-\frac{1}{2}\mathbf{y}_{N+1}^T \mathbf{C}_{N+1}^{-1} \mathbf{y}_{N+1})}{\sqrt{(2\pi)^{N+1}\det(\mathbf{C}_{N+1})}},$$

where $\mathbf{C}_{N+1}$ is the covariance matrix of a $(N+1)$-dimensional Gaussian distribution. The covariance matrix can be expressed as

$$\mathbf{C}_{N+1} = \begin{pmatrix} \mathbf{C}_N & \mathbf{k} \\ \mathbf{k}^T & \kappa \end{pmatrix},$$

where κ is the variance of the new point itself, $\mathbf{k}$ is is the vector of covariances between the new point and training data and $\mathbf{C}_N$ is the covariance of the Gaussian distribution corresponding to the N training data points [5].

Covariance functions provide prior information about the objective function and express the covariance between the function values of each two data points $\mathbf{x}_i$, $\mathbf{x}_j$ as $cov(f(\mathbf{x}_i), f(\mathbf{x}_j)) = k(\mathbf{x}_i, \mathbf{x}_j)$. Because the matrix $x_1, \ldots, x_N$ serves as a covariance matrix, it has to positive semidefinite.

2.1 Model Guided Sampling Optimization

MGSO has the ability to use a regression model for prediction and error estimation in order to get the probability of obtaining a better solution. It was inspired by two previously proposed methods in the field of black-box optimization. The first method, Estimation of Distribution Algorithms [9], creates a new set of solutions for the next generation using estimated probability distribution from previously selected candidate solutions. The second approach is surrogate modeling, described in Section 2.2.

MGSO was proposed as an alternative method for Jones' Efficient Global Optimization (EGO) [8]. Unlike EGO, MGSO produces not a single solution, but a whole population of solutions. The selection of candidate solutions is performed by sampling the probability of improvement (PoI) of the GP model, which serves as a measure of how promising the chosen point is for locating the optimum. PoI is determined by means of a chosen threshold T and the estimation of the objective function shape by the current GP model.

The crucial step in the MGSO algorithm is the sampling of PoI, which is determined by the predicted mean $\hat{f}(\mathbf{x}) = \hat{y}$ and the standard deviation $\hat{s}(\mathbf{x}) = s_y$ of the GP model $\hat{f}$ in any point $\mathbf{x}$ of the input space

$$\mathrm{PoI}_T(\mathbf{x}) = \Phi\left(\frac{T - \hat{f}(\mathbf{x})}{\hat{s}(\mathbf{x})}\right) = \mathrm{P}(\hat{f}(\mathbf{x} \leq T)),$$

which corresponds to the value of cumulative distribution function of the Gaussian for the value of threshold T. Although all the variables are sampled from Gaussian distribution, $\mathrm{PoI}(\mathbf{x})$ is not Gaussian-shaped as it depends on the threshold T and the modelled function f.

2.2 GP as Surrogate Model for CMA-ES

Surrogate modeling is a technique used in optimization in order to decrease the number of expensive function evaluations. A surrogate model, which is a regression model of suitable kind (in our case a GP), is constructed by training with known values of the objective function for some inputs first, and then it is used by the employed evolutionary optimization algorithm instead of the original objective function (in evolutionary optimization usually called fitness) during the search for the global optimum. Although, creating and training a model increases time complexity of the optimization algorithm, using a model instead of the original fitness function decreases the number of its evaluations, which is a crucial objective in expensive optimization.

Every regression model approximates the original fitness function with some error. To prevent the optimization from being mislead from such an erroneous approximation, it is necessary to use the original fitness function for some subset of evaluations. That subset is determined by the evolution control (EC) strategy [5].

An *individual-based* EC strategy consists in determining the subset of individuals evaluated by the original fitness function in each generation. The following description is illustrated by Figure 1. Denote λ to be the size of the population provided by CMA-ES. First, $\lambda'' = \alpha\lambda$ points are sampled from $N(m, \sigma^2\mathbf{C})$, where $\alpha \in [0,1]$, m is the mean, σ is the step-size and $\mathbf{C}$ stands for the covariance matrix (1). These λ'' points are evaluated by the original fitness function and included in training the model. Then, the extended population $\lambda' = \beta(\lambda - \lambda'')$, where $\beta \in [1, \infty)$, is sampled by a model using the same distribution (2). The extended population is required by the model for choosing promising points for re-evaluation by the original fitness function. Subsequently, $\gamma(\lambda - \lambda'')$ points, where $\gamma \in [0,1]$, are chosen according to some criterion from among the extended population, e.g. fitness value, and used in the evaluation by the original fitness function (3). The complement to λ points is gathered from the rest of the extended population by dividing it into $k = (1 - \gamma)(\lambda - \lambda'')$ clusters and selecting the best point from each cluster, which are also evaluated by the original fitness function and added to the final population (4) [3].

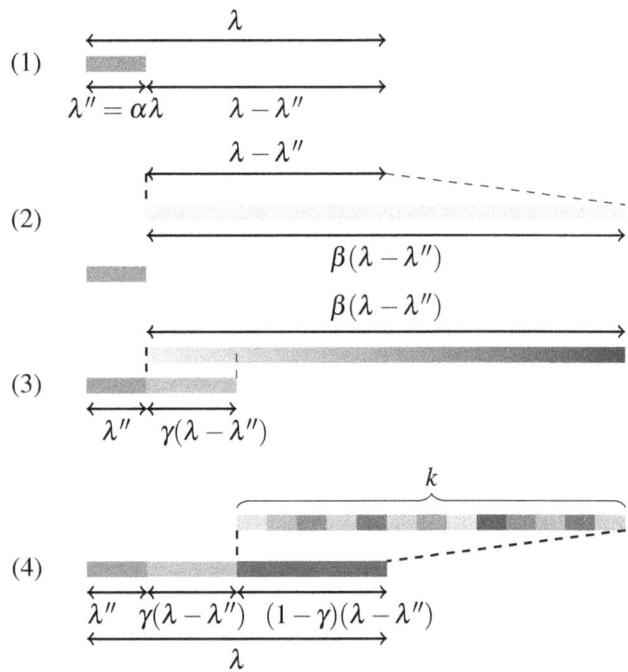

■ best points from each cluster
▨ best points from the extended population
□ extended population
▨ pre-sampled points

Figure 1: Individual-based EC

A *generation-based* EC strategy determines the number of model-evaluated generations between two generations evaluated by the original function. After a generation is evaluated by the original fitness function, the model is trained using the obtained values. The number of consecutive model-evaluated generations can be determined also dynamically, as introduced in so-called *adaptive* EC strategy [11], when the deviation between the original and the model fitness function is assessed and then it is decided whether to evaluate using the original fitness or the model.

Determining the most suitable EC parameters, however, is an open problem, which depends on the properties of the fitness function and the current performance of the surrogate model. Moreover, the most suitable parameters change during the optimization process.

3 Experimental Evaluation

Previous investigations compared the performance of MGSO [2] and CMA-ES with GP surrogate model [4] (denoted hereafter S-CMA-ES) with CMA-ES without a model, using the standard black-box optimization benchmarks [7]. In this work, we compare those methods using an additional set of 12 multimodal fitness functions from the CEC 2013 competition on niching methods for multimodal function optimization [10]:

- f1: Five-Uneven-Peak Trap (1D),
- f2: Equal Maxima (1D),
- f3: Uneven Decreasing Maxima (1D),
- f4: Himmelblau (2D),
- f5: Six-Hump Camel Back (2D),
- f6: Shubert (2D, 3D),
- f7: Vincent (2D, 3D),
- f8: Modified Rastrigin - All Global Optima (2D),
- f9: Composition Function 1 (2D),
- f10: Composition Function 2 (2D),
- f11: Composition Function 3 (2D, 3D, 5D, 10D),
- f12: Composition Function 4 (3D, 5D, 10D, 20D).

3.1 MGSO Performance

MGSO performance was examined using two covariance functions, isotropic squared exponential ($\mathbf{K}_{SE}^{iso}$) and squared exponential with automatic relevance determination ($\mathbf{K}_{SE}^{ard}$), with parameters shown in Table 1. The results in Tables 2 and 3 show the speed-up of MGSO with respect to CMA-ES. As can be seen, the $\mathbf{K}_{SE}^{iso}$ covariance function performed better among these two in more than the half of cases. Table 1 shows used parameter settings in our evaluations.

3.2 S-CMA-ES Performance

The speed-up results are shown in Tables 2 and 3. In performed evaluations, four covariance functions in the GP surrogate model were used, two types of the squared exponential covariance function, the isotropic version

$$\mathbf{K}_{SE}^{iso}(\mathbf{x}_i, \mathbf{x}_j) = \sigma_f^2 \exp\left(-\frac{1}{2\ell^2}(\mathbf{x}_i - \mathbf{x}_j)^\top(\mathbf{x}_i - \mathbf{x}_j)\right), \quad (1)$$

S-CMA-ES	
covariance functions	$cov \in$ $\{\mathbf{K}_{Matérn}^{v=\frac{5}{2}}, \mathbf{K}_{exp}, \mathbf{K}_{SE}^{iso}, \mathbf{K}_{SE}^{ard}\}$
starting values of (σ_f^2, ℓ)	$(0.1, 10)$ for $\mathbf{K}_{SE}^{iso}$ $(0.05 \times \boldsymbol{J}_{1,D}, 0.1)$ for $\mathbf{K}_{SE}^{ard}$ $(0.5, 2)$ otherwise
starting values of σ_n^2	0.01

MGSO	
covariance functions	$cov \in \{\mathbf{K}_{SE}^{iso}, \mathbf{K}_{SE}^{ard}\}$
starting values of (σ_f^2, ℓ)	$(0.1, 10)$ for $\mathbf{K}_{SE}^{iso}$ $(0.05 \times (\boldsymbol{J}_{1,D}), 0.1)$ for $\mathbf{K}_{SE}^{ard}$
starting values of σ_n^2	0.01

Table 1: Model parameter settings for S-CMA-ES and MGSO performance testing. The symbols $\mathbf{K}_{SE}^{iso}$, $\mathbf{K}_{SE}^{ard}$, $\mathbf{K}_{exp}$, $\mathbf{K}_{Matérn}^{v=\frac{5}{2}}$, denote, respectively, the isotropic squared exponential, squared exponential with automatic relevance determination, exponential and Matérn with parameter $v = \frac{5}{2}$ covariance functions. $\boldsymbol{J}_{1,D}$ denotes the vector of ones of length equal to the dimension D of the input space.

and the version using automatic relevance determination

$$\mathbf{K}_{SE}^{ard}(\mathbf{x}_i, \mathbf{x}_j) = \sigma_f^2 \exp\left(-\frac{1}{2}(\mathbf{x}_i - \mathbf{x}_j)^\top \lambda^{-2}(\mathbf{x}_i - \mathbf{x}_j)\right), \quad (2)$$

where λ stands for the characteristic length scale which measures the distance for being uncorrelated along x_i. The covariance matrices 1 and 2 differ only when λ is a diagonal matrix instead of a scalar. Two types of the Matérn covariance function were used,

$$\mathbf{K}_{Matérn}^{v=\frac{1}{2}}(r) = \exp\left(-\frac{r}{\ell}\right), \quad (3)$$

which is better known as exponential covariance function ($\mathbf{K}_{exp}$), and

$$\mathbf{K}_{Matérn}^{v=\frac{5}{2}}(r) = \sigma_f^2\left(1 + \frac{\sqrt{5}r}{\ell} + \frac{5r^2}{3\ell^2}\right)\exp\left(-\frac{\sqrt{5}r}{\ell}\right), \quad (4)$$

where $r = (\mathbf{x}_i - \mathbf{x}_j)$ and the parameter ℓ is the characteristic length-scale with which the distance of two considered data points is compared and σ_f^2 is the signal variance. The description of the listed covariance functions can be found in [12]. The considered covariance functions parameters are shown in Table 1.

In the performed experiments, different variants of the chosen EC strategies, described in Section 2.2, were examined, *generation-based* and *individual-based*. The result are discussed in the following sections.

3.3 Generation-Based EC Strategy

Apart from covariance function selection, *generation-based* EC strategy was determined by two other parameters, the number of model-evaluated generations and the multiplication factor of CMA-ES' step size σ, which is used in the original-evaluated generations in order to provide points for model training from a broader region of the input space. In the implementation, the first parameter was varied among the values 1, 2, 4 and 8 consecutive model-evaluated generations and the second parameter was varied among the values 1 and 2.

3.4 Individual-Based EC Strategy

Apart from covariance function selection, three other parameters, described in Section 2.2, were examined in the case of *individual-based* EC strategy. The first parameter $\alpha \in [0, 1]$ determines the proportion of the original population to be pre-sampled and evaluated by the original fitness function. The second parameter $\beta \in [1, \infty)$ is a multiplicator determining the size of extended population. The third parameter $\gamma \in [0, 1]$ determines the amount of points with the best model-fitness chosen from the extended population to be re-evaluated by the original fitness function to become a part of the final population. This parameter also determines the number of clusters, where the best point is chosen from each cluster and added to the final population.

In performed evaluations, the parameter α was varied among the values 0, 0.0625, 0.125, and 0.25, the parameter β was varied among the values 5 and 10 and γ was varied among the values 0, 0.1 and 0.2.

4 Results and Their Assessment

4.1 Result Tables

Tables 2 and 3 show the speed-up of S-CMA-ES and MGSO, compared to CMA-ES without a surrogate model. For the respective targets (distances to the true optimum Δf_{opt}), the speed-up of the expected running time (ERT) is shown. ERT is the number of function evaluations needed to reach the target divided by the ratio of the successful runs, which reached the target. Stopping criteria: the distance 10^{-8} to the true optimum and $100D$ original fitness function evaluations.

The first column in each box corresponds to the overall best settings (described in Section 4.2), the covariance function and EC settings of S-CMA-ES using the *generation-based* EC strategy in terms of the average speed-up. The second column corresponds to the best covariance function and *generation-based* EC settings for the respective function-dimension combination, if there was any better than the overall best observed settings. Analogously, the third and fourth columns show results for S-CMA-ES using *individual-based* EC strategy. Similarly, the last two columns in each box show the speed-up of the MGSO.

Signs "-" instead of the speed-up values mean that, unlike the CMA-ES, no run of the considered method (S-CMA-ES or MGSO) was able to reach that target. Signs

"+" mean that, unlike the employed method, no CMA-ES run was able to reach the target. Signs "*" mean that neither the considered method nor CMA-ES were able to reach the target. Speed-ups written in bold mark cases where the S-CMA-ES' or MGSO's median of the ERT is significantly lower than the median of the CMA-ES according to the one-sided Wilcoxon's test on the significance level $\alpha = 0.05$.

4.2 Observations

The MGSO method brought the highest speed-up in the case of $f2$, $f4$, 3D version of $f7$, 2D version of $f11$ and 5D version of $f12$. The worst results were observed in the case of $f1$, 2D version of $f10$ and 3D and 10D versions of $f12$. The best results were achieved using $\mathbf{K}_{\text{SE}}^{\text{iso}}$ covariance function, however, in the case of $f4$ and 2D version of $f7$ $\mathbf{K}_{\text{SE}}^{\text{ard}}$ covariance function brought much better results.

In the case of the *generation-based* EC, the overall best settings with respect to the median values are $(\mathbf{K}_{\text{SE}}^{\text{iso}}, 8, 1)$ – 8 consecutive model-evaluated generations with unmodified step size in combination with $\mathbf{K}_{\text{SE}}^{\text{iso}}$ covariance function. The overall best *generation-based* EC settings showed to be also the best *generation-based* EC settings of the respective functions, except for 3D version of $f12$, where S-CMA-ES performed better using larger step size. Using different covariance functions didn't bring much better results than the overall best covariance function.

The *individual-based* EC strategy achieved the best results with the overall settings $(\mathbf{K}_{\text{SE}}^{\text{ard}}, 0, 5, 0.1)$ – squared exponential covariance matrix with automatic relevance determination, no pre-sampling before training the model, 5 as the multiplicator determining the size of extended population and 0.1 as a multiplicator determining the amount of best points chosen from the extended population. The best results using described parameters were achieved in the case of functions $f3$ and 2D and 3D version of $f6$. However, the overall performance of the *individual-based* EC strategy lags far behind the *generation-based* EC strategy, MGSO and even CMA-ES itself.

4.3 Best-Fitness Progress Diagrams

Figure 2 shows examples of the best-fitness progress with the best observed settings (see Table 2 for details). Medians and the first and third quartiles of the best fitness reached are shown; medians and quartiles measured for MGSO and S-CMA-ES on 15 and 10 independent runs (for both EC strategies), respectively.

The optimization progress of the *individual-based* EC strategy showed to be the slowest in comparison to other methods. MGSO outperformed CMA-ES in most cases and the highest speed-up was achieved in the later phase of the optimization process. The *generation-based* EC strategy achieved the highest speed-up in the middle phase of the optimization process. However, the *generation-based*

f1 (1–D)

Δf_{opt}	S-CMA-ES - generation EC		S-CMA-ES - individual EC		MGSO	
1e1	00.90	01.00	01.00		00.17	00.14
1e0	00.90	01.00	01.00		00.06	00.14
1e-1	00.90	01.00	01.00		00.06	00.08
1e-2	00.90	01.00	01.00		-	00.05
1e-3	00.90	01.00	01.00		-	00.05
1e-4	00.90	01.00	01.00		-	-
1e-5	00.90	01.00	01.00		-	-
1e-6	00.90	01.00	01.00		-	-
1e-7	00.90	01.00	01.00		-	-
1e-8	00.90	01.00	01.00		-	-
param:	$(\mathbf{K}_{SE}^{iso},8,1)$	$(\mathbf{K}_{SE}^{ard},8,1)$	$(\mathbf{K}_{SE}^{ard},0,5,0.1)$		$\mathbf{K}_{SE}^{iso}$	$\mathbf{K}_{SE}^{ard}$

f2 (1–D)

Δf_{opt}	S-CMA-ES - generation EC		S-CMA-ES - individual EC		MGSO
1e-1	01.00	**01.44**	00.76	00.13	01.30
1e-2	**01.41**	**02.54**	00.30	00.10	**03.30**
1e-3	**01.85**	**02.54**	00.12	00.86	01.08
1e-4	**03.11**	**03.75**	-	01.73	**01.73**
1e-5	**04.04**	**04.49**	-	-	**02.25**
1e-6	**04.34**	**04.83**	-	-	**02.37**
1e-7	**12.95**	**13.57**	-	-	**08.05**
1e-8	12.21	15.23	-	-	16.78
param:	$(\mathbf{K}_{SE}^{iso},8,1)$	$(\mathbf{K}_{SE}^{ard},8,1)$	$(\mathbf{K}_{SE}^{ard},0,5,0.1)$	$(\mathbf{K}_{Matérn}^{v=\frac{5}{2}},2^{-2},10,0.2)$	$\mathbf{K}_{SE}^{iso}$

f3 (1–D)

Δf_{opt}	S-CMA-ES - generation EC	S-CMA-ES - individual EC		MGSO
1e-1	**02.44**	00.52	00.45	01.83
1e-2	05.71	01.44	01.51	02.13
1e-3	21.36	07.12	17.80	09.49
1e-4	18.41	07.12	-	08.63
1e-5	20.07	07.76	-	09.41
1e-6	20.07	07.76	-	09.41
1e-7	*	*	*	*
1e-8	*	*	*	*
param:	$(\mathbf{K}_{SE}^{iso},8,1)$	$(\mathbf{K}_{SE}^{ard},0,5,0.1)$	$(\mathbf{K}_{exp},0,5,0.2)$	$\mathbf{K}_{SE}^{iso}$

f4 (2–D)

Δf_{opt}	S-CMA-ES - generation EC		S-CMA-ES - individual EC		MGSO	
1e2	00.54	01.00	00.16	00.11	00.35	00.35
1e1	01.24	**01.63**	00.17	00.44	**01.55**	01.55
1e0	**02.59**	**03.52**	-	01.26	**02.20**	04.00
1e-1	**02.74**	**03.47**	-	-	**01.97**	02.62
1e-2	**02.99**	**03.66**	-	-	**02.44**	03.25
1e-3	**03.91**	**05.00**	-	-	**03.58**	04.57
1e-4	**04.45**	**05.22**	-	-	**04.52**	04.68
1e-5	**13.44**	**14.83**	-	-	00.67	13.23
1e-6	**17.63**	**20.10**	-	-	-	18.12
1e-7	+	+	*	*	*	+
1e-8	+	+	*	*	*	+
param:	$(\mathbf{K}_{SE}^{iso},8,1)$	$(\mathbf{K}_{exp},8,1)$	$(\mathbf{K}_{SE}^{ard},0,5,0.1)$	$(\mathbf{K}_{exp},2^{-2},5,0.1)$	$\mathbf{K}_{SE}^{iso}$	$\mathbf{K}_{SE}^{ard}$

f5 (2–D)

Δf_{opt}	S-CMA-ES - generation EC	S-CMA-ES - individual EC		MGSO	
1e0	01.00	00.12	00.09	00.65	00.65
1e-1	**01.72**	00.05	00.08	00.80	00.80
1e-2	**03.28**	-	00.70	**01.52**	01.52
1e-3	**02.95**	-	-	01.51	01.92
1e-4	**03.27**	-	-	**01.96**	02.48
1e-5	**03.74**	-	-	**02.55**	03.23
1e-6	*	*	*	*	*
1e-7	*	*	*	*	*
1e-8	*	*	*	*	*
param:	$(\mathbf{K}_{SE}^{iso},8,1)$	$(\mathbf{K}_{SE}^{ard},0,5,0.1)$	$(\mathbf{K}_{Matérn}^{v=\frac{5}{2}},0,10,0.2)$	$\mathbf{K}_{SE}^{iso}$	$\mathbf{K}_{SE}^{ard}$

f6 (2–D)

Δf_{opt}	S-CMA-ES - generation EC		S-CMA-ES - individual EC		MGSO	
1e2	**01.18**	01.18	-	01.74	**03.04**	02.03
1e1	04.97	04.20	-	-	00.36	02.21
1e0	07.16	08.40	-	-	-	01.80
1e-1	+	+	*	*	*	+
1e-2	+	+	*	*	*	+
1e-3	+	+	*	*	*	+
1e-4	+	+	*	*	*	+
1e-5	*	*	*	*	*	*
1e-6	*	*	*	*	*	*
1e-7	*	*	*	*	*	*
1e-8	*	*	*	*	*	*
param:	$(\mathbf{K}_{SE}^{iso},8,1)$	$(\mathbf{K}_{SE}^{ard},4,1)$	$(\mathbf{K}_{SE}^{ard},0,5,0.1)$	$(\mathbf{K}_{SE}^{ard},0,10,0)$	$\mathbf{K}_{SE}^{iso}$	$\mathbf{K}_{SE}^{ard}$

f6 (3–D)

Δf_{opt}	S-CMA-ES - generation EC	S-CMA-ES - individual EC		MGSO
1e4	01.00	01.00	01.00	00.27
1e3	17.55	-	07.42	04.53
1e2	07.72	-	-	04.06
1e1	05.96	-	-	-
1e0	+	*	*	*
1e-1	+	*	*	*
1e-2	+	*	*	*
1e-3	+	*	*	*
1e-4	+	*	*	*
1e-5	+	*	*	*
1e-6	+	*	*	*
1e-7	+	*	*	*
param:	$(\mathbf{K}_{SE}^{iso},8,1)$	$(\mathbf{K}_{SE}^{ard},0,5,0.1)$	$(\mathbf{K}_{SE}^{ard},0,10,0.1)$	$\mathbf{K}_{SE}^{iso}$

f7 (2–D)

Δf_{opt}	S-CMA-ES - generation EC		S-CMA-ES - individual EC		MGSO	
1e-2	**03.28**	**02.93**	00.05	00.43	00.14	01.02
1e-3	**04.42**	**04.81**	-	-	00.27	00.54
1e-4	**05.63**	**06.09**	-	-	-	00.52
1e-5	**07.91**	**07.91**	-	-	-	01.45
1e-6	**16.24**	**16.24**	-	-	-	03.39
1e-7	65.25	68.62	-	-	-	11.29
1e-8	+	+	*	*	*	+
param:	$(\mathbf{K}_{SE}^{iso},8,1)$	$(\mathbf{K}_{exp},8,1)$	$(\mathbf{K}_{SE}^{ard},0,5,0.1)$	$(\mathbf{K}_{SE}^{ard},2^{-4},10,0)$	$\mathbf{K}_{SE}^{iso}$	$\mathbf{K}_{SE}^{ard}$

f7 (3–D)

Δf_{opt}	S-CMA-ES - generation EC	S-CMA-ES - individual EC		MGSO
1e-1	03.15	00.03	01.28	01.14
1e-2	**05.26**	-	-	**01.56**
1e-3	**10.36**	-	-	**03.16**
1e-4	**13.33**	-	-	**04.95**
1e-5	**64.47**	-	-	22.83
1e-6	+	*	*	+
1e-7	+	*	*	+
1e-8	+	*	*	+
param:	$(\mathbf{K}_{SE}^{iso},8,1)$	$(\mathbf{K}_{SE}^{ard},0,5,0.1)$	$(\mathbf{K}_{Matérn}^{v=\frac{5}{2}},0,10,0)$	$\mathbf{K}_{SE}^{iso}$

f8 (2–D)

Δf_{opt}	S-CMA-ES - generation EC		S-CMA-ES - individual EC		MGSO	
1e0	**02.68**	**02.68**	-	00.96	**01.68**	01.68
1e-1	**03.91**	**03.91**	-	-	**02.22**	02.22
1e-2	**04.46**	**04.79**	-	-	**03.20**	03.20
1e-3	**07.15**	**07.65**	-	-	**03.29**	03.29
1e-4	**12.95**	**13.74**	-	-	**05.61**	06.73
1e-5	**22.40**	**24.85**	-	-	**11.39**	13.53
1e-6	+	+	*	*	+	+
1e-7	+	+	*	*	+	+
1e-8	+	+	*	*	+	+
param:	$(\mathbf{K}_{SE}^{iso},8,1)$	$(\mathbf{K}_{Matérn}^{v=\frac{5}{2}},8,1)$	$(\mathbf{K}_{SE}^{ard},0,5,0.1)$	$(\mathbf{K}_{exp},2^{-4},5,0)$	$\mathbf{K}_{SE}^{iso}$	$\mathbf{K}_{SE}^{ard}$

Table 2: Speed-up of S-CMA-ES using *individual-* and *generation-based* strategies and MGSO, compared to CMA-ES without a surrogate model – functions $f1 - f10$ (see Section 4.1 for details). Empty columns signify that the best observed settings for the respective function-dimension combination are identical with the overall best observed settings.

f9 (2–D)

Δf_{opt}	S-CMA-ES - generation EC		S-CMA-ES - individual EC		MGSO
1e1	**03.40**	**02.74**	-	00.10	01.06
1e0	**03.38**	**03.11**	-	00.07	**01.15**
1e-1	**03.82**	**03.82**	-	-	**01.53**
1e-2	**04.35**	**04.35**	-	-	**01.52**
1e-3	**07.80**	**07.80**	-	-	**02.68**
1e-4	**22.40**	**23.56**	-	-	**08.54**
1e-5	+	+	*	*	+
1e-6	+	+	*	*	+
1e-7	+	+	*	*	+
1e-8	+	+	*	*	*
param:	$(\mathbf{K}_{SE}^{iso}, 8, 1)$	$(\mathbf{K}_{SE}^{ard}, 8, 1)$	$(\mathbf{K}_{SE}^{ard}, 0, 5, 0.1)$	$(\mathbf{K}_{SE}^{ard}, 2^{-4}, 5, 0)$	$\mathbf{K}_{SE}^{iso}$

f10 (2–D)

Δf_{opt}	S-CMA-ES - generation EC		S-CMA-ES - individual EC		MGSO
1e2	**01.98**	**01.76**	00.03	00.26	00.39
1e1	**02.29**	**02.62**	-	-	00.46
1e0	**03.01**	**03.21**	-	-	00.07
1e-1	**03.71**	**04.00**	-	-	-
1e-2	**06.15**	**07.02**	-	-	-
1e-3	18.28	21.88	-	-	-
1e-4	+	+	*	*	*
1e-5	+	+	*	*	*
1e-6	+	+	*	*	*
1e-7	+	+	*	*	*
1e-8	+	+	*	*	*
param:	$(\mathbf{K}_{SE}^{iso}, 8, 1)$	$(\mathbf{K}_{Matérn}^{v=\frac{5}{2}}, 8, 1)$	$(\mathbf{K}_{SE}^{ard}, 0, 5, 0.1)$	$(\mathbf{K}_{SE}^{ard}, 0, 10, 0)$	$\mathbf{K}_{SE}^{iso}$

f11 (2–D)

Δf_{opt}	S-CMA-ES - generation EC		S-CMA-ES - individual EC		MGSO
1e2	**03.92**	**03.38**	00.04	01.06	**01.24**
1e1	**03.69**	**03.69**	-	-	**01.43**
1e0	**04.63**	**05.01**	-	-	**02.32**
1e-1	**05.19**	**05.19**	-	-	**03.18**
1e-2	**08.70**	**09.23**	-	-	**02.51**
1e-3	**12.06**	**14.04**	-	-	**03.72**
1e-4	59.40	65.25	-	-	18.46
1e-5	+	+	*	*	+
1e-6	+	+	*	*	*
1e-7	+	+	*	*	*
1e-8	+	+	*	*	*
param:	$(\mathbf{K}_{SE}^{iso}, 8, 1)$	$(\mathbf{K}_{SE}^{ard}, 8, 1)$	$(\mathbf{K}_{SE}^{ard}, 0, 5, 0.1)$	$(\mathbf{K}_{SE}^{iso}, 2^{-4}, 10, 0)$	$\mathbf{K}_{SE}^{iso}$

f11 (3–D)

Δf_{opt}	S-CMA-ES - generation EC		S-CMA-ES - individual EC		MGSO
1e3	00.53	01.00	01.00	01.00	00.27
1e2	**02.51**	**03.04**	-	-	**01.33**
1e1	**03.04**	**03.38**	-	-	**01.41**
1e0	**03.12**	**03.47**	-	-	**01.85**
1e-1	**03.90**	**04.33**	-	-	**01.71**
1e-2	**05.95**	**07.21**	-	-	**02.19**
1e-3	27.83	32.15	-	-	11.06
1e-4	+	+	*	*	+
1e-5	+	+	*	*	+
1e-6	+	+	*	*	+
1e-7	+	+	*	*	*
1e-8	+	+	*	*	*
param:	$(\mathbf{K}_{SE}^{iso}, 8, 1)$	$(\mathbf{K}_{exp}, 8, 1)$	$(\mathbf{K}_{SE}^{ard}, 0, 5, 0.1)$	$(\mathbf{K}_{SE}^{ard}, 0, 5, 0)$	$\mathbf{K}_{SE}^{iso}$

f12 (3–D)

Δf_{opt}	S-CMA-ES - generation EC		S-CMA-ES - individual EC		MGSO	
1e2	**02.05**	01.72	00.37	00.27	00.91	00.46
1e1	**01.42**	02.27	-	00.67	00.31	00.63
1e0	-	**02.28**	-	-	00.48	00.65
1e-1	-	**02.76**	-	-	-	-
1e-2	-	**03.78**	-	-	-	-
1e-3	-	19.56	-	-	-	-
1e-4	*	+	*	*	*	*
1e-5	*	+	*	*	*	*
1e-6	*	+	*	*	*	*
1e-7	*	+	*	*	*	*
1e-8	*	+	*	*	*	*
param:	$(\mathbf{K}_{SE}^{iso}, 8, 1)$	$(\mathbf{K}_{SE}^{ard}, 8, 2)$	$(\mathbf{K}_{SE}^{ard}, 0, 5, 0.1)$	$(\mathbf{K}_{Matérn}^{v=\frac{5}{2}}, 0, 10, 0.1)$	$\mathbf{K}_{SE}^{iso}$	$\mathbf{K}_{SE}^{ard}$

f11 (5–D)

Δf_{opt}	S-CMA-ES - generation EC	S-CMA-ES - individual EC	MGSO
1e2	**03.02**	-	**01.36**
1e1	**03.24**	-	**01.51**
1e0	**04.37**	-	**01.96**
1e-1	**09.79**	-	03.12
1e-2	27.12	-	07.51
1e-3	47.82	-	13.17
1e-4	+	*	+
1e-5	+	*	*
1e-6	+	*	*
1e-7	+	*	*
1e-8	+	*	*
param:	$(\mathbf{K}_{SE}^{iso}, 8, 1)$	$(\mathbf{K}_{SE}^{ard}, 0, 5, 0.1)$	$\mathbf{K}_{SE}^{iso}$

f11 (10–D)

Δf_{opt}	S-CMA-ES - generation EC		S-CMA-ES - individual EC		MGSO
1e3	00.89	01.14	00.17	00.64	00.27
1e2	**03.65**	**04.11**	-	-	01.05
1e1	**08.51**	**09.35**	-	-	02.61
1e0	**09.78**	**10.59**	-	-	02.10
1e-1	**20.02**	**21.50**	-	-	-
1e-2	+	+	*	*	*
1e-3	+	+	*	*	*
1e-4	+	+	*	*	*
1e-5	+	+	*	*	*
1e-6	+	+	*	*	*
1e-7	*	+	*	*	*
1e-8	*	*	*	*	*
param:	$(\mathbf{K}_{SE}^{iso}, 8, 1)$	$(\mathbf{K}_{Matérn}^{v=\frac{5}{2}}, 8, 1)$	$(\mathbf{K}_{SE}^{ard}, 0, 5, 0.1)$	$(\mathbf{K}_{SE}^{iso}, 2^{-4}, 5, 0)$	$\mathbf{K}_{SE}^{iso}$

f12 (5–D)

Δf_{opt}	S-CMA-ES - generation EC		S-CMA-ES - individual EC		MGSO
1e2	**01.64**	**01.70**	-	00.04	00.78
1e1	**05.60**	**06.61**	-	-	**04.73**
1e0	19.24	33.01	-	-	28.86
1e-1	+	+	*	*	+
1e-2	+	+	*	*	+
1e-3	+	+	*	*	+
1e-4	+	+	*	*	*
1e-5	+	+	*	*	*
1e-6	+	+	*	*	*
1e-7	*	+	*	*	*
1e-8	*	*	*	*	*
param:	$(\mathbf{K}_{SE}^{iso}, 8, 1)$	$(\mathbf{K}_{SE}^{ard}, 4, 1)$	$(\mathbf{K}_{SE}^{ard}, 0, 5, 0.1)$	$(\mathbf{K}_{SE}^{iso}, 2^{-2}, 5, 0)$	$\mathbf{K}_{SE}^{iso}$

f12 (10–D)

Δf_{opt}	S-CMA-ES - generation EC	S-CMA-ES - individual EC	MGSO	
1e2	**08.98**	-	**01.88**	02.02
1e1	**06.83**	-	-	01.05
1e0	17.61	-	-	-
1e-1	+	*	*	*
1e-2	+	*	*	*
1e-3	+	*	*	*
1e-4	+	*	*	*
1e-5	+	*	*	*
1e-6	*	*	*	*
1e-7	*	*	*	*
1e-8	*	*	*	*
param:	$(\mathbf{K}_{SE}^{iso}, 8, 1)$	$(\mathbf{K}_{SE}^{ard}, 0, 5, 0.1)$	$\mathbf{K}_{SE}^{iso}$	$\mathbf{K}_{SE}^{ard}$

f12 (20–D)

Δf_{opt}	S-CMA-ES - generation EC	S-CMA-ES - individual EC		MGSO
1e3	01.39	00.05	00.18	00.42
1e2	**02.73**	-	-	00.03
1e1	**07.24**	-	-	-
1e0	59.06	-	-	-
1e-1	+	*	*	*
1e-2	+	*	*	*
1e-3	+	*	*	*
1e-4	+	*	*	*
1e-5	+	*	*	*
1e-6	*	*	*	*
1e-7	*	*	*	*
1e-8	*	*	*	*
param:	$(\mathbf{K}_{SE}^{iso}, 8, 1)$	$(\mathbf{K}_{SE}^{ard}, 0, 5, 0.1)$	$(\mathbf{K}_{SE}^{iso}, 0, 10, 0)$	$\mathbf{K}_{SE}^{iso}$

Table 3: Speed-up of S-CMA-ES using *individual-* and *generation-based* strategies and MGSO, compared to CMA-ES without a surrogate model – functions $f11 - f20$ (see Section 4.1 for details). Empty columns signify that the best observed settings for the respective function-dimension combination are identical with the overall best observed settings.

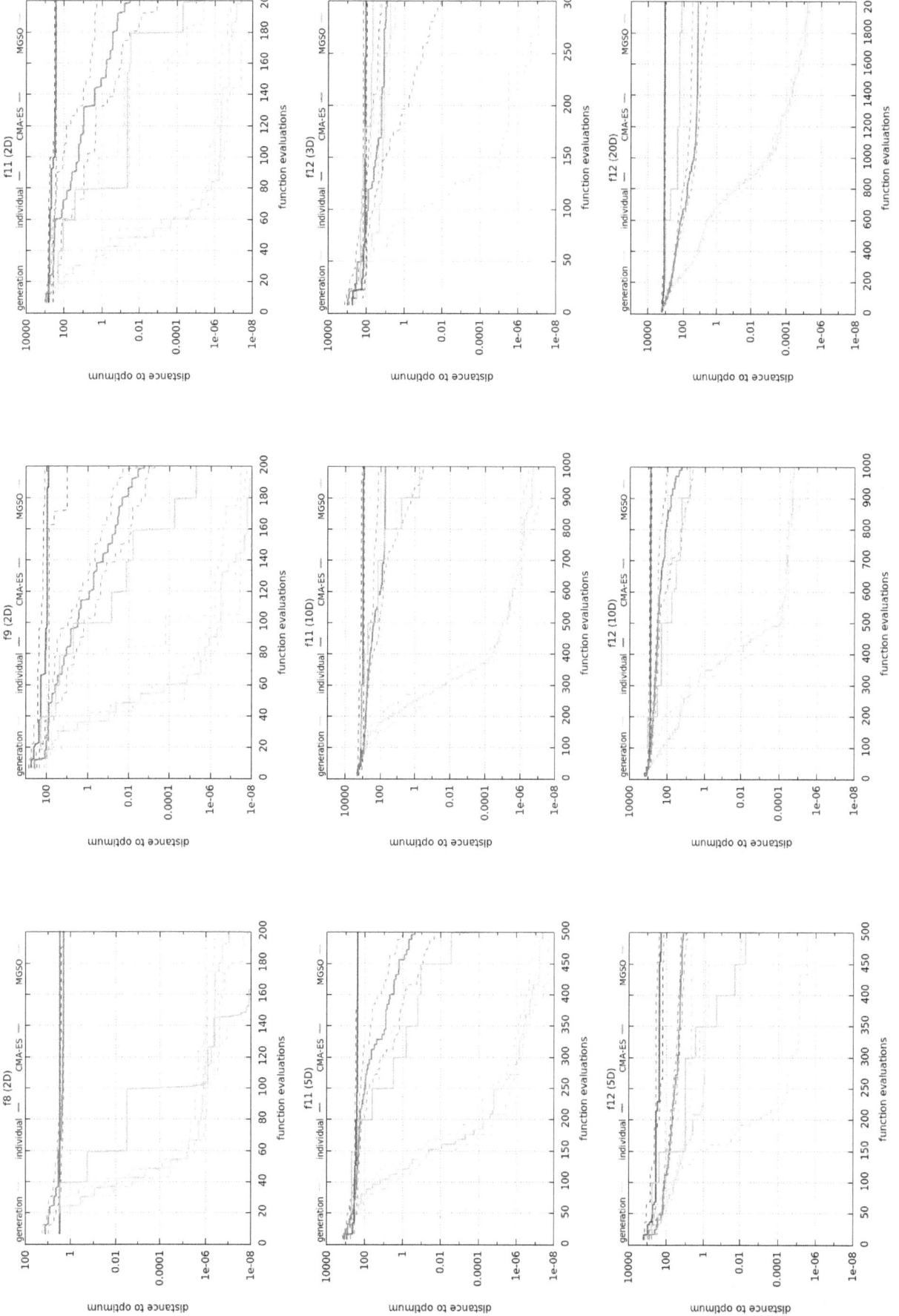

Figure 2: Examples of the best-fitness progress with the best observed settings, solid line: median, dashed lines: quartiles.

EC strategy generally brought the best results, it was outperformed by MGSO in the case of 2D version of $f8$ and 5D version of $f12$ in the later phase of the optimization process.

5 Conclusion

In this paper, two optimization approaches based on Gaussian processes were tested on the set of multimodal fitness functions from the CEC 2013 competition [10], and were compared to the state-of-the-art evolutionary approach in black-box optimization, CMA-ES. One of them is Model Guided Sampling Optimization [2], the other approach, S-CMA-ES [5], consists in using GP as a surrogate model for CMA-ES. The performance of the methods was compared with respect to the number of function evaluations.

In the case of S-CMA-ES, two evolution control strategies were used, the *individual-* and *generation-based*. Although S-CMA-ES using *generation-based* EC strategy outperformed MGSO, both methods showed the performance improvement in most cases. On the other hand, the *individual-based* EC strategy brought the worst results of all considered methods. We also observed, that S-CMA-ES performs better using *generation-based* EC setting with more consecutive model-evaluated generations. Isotropic squared exponential covariance function showed to be the most suitable for the optimization from all tested covariance functions.

This is a work in progress that is a part of a broader ongoing research. Terefore, it would be premature to draw deeper conclusions at this stage. We hope to be able to draw such conclusions after further investigations will be performed in the future.

Acknowledgement

The research reported in this paper has been supported by the Czech Science Foundation (GAČR) grant 13-17187S.

References

[1] Nik, M. A., Fayazbakhsh, K., Pasini, D., Lessard, L.: A comparative study of metamodeling methods for the design optimization of variable stiffness composites. Composite Structures **107** (2014), 494–501

[2] Bajer, L., Charypar, V., Holeňa, M.: Model guided sampling optimization with gaussian processes for expensive black-box optimization. In: Blum, C. (ed.), GECCO Companion'13, 2013, New York: ACM

[3] Bajer, L., Holeňa, M.: Two gaussian approaches to black-box optomization. CoRR, abs/1411.7806, 2014

[4] Bajer, L., Pitra, Z., Holeňa, M.: Benchmarking Gaussian processes and random forests surrogate models on the BBOB noiseless testbed. In: GECCO'15, Madrid, Spain, 2015, ACM

[5] Bajer, L., Pitra, Z., Holeňa, M.: Investigation of Gaussian processes and random forests as surrogate models for evolutionary black-box optimization. In: GECCO'15, Poster Abstracts, Madrid, Spain, 2015, ACM, TBA

[6] Chaput, J. C., Szostak, J. W.: Evolutionary optimization of a nonbiological ATP binding protein for improved folding stability. Chemistry & Biology bf 11(6) (June 2004), 865–874

[7] Hansen, N., Finck, S., Ros, R., Auger, A.: Real-parameter black-box optimization benchmarking 2012: experimental setup. Technical Report, INRIA, 2012

[8] Jones, D. R., Schonlau, M., Welch, W. J.: Efficient Global Optimization of expensive black-box functions. J. of Global Optimization **13(4)** (Dec. 1998), 455–492

[9] Larranaga, P., Lozano, J.: Estimation of distribution algorithms: a new tool for evolutionary computation. Kluwer Academic Pub, 2002

[10] Li, X., Engelbrecht, A., Epitropakis, M. G.: Benchmark functions for CEC'2013 special session and competition on niching methods for multimodal function optimization'. 2013.

[11] Loshchilov, I., Schoenauer, M., Sebag, M.: Self-adaptive surrogate-assisted covariance matrix adaptation evolution strategy. In: Proceedings of the Fourteenth International Conference On Genetic And Evolutionary Computation Conference, GECCO'12, 321–328, New York, NY, USA, 2012, ACM

[12] Rasmussen, C., Williams, C.: Gaussian processes for machine learning. Adaptive Computation and Machine Learning, MIT Press, Cambridge, MA, USA, Jan. 2006

[13] Tesauro, G., Jong, N. K., Das, R., Bennani, M. N.: On the use of hybrid reinforcement learning for autonomic resource allocation. Cluster Computing **10(3)** (2007), 287–299

J. Yaghob (Ed.): ITAT 2015 pp. 167–171
Charles University in Prague, Prague, 2015

ITAT

Limitations of One-Hidden-Layer Perceptron Networks

Věra Kůrková

Institute of Computer Science, Czech Academy of Sciences,
vera@cs.cas.cz,
WWW home page: http://www.cs.cas.cz/~vera

Abstract: Limitations of one-hidden-layer perceptron networks to represent efficiently finite mappings is investigated. It is shown that almost any uniformly randomly chosen mapping on a sufficiently large finite domain cannot be tractably represented by a one-hidden-layer perceptron network. This existential probabilistic result is complemented by a concrete example of a class of functions constructed using quasi-random sequences. Analogies with central paradox of coding theory and no free lunch theorem are discussed.

1 Introduction

A widely-used type of a neural-network architecture is a network with one-hidden-layer of computational units (such as perceptrons, radial or kernel units) and one linear output unit. Recently, new hybrid learning algorithms for feedforward networks with two or more hidden layers, called deep networks [9, 3], were successfully applied to various pattern recognition tasks. Thus a theoretical analysis identifying tasks for which shallow networks require considerably larger model complexities than deep ones is needed. In [4, 5], Bengio et al. suggested that a cause of large model complexities of shallow networks with one hidden layer might be in the "amount of variations" of functions to be computed and they illustrated their suggestion by an example of representation of d-dimensional parities by Gaussian SVM.

In practical applications, feedforward networks compute functions on finite domains in $\mathbb{R}^d$ representing, e.g., scattered empirical data or pixels of images. It is well-known that shallow networks with many types of computational units have the "universal representation property", i.e., they can exactly represent any real-valued function on a finite domain. This property holds, e.g., for networks with perceptrons with any sigmoidal activation function [10] and for networks with Gaussian radial units [15]. However, proofs of universal representation capabilities assume that networks have numbers of hidden units equal to sizes of domains of functions to be computed. For large domains, this can be a factor limiting practical implementations. Upper bounds on rates of approximation of multivariable functions by shallow networks with increasing numbers of units were studied in terms of variational norms tailored to types of network units (see, e.g., [11] and references therein).

In this paper, we employ these norms to derive lower bounds on model complexities of shallow networks representing finite mappings. Using geometrical properties of high-dimensional spaces we show that a representation of almost any uniformly randomly chosen function on a "large" finite domain by a shallow perceptron networks requires "large" number of units or "large" sizes of output weights. We illustrate this existential probabilistic result by a concrete construction of a class of functions based on Hadamard and quasi-noise matrices. We discuss analogies with central paradox of coding theory and no free lunch theorem.

The paper is organized as follows. Section 2 contains basic concepts and notations on shallow networks and dictionaries of computational units. Section 3 reviews variational norms as tools for investigation of network complexity. In Section 3, estimates of probabilistic distributions of sizes of variational norms are proven. In section 4, concrete examples of functions which cannot be tractably represented by perceptron networks are constructed using Hadamard and pseudo-noise matrices. Section 5 is a brief discussion.

2 Preliminaries

One-hidden-layer networks with single linear outputs (*shallow networks*) compute input-output functions from sets of the form

$$\text{span}_n G := \left\{ \sum_{i=1}^{n} w_i g_i \,\middle|\, w_i \in \mathbb{R}, g_i \in G \right\},$$

where G, called a *dictionary*, is a set of functions computable by a given type of units, the coefficients w_i are called output weights, and n is the number of hidden units. This number is sometimes used as a measure of *model complexity*.

In this paper, we focus on representations of functions on finite domains $X \subset \mathbb{R}^d$. We denote by

$$\mathscr{F}(X) := \{ f \,|\, f : X \to \mathbb{R} \}$$

the *set of all real-valued functions on X*. On $\mathscr{F}(X)$ we have the Euclidean inner product defined as

$$\langle f, g \rangle := \sum_{u \in X} f(u) g(u)$$

and the Euclidean norm

$$\|f\| := \sqrt{\langle f, f \rangle}.$$

To distinguish the inner product $\langle .,.\rangle$ on $\mathcal{F}(X)$ from the inner product on $X \subset \mathbb{R}^d$, we denote it $\cdot$, i.e., for $u, v \in X$,

$$u \cdot v := \sum_{i=1}^{d} u_i v_i.$$

We investigate networks with units from the dictionary of *signum perceptrons*

$$P_d(X) := \{\operatorname{sgn}(v \cdot . + b) : X \to \{-1, 1\} \mid v \in \mathbb{R}^d, b \in \mathbb{R}\}$$

where $\operatorname{sgn}(t) := -1$ for $t < 0$ and $\operatorname{sign}(t) := 1$ for $t \geq 0$. Note that from the point of view of model complexity, there is only a minor difference between networks with signum perceptrons and those with Heaviside perceptrons as

$$\operatorname{sgn}(t) = 2\vartheta(t) - 1$$

and

$$\vartheta(t) := \frac{\operatorname{sgn}(t) + 1}{2},$$

where $\vartheta(t) := 0$ for $t < 0$ and $\vartheta(t) = 1$ for $t \geq 0$.

3 Model Complexity and Variational Norms

A useful tool for derivation of estimates of numbers of units and sizes of output weights in shallow networks is the concept of a variational norm tailored to network units introduced in [12] as an extension of a concept of variation with respect to half-spaces from [2]. For a subset G of a normed linear space $(\mathcal{X}, \|.\|_{\mathcal{X}})$, *G-variation (variation with respect to the set G)*, denoted by $\|.\|_G$, is defined as

$$\|f\|_G := \inf \{c \in \mathbb{R}_+ \mid f/c \in \operatorname{cl}_{\mathcal{X}} \operatorname{conv}(G \cup -G)\},$$

where $\operatorname{cl}_{\mathcal{X}}$ denotes the closure with respect to the norm $\|\cdot\|_{\mathcal{X}}$ on $\mathcal{X}$, $-G := \{-g \mid g \in G\}$, and

$$\operatorname{conv} G := \left\{ \sum_{i=1}^{k} a_i g_i \mid a_i \in [0, 1], \sum_{i=1}^{k} a_i = 1, g_i \in G, k \in \mathbb{N} \right\}$$

is the convex hull of G. The following straightforward consequence of the definition of G-variation shows that in all representations of a function with "large" G-variation by shallow networks with units from the dictionary G, the number of units must be "large" or absolute values of some output weights must be "large".

Proposition 1. *Let G be a finite subset of a normed linear space $(\mathcal{X}, \|.\|_{\mathcal{X}})$, then for every $f \in \mathcal{X}$,*

$$\|f\|_G = \min \left\{ \sum_{i=1}^{k} |w_i| \,\middle|\, f = \sum_{i=1}^{k} w_i g_i, w_i \in \mathbb{R}, g_i \in G \right\}.$$

Note that classes of functions defined by constraints on their variational norms represent a similar type of a concept as classes of functions defined by constraints on both

numbers of gates and sizes of output weights studied in theory of circuit complexity [16].

To derive lower bounds on variational norms, we use the following theorem from [13] showing that functions which are "not correlated" to any element of the dictionary G have large variations.

Theorem 2. *Let $(\mathcal{X}, \|.\|_{\mathcal{X}})$ be a Hilbert space with inner product $\langle .,.\rangle_{\mathcal{X}}$ and G its bounded subset. Then for every $f \in \mathcal{X} - G^{\perp}$,*

$$\|f\|_G \geq \frac{\|f\|^2}{\sup_{g \in G} |\langle f, g \rangle_{\mathcal{X}}|}.$$

The following theorem shows that when a dictionary $G(X)$ is not "too large", then for a "large" domain X, almost any randomly chosen function has large $G(X)$-variation. We denote by

$$S_r(X) := \{f \in \mathcal{F}(X) \mid \|f\| = r\}$$

the sphere of radius r in $\mathcal{F}(X)$ and for $f \in \mathcal{F}(X)$, $f^o := \frac{f}{\|f\|}$. The proof of the theorem is based on geometry of spheres in high-dimensional Euclidean spaces. In large dimensions, most of areas of spheres lie very close to their "equators" [1].

Theorem 3. *Let d be a positive integer, $X \subset \mathbb{R}^d$ with $\operatorname{card} X = m$, $G(X)$ a subset of $\mathcal{F}(X)$ with $\operatorname{card} G(X) = n$ such that for all $g \in G(X)$, $\|f\| \leq r$, μ be a uniform probability measure on $S_r(X)$, and $b > 0$. Then*

$$\mu(\{f \in S_r(X) \mid \|f\|_{G(X)} \geq b\}) \geq 1 - 2n e^{-\frac{m}{2b^2}}.$$

Proof. Denote for $g \in S_r(X)$ and $\varepsilon \in (0, 1)$,

$$C(g, \varepsilon) := \{h \in S_r^{m-1} \mid |\langle h^o, g^o \rangle| \geq \varepsilon\}.$$

As $C(g, \varepsilon)$ is equivalent to a polar cap in $\mathbb{R}^{\operatorname{card} X}$, whose measure is exponentially decreasing with the dimension m, we have

$$\mu(C(g, \varepsilon)) \leq e^{-\frac{m\varepsilon^2}{2}}$$

(see, e.g., [1]). By Theorem 2,

$$\{f \in S_r(X) \mid \|f\|_{G(X)} \geq b\} = S_r(X) - \bigcup_{g \in G} C(g, 1/b).$$

Hence the statement follows. $\qquad\square$

Theorem 3 can be applied to dictionaries $G(X)$ on domains $X \subset \mathbb{R}^d$ with $\operatorname{card} X = m$, which are "relatively small". In particular, dictionaries of signum and Heaviside perceptrons are relatively "small". Estimates of their sizes can be obtained from bounds on numbers of linearly separable dichotomies to which finite subsets of $\mathbb{R}^d$ can be partitioned. Various estimates of numbers of dichotomies have been derived by several authors starting from results by Schläfli [17]. The next bound is obtained by combining a theorem from [7, p.330] with an upper bound on partial sum of binomials.

Theorem 4. *For every d and every $X \subset \mathbb{R}^d$ such that* $\operatorname{card} X = m$,

$$\operatorname{card} P_d(X) \leq 2 \sum_{i=0}^{d} \binom{m-1}{i} \leq 2 \frac{m^d}{d!}.$$

Combining Theorems 3 and 4, we obtain a lower bound on measures of sets of functions having variations with respect to signum perceptrons bounded from below by a given bound b.

Corollary 1. *Let d be a positive integer, $X \subset \mathbb{R}^d$ with* $\operatorname{card} X = m$, μ *a uniform probability measure on* $S_{\sqrt{m}}(X)$, *and $b > 0$. Then*

$$\mu(\{ f \in S_{\sqrt{m}}(X) \mid \|f\|_{P_d(X)} \geq b \}) \geq 1 - 4 \frac{m^d}{d!} e^{-\frac{m}{2b^2}}.$$

For example, for the domain $X = \{0,1\}^d$ and $b = 2^{\frac{d}{4}}$, we obtain from Corollary 1 a lower bound

$$1 - \frac{2^{d^2 + 2} e^{-(2^{\frac{d}{2}} - 1)}}{d!}$$

on the probability that a function on $\{0,1\}^d$ with the norm $2^{d/2}$ has variation with respect to signum perceptrons greater or equal to $2^{\frac{d}{4}}$. Thus for large d almost any uniformly randomly chosen function on the d-dimensional Boolean cube $\{0,1\}^d$ of the same norm $2^{d/2}$ as signum perceptrons, has variation with respect to signum perceptrons depending on d exponentially.

4 Construction of Functions with Large Variations

The results derived in the previous section are existential. In this section, we construct a class of functions, which cannot be represented by shallow perceptron networks of low model complexities. We construct such functions using Hadamard matrices. We show that the class of Hadamard matrices contains circulant matrices with rows being segments of pseudo-noise sequences which mimic some properties of random sequences.

Recall that a *Hadamard matrix* of order m is an $m \times m$ square matrix M with entries in $\{-1,1\}$ such that any two distinct rows (or equivalently columns) of M are orthogonal. Note that this property is invariant under permutating rows or columns and under sign flipping all entries in a column or a row. Two distinct rows of a Hadamard matrix differ in exactly $m/2$ positions.

The next theorem gives a lower bound on variation with respect to signum perceptrons of a $\{-1,1\}$-valued function constructed using a Hadamard matrix.

Theorem 5. *Let M be an $m \times m$ Hadamard matrix,* $\{x_i \mid i = 1, \ldots, m\} \subset \mathbb{R}^d$, $\{y_j \mid j = 1, \ldots, m\} \subset \mathbb{R}^d$, $X = \{x_i \mid i = 1, \ldots, m\} \times \{y_j \mid j = 1, \ldots, m\} \subset \mathbb{R}^{2d}$, *and $f_M : X \to \{-1,1\}$ be defined as $f_M(x_i, y_j) =: M_{i,j}$. Then*

$$\|f_M\|_{P_d(X)} \geq \frac{\sqrt{m}}{log_2 m}.$$

Proof. By Theorem 2,

$$\|f_M\|_{P_d(X)} \geq \frac{\|f_M\|^2}{\sup_{g \in P_d(X)} \langle f_M, g \rangle} = \frac{m^2}{\sup_{g \in P_d(X)} \langle f_M, g \rangle}.$$

For each $g \in P_d(X)$, let $M(g)$ be an $m \times m$ matrix defined as $M(g)_{i,j} = g(x_i, y_j)$. It is easy to see that

$$\langle f_M, g \rangle = \sum_{i,j} M_{i,j} M(g)_{i,j}.$$

Using suitable permutations, we reorder rows and columns of both matrices $M(g)$ and M in such a way that each row and each column of the reordered matrix $\bar{M}(g)$ starts with a (possibly empty) initial segment of -1's followed by a (possibly empty) segment of 1's. Denoting $\bar{M}$ the reordered matrix M we have

$$\langle f_M, g \rangle = \sum_{i,j} M_{i,j} M(g)_{i,j} = \sum_{i,j} \bar{M}_{i,j} \bar{M}(g)_{i,j}.$$

As the property of being a Hadamard matrix is invariant under permutations of rows and columns, we can apply Lindsay lemma [8, p.88] to submatrices of the Hadamard matrix $\bar{M}$ on which all entries of the matrix $\bar{M}(g)$ are either -1 or 1. Thus we obtain an upper bound $m\sqrt{m}$ on the differences of $+1$s and -1s in suitable submatrices of $\bar{M}$. Iterating the procedure at most $\log_2 m$-times, we obtain an upper bound $m\sqrt{m} \log_2 m$ on $\sum_{i,j} \bar{M}_{i,j} \bar{M}(g)_{i,j} = \langle f_M, g \rangle$. Thus

$$\|f_M\|_{P_d(X)} \geq \frac{m^2}{m\sqrt{m} \log_2 m} = \frac{\sqrt{m}}{\log_2 m}.$$

$\square$

Theorem 5 shows that functions whose representations by shallow perceptron networks require numbers of units or sizes of output weights bounded from below by $\frac{\sqrt{m}}{\log_2 m}$ can be constructed using Hadamard matrices. In particular, when the domain is d-dimensional Boolean cube $\{0,1\}^d$, where d is even, the lower bound is $\frac{2^{d/4}}{d/2}$. So the lower bounds grows with d exponentially.

Recall that if a Hadamard matrix of order m exists, then $m = 1$ or $m = 2$ or m is divisible by 4 [14, p.44]. It is conjectured that there exists a Hadamard matrix of every order divisible by 4. Listings of Hadamard matrices of various orders can be found at Neil Sloane's library of Hadamard matrices.

We show that suitable Hadamard matrices can be obtained from pseudo-noise sequences. An infinite sequence

$a_0, a_1, \ldots, a_i, \ldots$ of elements of $\{0,1\}$ is called *k-th order linear recurring sequence* if for some $h_0, \ldots, h_k \in \{0,1\}$

$$a_i = \sum_{j=1}^{k} a_{i-j} h_{k-j} \mod 2$$

for all $i \geq k$. It is called *k-th order pseudo-noise (PN) sequence* (or *pseudo-random sequence*) if it is *k*-th order linear recurring sequence with minimal period $2^k - 1$.

A $2^k \times 2^k$ matrix L is called *pseudo-noise* if for all $i = 1, \ldots, 2^k$, $L_{1,i} = 0$ and $L_{i,1} = 0$ and for all $i = 2, \ldots, 2^k$ and $j = 2, \ldots, 2^k$

$$L_{i,j} = \bar{L}_{i-1,j-1}$$

where the $(2^k - 1) \times (2^k - 1)$ matrix $\bar{L}$ is a circulant matrix with rows formed by shifted segments of length $2^k - 1$ of a *k*-th order pseudo-noise sequence.

PN sequences have many useful applications because some of their properties mimic those of random sequences. A run is a string of consecutive 1's or a string of consecutive 0's. In any segment of length $2^k - 1$ of *k*-th order PN-sequence, one-half of the runs have length 1, one quarter have length 2, one-eighth have length 3, and so on. In particular, there is one run of length k of 1's, one run of length $k - 1$ of 0's. Thus every segment of length $2^k - 1$ contains $2^{k/2}$ ones and $2^{k/2} - 1$ zeros [14, p.410].

Let $\tau : \{0,1\} \to \{-1,1\}$ be defined as $\tau(x) = -1^x$ (i.e., $\tau(0) = 1$ and $\tau(1) = -1$). The following theorem states that a matrix obtained by applying τ to entries of a pseudo-noise matrix is a Hadamard matrix.

Theorem 6. *Let L be a $2^k \times 2^k$ pseudo-noise matrix and L_τ be the $2^k \times 2^k$ matrix with entries in $\{-1,1\}$ obtained from L by applying τ to all its entries. Then L_τ is a Hadamard matrix.*

Proof. We show that inner product of any two rows of L_τ is equal to zero. The autocorrelation of a sequence $a_0, a_1, \ldots, a_i, \ldots$ of elements of $\{0,1\}$ with period $2^k - 1$ is defined as

$$\rho(t) = \frac{1}{2^k - 1} \sum_{j=0}^{2^k - 1} -1^{a_j + a_{j+t}}.$$

For every pseudo-noise sequence,

$$\rho(t) = -\frac{1}{2^k - 1}$$

for every $t = 1, \ldots, 2^k - 2$ [14, p. 411]. Thus the inner product of every two rows of the matrix $\bar{L}_\tau$ is equal to -1. As all elements of the first column of L_τ are equal to 1, inner product of every pair of its rows is equal to zero. $\square$

Theorem 5 implies that for every pseudo-noise matrix L of order 2^k and $X \subset \mathbb{R}^d$ such that $\text{card} X = 2^k \times 2^k$, there exists a function $f_{L_\tau} : X \to \{-1,1\}$ induced by the matrix L_τ obtained from L by replacing 0's with 1's and 1's with -1's such that

$$\|f_{L_\tau}\|_{P_d(X)} \geq \frac{2^{k/2}}{k}.$$

So the variation of f_{L_τ} with respect to signum perceptrons depends on k exponentially. In particular, setting $X = \{0,1\}^d$, where $d = 2k$ is even, we obtain a function of d variables with variation with respect to signum perceptrons growing with d exponentially as

$$\|f_{L_\tau}\|_{P_d(X)} \geq \frac{2^{d/4}}{d/2}.$$

Representation of this function by a shallow perceptron network requires number of units or sizes of some output weights depending on d exponentially.

It is easy to show that for each even integer d, the function induced by Sylvester-Hadamard matrix

$$M_{u,v} = -1^{u \cdot v},$$

where $u, v \in \{0,1\}^{d/2}$, can be represented by a two-hidden-layer network with $d/2$ units in each hidden layer.

5 Discussion

We proved that almost any uniformly randomly chosen function on a sufficiently large finite set in $\mathbb{R}^d$ has large variation with respect to signum perceptrons and thus it cannot be tractably represented by a shallow perceptron network.

It seems to be a paradox that although representations of almost all functions by shallow perceptron networks are "untractable", it is difficult to construct such functions. The situation can be rephrased in analogy with the title of an article from coding theory "Any code of which we cannot think is good" [6] as "representation of almost any function of which we cannot think by shallow perceptron networks is untractable". A central paradox of coding theory concerns the existence and construction of the best codes. Virtually every linear code is good (in the sense that it meets the Gilbert-Varshamov bound on distance versus redundancy), however despite the sophisticated constructions for codes derived over the years, no one has succeeded in demonstrating a constructive procedure that yields such good codes.

The only class of functions having "large" variations which we succeeded to construct is the class described in section 4 based on Hadamard matrices. Among these matrices belong quasi-noise (quasi-random) matrices with rows obtained as shifts of segments of quasi-noise sequences. These sequences have been used in construction of codes, interplanetary satellite picture transmission, precision measurements, acoustics, radar camouflage, and light diffusers. Pseudo-noise sequences permit design of surfaces that scatter incoming signals very broadly making reflected energy "invisible" or "inaudible".

It should be emphasized that similarly as "no free lunch theorem" [18], our results assume uniform distributions of functions to be represented. However, probability distributions of functions modeling some practical tasks of interest (such as colors of pixels in a photograph) might be highly non uniform.

Acknowledgments. This work was partially supported by the grant COST LD13002 of the Ministry of Education of the Czech Republic and institutional support of the Institute of Computer Science RVO 67985807.

References

[1] Ball, K.: An elementary introduction to modern convex geometry. In: S. Levy, (ed.), Falvors of Geometry, 1–58, Cambridge University Press, 1997

[2] Barron, A. R.: Neural net approximation. In: K. S. Narendra, (ed.), Proc. 7th Yale Workshop on Adaptive and Learning Systems, 69–72, Yale University Press, 1992

[3] Bengio, Y.: Learning deep architectures for AI. Foundations and Trends in Machine Learning **2** (2009), 1–127

[4] Bengio, Y., Delalleau, O., Le Roux, N.: The curse of dimensionality for local kernel machines. Technical Report 1258, Département d'Informatique et Recherche Opérationnelle, Université de Montréal, 2005

[5] Bengio, Y., Delalleau, O., Le Roux, N.: The curse of highly variable functions for local kernel machines. In: Advances in Neural Information Processing Systems 18, 107–114, MIT Press, 2006

[6] Coffey, J. T., Goodman, R. M.: Any code of which we cannot think is good. IEEE Transactions on Information Theory **36** (1990), 1453–1461

[7] Cover, T.: Geometrical and statistical properties of systems of linear inequalities with applictions in pattern recognition. IEEE Trans. on Electronic Computers **14** (1965), 326–334

[8] Erdös, P., Spencer, J. H.: Probabilistic Methods in Combinatorics. Academic Press, 1974

[9] Hinton, G. E., Osindero, S., Teh, Y. W.: A fast learning algorithm for deep belief nets. Neural Computation **18** (2006), 1527–1554

[10] Ito, Y.: Finite mapping by neural networks and truth functions. Mathematical Scientist **17** (1992), 69–77

[11] Kainen, P. C., Kůrková, V., Sanguineti, M.: Dependence of computational models on input dimension: Tractability of approximation and optimization tasks. IEEE Trans. on Information Theory **58** (2012), 1203–1214

[12] Kůrková, V.: Dimension-independent rates of approximation by neural networks. In: K. Warwick and M. Kárný, (eds.), Computer-Intensive Methods in Control and Signal Processing. The Curse of Dimensionality, 261–270, Birkhäuser, Boston, MA, 1997

[13] Kůrková, V., Savický, P., Hlaváčková, K.: Representations and rates of approximation of real-valued Boolean functions by neural networks. Neural Networks **11** (1998), 651–659

[14] MacWilliams, F. J., Sloane, N. J. A.: The theory of error-correcting codes. North Holland, New York, 1977

[15] Micchelli, C. A.: Interpolation of scattered data: Distance matrices and conditionally positive definite functions. Constructive Approximation **2** (1986), 11–22

[16] Roychowdhury, V., Siu, K.-Y., Orlitsky, A.: Neural models and spectral methods. In: V. Roychowdhury, K. Siu, and A. Orlitsky, (eds.), Theoretical Advances in Neural Computation and Learning, 3–36, Springer, New York, 1994

[17] Schläfli, L.: Theorie der vielfachen Kontinuität. Zürcher & Furrer, Zürich, 1901

[18] Wolpert, D. H., Macready, W. G.: No free lunch theorems for optimization. IEEE Transactions on Evolutionary Computation **1(1)** (1997), 67–82

J. Yaghob (Ed.): ITAT 2015 pp. 172–178
Charles University in Prague, Prague, 2015

ÍTAT

Benchmarking Classifier Performance with Sparse Measurements

Jan Motl[1]

Czech Technical University, Prague, Czech Republic,
jan.motl@fit.cvut.cz,
WWW home page: http://relational.cvut.cz

Abstract: The presented paper describes a methodology, how to perform benchmarking, when classifier performance measurements are sparse. The described methodology is based on missing value imputation and was demonstrated to work, even when 80% of measurements are missing, for example because of unavailable algorithm implementations or unavailable datasets. The methodology was then applied on 29 relational classifiers & propositional tools and 15 datasets, making it the biggest meta-analysis in relational classification up to date.

1 Introduction

You can't improve what you can't measure. However, in some fields the comparison of different approaches is demanding. For example, in the field of relational classifiers, essentially each classifier uses different syntax and requires data in different format, making the comparison of relational classifiers difficult. Despite these obstacles, each author of a new relational classifier attempts to prove that his algorithm is better than some previous algorithm and takes the burden of comparing his algorithm to a small set of algorithms on a limited set of datasets. But how can we compare the algorithms, if they are not evaluated on the same set of datasets?

1.1 Literature Review

The biggest meta-analysis of relational classifiers (to our best knowledge) is "Is mutagenesis still challenging?" [27], where 19 algorithms are examined. While this analysis is highly interesting, it limits itself on comparison of the classifiers on a single dataset.

The biggest analysis in the regard of used datasets is from Dhafer [21], where a single algorithm is tested on 10 datasets and 20 tasks (some datasets have multiple targets).

If we are interested into comparison of multiple algorithms on multiple datasets, the counts are comparably smaller. For example, in the article from Bina [2] 6 algorithms on 5 datasets are examined.

This meta-analysis presents comparison of 29 algorithms on 15 datasets.

1.2 Baseline

Traditionally, a set of algorithms is evaluated on a set of datasets. And then the algorithms are ordered with one (or

	Algo. 1	Algo. 2	Algo. 3
Dataset A	0.55	0.5	0.45
Dataset B	0.65	0.6	0.55
Dataset C	0.95	1	0.9
Average accuracy	0.72	0.7	0.63
Average ranking	1.33	1.67	3
# Wins	2	1	0

Table 1: Hypothetical evaluation of classifiers based on accuracy (bigger is better) with three ordering methods. In this scenario, all the methods are in agreement (Algorithm 1 is always the best).

	Algo. 1	Algo. 2	Algo. 3
Dataset A	0.55	0.5	-
Dataset B	0.65	0.6	-
Dataset C	-	1	0.9
Average accuracy	0.6	0.7	0.9
Average ranking	1	1.67	2
# Wins	2	1	0

Table 2: With sparse measurements, average measure predicts that Algorithm 3 is the best while the rest of the methods predict that Algorithm 1 is the best.

multiple) of the following methods:

- Average measure
- Average ranking
- Count of wins

The different ordering methods [6] are illustrated on an example in Table 1. In this hypothetical scenario 3 algorithms are evaluated on 3 datasets with accuracy. Based on each ordering method, the first algorithm is the best and the third algorithm is the worst.

But what if not all the measures are available? With the same data, but some missing, we can get different results (Table 2). Based on average accuracy the third algorithm is the best. But we are getting this result only because the third algorithm was evaluated on the datasets with high average accuracy (i.e. easy datasets), while the rest of the algorithms were evaluated on datasets with lower average accuracy (i.e. hard datasets).

Average ranking and count of wins are more robust to missing values. However, neither of them is infallible. Imagine that someone publishes an algorithm and its weaker version and measures the accuracy not only on the common datasets but also on thousands of randomly generated datasets that are never ever going to be classified by any other algorithm. Then the stronger version of the classifier is going to score at least a thousand wins and place on the first position on the leaderboard regardless of the score on the few common datasets.

If all the algorithms were evaluated on at least one common dataset, we could also order the algorithms just based on the common datasets. But if there isn't any dataset on which all the algorithms are evaluated, we have to come out with another solution.

The solution is to perform missing value imputation and convert the problem to the problem we can already solve.

2 Imputation

The proposed missing value imputation iteratively approximates:

$$acc \approx \overrightarrow{alg} * \overrightarrow{dat} \tag{1}$$

with following pseudocode:

```
acc = pivot(input, @mean)
alg = rowmean(acc)
dat = ones(1, ncol(acc))
for i = 1:nit
    dat = dat + colmean(acc - alg*dat)
    alg = alg + rowmean(acc - alg*dat)
end
```

Where:

input: Matrix with three columns: {algorithm name, dataset name, measured accuracy}.

acc: Matrix with accuracies, where algorithms are in rows and datasets in columns.

alg: Column vector with average accuracy of the algorithms over all the datasets. Initialized to average algorithm accuracy.

dat: Row vector with relative difficulty of the datasets. Initialized to a vector of ones.

nit: Parameter describing the count of iterations. 10 iterations are commonly sufficient.

2.1 Evaluation on a Dense Dataset

To assess the ability of the proposed imputation to properly order relational classifiers, a test on a related task was performed. Arguably the closest task to relational classification, which is well benchmarkable, is propositional classification - the most common type of classification, where a single table is classified.

Conveniently, accuracies of 179 propositional classifiers on 121 datasets were published in a recent study by Fernandez-Delgado [8]. Since not all the examined algorithms always finished successfully, e.g. due to colineality of data, a dense submatrix of 179 algorithms on 14 datasets

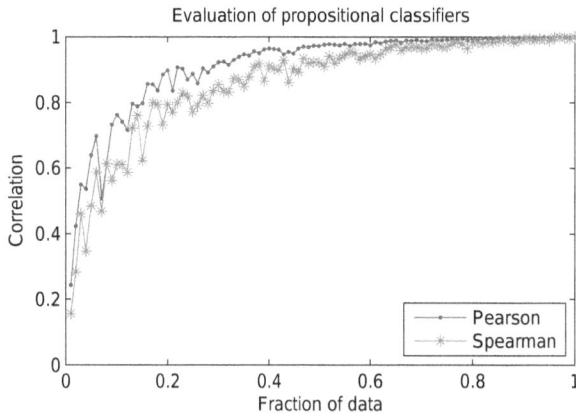

Figure 1: Correlation of the predicted algorithm order with the ground truth based on the proportion of missing data.

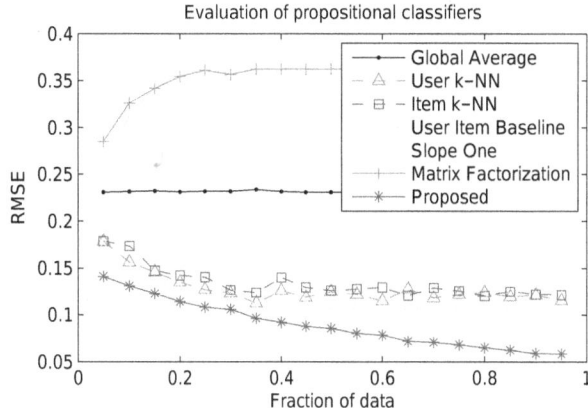

Figure 2: RMSE of the sampled and imputed submatrix with the dense submatrix.

was extracted. The dense submatrix was then randomly sampled with a variable count of measurements. The missing values were then imputed. The resulting learning curve, depicted in figure 1, suggests, that once 20% of all combinations algorithm × dataset are used, a fairly good estimate of the actual ordering can be estimated.

A comparison of the proposed imputation method to other imputation methods in regard to Root Mean Square Error (RMSE) is in figure 2. The reference methods are from RapidMiner and their parameters were optimized with grid search.

2.2 Theoretical Evaluation

According to the No-Free-Lunch theorem [41], the best classifier will not be the same for all the data sets. Hence we shouldn't even attempt to measure and average classifier's performance over a wide set of datasets. But merely describe strengths and weaknesses of different classifiers. In this respect, the selected imputation method fails because it is not able to model interactions between datasets and algorithms. Nevertheless, in the practice, some classifiers appear to be systematically better then other [8]. If all we want is to order classifiers based on their expected accuracy on a set of datasets, the absence of ability to model interactions is irrelevant.

Another property of the used methodology is that it doesn't permit mixture of measures. That is unfortunate since some articles [18] report only accuracy, while other articles [10] report only precision, recall and F-measure.

A special attention is necessary when we are comparing results from several authors, because not only the evaluation methodology can differ (e.g. 10-fold cross-validation vs. single hold-out sample), but also datasets can differ despite the common name. For example, the canonical version of East-West dataset (further abbreviated as *Trains*) contains 10 instances [30]. However, some authors prefer an extended version of the dataset with 20 instances [35]. Nevertheless, data quality is the pitfall common to all analyses. To alleviate the problem with different datasets in the future, a new (and the first) relational repository was based at `relational.fit.cvut.cz`. Further discussion about the collected data is provided in the next section.

The final limitation of the method is that it doesn't provide trustworthy confidence intervals. The first reason is that measures for the same algorithm and dataset are averaged and treated as a single measure. The second reason is that the algorithm performs missing value imputation, violating the assumption of sample independence.

A list summarizing the advantages and constrains of the proposed method follows:

Advantages:

- Permits benchmarking with sparse measures.
- Respects that some datasets are tougher than others.
- Allows conflicting measurements (for example, by different authors).

Disadvantages:

- Neglects interactions between datasets and algorithms.
- Requires one common measure (e.g. we cannot mix accuracy and F-measure).
- Requires comparably prepared datasets (e.g. using the same instance count).
- Doesn't provide confidence intervals.

3 Classification of Relational Data

The proposed methodology how to benchmark with sparse measurements is applied on relational classifiers, including propositional tools. In the following paragraphs description of the collected measures, benchmarked algorithms and datasets follow. The collected data can be downloaded from `motl.us\benchmarking`.

Algorithm	Algorithm type	Reference
Aleph	ILP	[35]
CILP++*	Neural Network	[9]
CrossMine	ILP	[42]
E-NB*	Probabilistic	[37]
FOIL	ILP	[23]
FORF-NA	Decision Tree	[40]
Graph-NB	Probabilistic	[26]
HNBC*	Probabilistic	[37]
kFOIL	Kernel	[23]
Lynx-RSM*	Propositionalization	[29]
MLN	Probabilistic	[36]
MRDTL-2	Decision Tree	[25]
MRNBC*	Probabilistic	[37]
MVC*	Multi-View	[11]
MVC-IM*	Multi-View	[12]
MulSVM*	Propositionalization	[43]
nFOIL*	Probabilistic	[22]
PIC*	Probabilistic	[37]
RELAGGS	Propositionalization	[17]
RPT*	Probabilistic	[28]
RSD	Propositionalization	[18]
RollUp	Propositionalization	[16]
SINUS	Propositionalization	[18]
SimFlat*	Propositionalization	[12]
SDF*	Decision Tree	[2]
TILDE	Decision Tree	[37]
TreeLiker-Poly	ILP	[20]
TreeLiker-RelF	ILP	[20]
Wordification*	Propositionalization	[35]

Table 3: List of 29 relational classifiers and propositional algorithms used in the meta-analysis. A star by the algorithm name marks algorithms, for which measurements by someone else than by the algorithm authors was not found.

3.1 Measure Selection

Since almost all relational classifiers in the literature are evaluated on classification accuracy (with exceptions like CLAMF [10], ACORA [34] or SAYU [4], which are evaluated in the literature only with measures based on precision & recall) but only a few were evaluated with a different measure (like precision & recall, F-measure, AUC or AUC-PR), the meta-analysis limits itself to classification accuracy. The methods how to measure accuracy may differ, but only testing accuracies (not training) were collected.

Other interesting measures, like runtime or memory consumption, were not evaluated, as they are rarely published. And even if they were published, they would be hardly comparable as the measurements are platform dependent.

Dataset	Target	#Instances	Reference
Alzheimer	acetyl	1326	[15]
Alzheimer	amine	686	[15]
Alzheimer	memory	642	[15]
Alzheimer	toxicity	886	[15]
Carcinogenesis	carcinogenic	329	[38]
ECML	insurance	7329	[19]
Financial	loan status	682	[1]
Hepatitis	biopsy	690	[35]
IMDb	ratings	281449	[2]
KRK	depth-of-win	1000	[18]
Mondial	religion	185	[39]
MovieLens	age	941	[2]
Musk-small	musk	92	[7]
Musk-large	musk	102	[7]
Mutagenesis	mutagenic	188	[5]
Thrombosis	degree	770	[3]
Trains	direction	10	[30]
UW-CSE	advisedBy	339	[36]

Table 4: List of 15 datasets with their targets (Alzheimer dataset has multiple targets) used in the meta-analysis.

3.2 Algorithm Selection

The selection of relational classifiers and propositionalization tools was restricted to algorithms, which:

- Were published in conference or journal paper.
- Were benchmarked on at least four datasets.
- Were evaluated on classification accuracy.

The list of compared algorithms is in table 3.

3.3 Dataset Selection

Datasets were selected based on the following criteria:

- The dataset has a defined classification target.
- The dataset consists of at least two tables.
- The dataset is used by at least four algorithms.

The used relational datasets are listed in table 4.

4 Results

Box plot in Figure 3 depicts estimated average classification accuracies of 29 algorithms on 15 datasets (18 tasks). The input data consists of 26% of all combinations algorithm × dataset, making the estimates solid (recall figure 1). The accuracies were estimated with 1000 bootstrap samples. Whiskers depict 1.5 IQR.

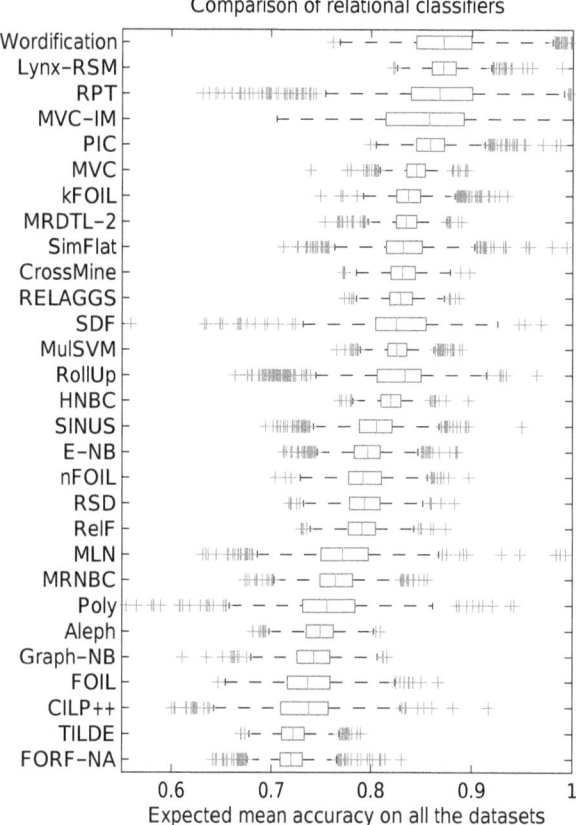

Figure 3: Box plot with expected accuracies.

4.1 Validation

The ordering of algorithms from the meta-analysis should roughly correspond to the ordering of the algorithms in the individual articles. Differences in the orderings are evaluated with Spearman correlation in table 5.

As we can see, the orderings in the literature can contradict – once RELLAGS is better than CrossMine, once CrossMine is better than RELLAGS.

5 Discussion

Summary of figure 3 based on algorithm type is in figure 4. Interestingly, kernel and multi-view approaches are averagely the most accurate algorithms. But some propositionalization algorithms, namely Wordification, Lynx-RSM and RPT (Relational Probabilistic Tree) beat them. Nevertheless, note that propositionalization algorithms are overrepresented in the meta-analysis, making it more likely that some of them place at extreme positions.

Also note, that performance of algorithms is influenced by their setting. While algorithms in figure 3 are ordered based on the average accuracy per combination of algorithm × dataset, algorithms in figure 5 are ordered based

Ordering	Spearman Correlation	Reference
TILDE < FORF-NA < Graph-NB < RelF < Poly < SDF	0.89	[2]
MRNBC < TILDE < E-NB < HNBC < PIC	0.9	[37]
TILDE < RELLAGS < CrossMine < MVC-IM	1.0	[12]
FOIL < TILDE < CrossMine < RELLAGS < MVC	0.9	[31]

Table 5: Comparison of algorithm ordering in the literature with ordering from the imputation.

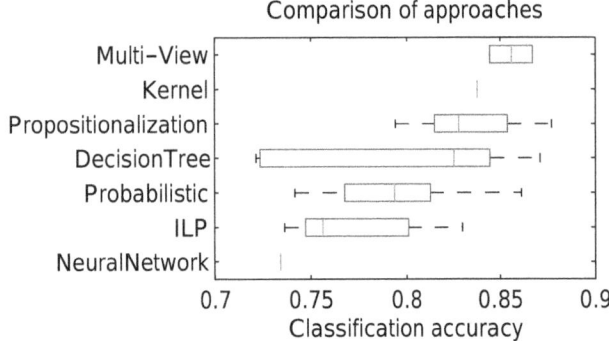

Figure 4: Estimated accuracies by algorithm type.

on the maximal known accuracy per combination of algorithm × dataset. Notably, with the right setting, RSD improves its ranking by 5 positions.

Finally, it is trusted that each individual author measured accuracy correctly. For example, in the article from 2014 [24] authors of Wordification applied cross-validation only on the propositional classifiers, leaving discretization and propositionalization out of the cross-validation. This design can lead to overly optimistic estimates of the accuracy. In the follow up article from 2015 [35] the authors of Wordification are already applying cross-validating on the whole process flow. Wordification accuracies used in this article are exclusively from [35].

5.1 Fairness of Comparison

The comparison solely based on the classification accuracy can be unfair. For example, CLAMF* [10] and Co-MoVi* [32] are designed to work with temporal datasets. Hence in Financial dataset they estimate probability of a loan default only from the data before the time of a loan application, while classifiers designed for the statical datasets also use data at and after the time of the loan application. All classifiers used in the meta-analysis treat all the datasets as if they were statical.

Validation of temporal datasets can be furthermore complicated by repeated target events. For example, a customer may apply for a loan many times. And now there are two perfectly plausible goals - we may want to calculate probability of default of a current customer with history of

loans or probability of default of a new customer. Generally, the second task is tougher because one of the best predictors of customer's behavior is the customer's past behavior. Nevertheless, all datasets in the meta-analysis have exactly one target value per classified object (including Financial dataset). Note that difference between within/across classification in IMDb and MovieLens datasets is another issue [33].

Also the goals of modeling can differ. For example Markov Logic Network* in [14] is evaluated as generative model, so accuracies reported are over all predicates, not just the target one. And accuracies can vary substantially with respect to the chosen target predicate. All the algorithms in the meta-analysis are evaluated in a discriminative setting.

Additionally, not all classifiers are designed to perform well on a wide spectrum of datasets. Indeed, there are algorithms like MOLFEA [13] that are designed to work only on a narrow subset of datasets. A possible specialization of the algorithms in the meta-analysis is not taken in the consideration.

At last different authors may have different ethos. Algorithms that were evaluated only by the algorithm authors (to our best knowledge) were marked with a star in table 3. And many algorithms that place at the top of the ranking are stared algorithms. However, this trend can also be explained with following hypotheses:

- Recent algorithms tend to be better than the old algorithms. And recent algorithms (like Wordification [35]) did not have enough time to accumulate references.

- New algorithms tend to look better in comparison to mediocre algorithms than in comparison to the best algorithm in the field. Hence authors prefer to compare their algorithms against mediocre algorithms.

- Third-party evaluators do not have the knowledge and resources to find the best algorithm setting. Hence popular algorithms have, on average, low accuracy. This problem is partially mitigated by considering only the best reported accuracies in figure 5.

Overall, comparison of measurements from several sources is not a simple task at all.

---

*The algorithm is not included in the meta-analysis because it's accuracy wasn't measured on enough datasets.

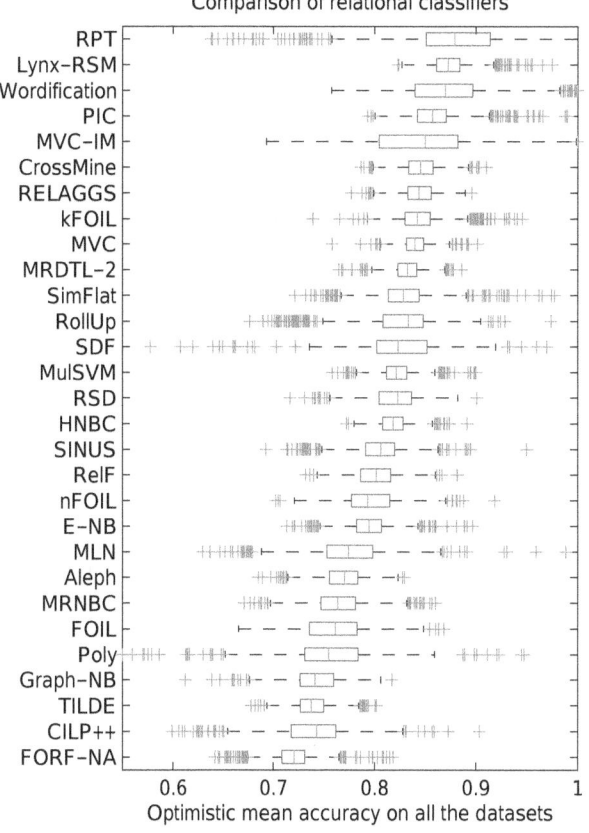

Figure 5: Box plot with optimistic accuracies.

6 Conclusion

Based on the performed analysis, Wordification, Lynx−RSM and Relational Probabilistic Tree on average outperform other 26 algorithms for relational classifications. Other promising categories of relational classifiers are multi-view and kernel based approaches.

Acknowledgement

The research reported in this paper has been supported by the Czech Science Foundation (GAČR) grant 13-17187S.

References

[1] Berka, P.: Workshop notes on Discovery Challenge PKDD'99, 1999

[2] Bina, B., Schulte, O., Crawford, B., Qian, Z., Xiong, Y.: Simple decision forests for multi-relational classification. Decision Support Systems 54(3) 2013, 1269–1279

[3] Coursac, I., Duteil, N.: PKDD 2001 Discovery Challenge – Medical Domain, 2001

[4] Davis, J., Burnside, E., Page, D.: View learning extended: inventing new tables for statistical relational learning. ICML Workshop on Open Problems in Statistical Relational Learning, 2006

[5] Debnath, A. K., Lopez de Compadre, R. L., Debnath, G., Shusterman, A. J., Hansch, C.: Structure-activity relationship of mutagenic aromatic and heteroaromatic nitro compounds. Correlation with molecular orbital energies and hydrophobicity. Journal of Medicinal Chemistry 34(2) (1991), 786–797

[6] Demšar, J.: Statistical comparisons of classifiers over multiple data sets. The Journal of Machine Learning Research 7 (2006), 1–30

[7] Dietterich, T.: Solving the multiple instance problem with axis-parallel rectangles. Artificial Intelligence 89(1–2) (1997), 31–71

[8] Fernández-Delgado, M., Cernadas, E., Barro, S., Amorim, D.: Do we need hundreds of classifiers to solve real world classification problems? Journal of Machine Learning Research 15 (2014), 3133–3181

[9] França, M. V. M., Zaverucha, G., D'Avila Garcez, A.: Fast relational learning using bottom clause propositionalization with artificial neural networks. Machine Learning 94(1) (2014), 81–104

[10] Frank, R., Moser, F., Ester, M.: A method for multi-relational classification using single and multi-feature aggregation functions. Lecture Notes in Computer Science 4702 (2007), 430–437

[11] Guo, H., Viktor, H. L.: Mining relational databases with multi-view learning. Proceedings of the 4th International Workshop on Multi-Relational Mining – MRDM'05, 2005, 15–24

[12] Guo, H., Viktor, H. L.: Learning from skewed class multi-relational databases. Fundamenta Informaticae 89(1) (2008), 69–94

[13] Helma, C., Kramer, S., De Raedt, L.: The molecular feature miner MOLFEA. In: Proceedings of the Beilstein Workshop 2002: Molecular Informatics: Confronting Complexity; Beilstein Institut, 2002, 1–16

[14] Khosravi, H., Schulte, O., Hu, J., Gao, T.: Learning compact Markov logic networks with decision trees. Machine Learning 89(3) (2012), 257–277

[15] King, R. D., Sternberg, M., Srinivasan, A.: Relating chemical activity to structure: an examination of ILP successes. New Generation Computing 13 (3–4) (1995), 411–433

[16] Knobbe, A., J., De Haas, M., Siebes, A.: Propositionalisation and aggregates. Lecture Notes in Computer Science 2168 (2001), 277–288

[17] Krogel, M.-A.: On propositionalization for knowledge discovery in relational databases. PhD Thesis, Otto-von-Guericke-Universität Magdeburg, 2005

[18] Krogel, M.-A., Rawles, S., Železný, F., Flach, P. A., Lavrač, N., Wrobel, S.: Comparative evaluation of approaches to propositionalization. In: Proceedings of the 13th International Conference on Inductive Logic Programming, volume 2835, 194–217, 2003

[19] Krogel, M.-A., Wrobel, S.: Facets of aggregation approaches to propositionalization. In: Inductive Logic Programming: 13th International Conference, 30–39, Springer, Berlin, 2003

[20] Kuželka, O.: Fast construction of relational features for machine learning. PhD Thesis, Czech Technical University, 2013

[21] Lahbib, D., Boullé, M., Laurent, D.: Itemset-based variable construction in multi-relational supervised learning. Lecture Notes in Computer Science (including subseries Lecture Notes in Artificial Intelligence and Lecture Notes in Bioinformatics), 7842 LNAI:130–150, 2013

[22] Landwehr, N.: Integrating naive bayes and FOIL. Journal of Machine Learning Research **8** (2007), 481–507

[23] Landwehr, N., Passerini, A., De Raedt, L., Frasconi, P.: kFOIL: Learning simple relational kernels. Aaai **6** (2006), 389–394

[24] Lavrač, N., Perovšek, M., Vavpetič, A.: Propositionalization online. In: ECML PKDD 2014, 456–459, Springer-Verlag, 2014

[25] Leiva, H. A., Atramentov, A., Honavar, V.: A multi-relational decision tree learning algorithm. Proceedings of the 13th International Conference on Inductive Logic Programming, 2002, 38–56

[26] Liu, H., Yin, X., Han, J.: An efficient multi-relational Naïve Bayesian classifier based on semantic relationship graph. Proceedings of the 4th International Workshop on Multi-Relational Mining – MRDM'05, Eleventh ACM SIGKDD International Conference on Knowledge Discovery and Data Mining (KDD-2005), 2005, 39–48

[27] Lodhi, H., Muggleton, S.: Is mutagenesis still challenging? Proceedings of the 15th International Conference on Inductive Logic Programming, ILP 2005, Late-Breaking Papers., 2005, 35–40

[28] Macskassy, S. A., Provost, F.: A simple relational classifier. Technical Report, Stern New York University, 2003

[29] Di Mauro, N., Esposito, F.: Ensemble relational learning based on selective propositionalization. CoRR (2013), 1–10

[30] Michie, D., Muggleton, S., Page, D., Srinivasan, A.: To the international computing community: a new east-west challenge. Technical Report, Oxford University Computing Laboratory, Oxford, 1994

[31] Modi, S.: Relational classification using multiple view approach with voting. International Journal of Computer Applications **70(16)** (2013), 31–36

[32] Neto, R. O., Adeodato, P. J. L., Salgado, A. C., Filho, D. R., Machado, G. R.: CoMoVi: a framework for data transformation in credit behavioral scoring applications using model driven architecture. SEKE 2014 (2014), 286–291

[33] Neville, J., Gallagher, B., Eliassi-Rad, T., Wang, T.: Correcting evaluation bias of relational classifiers with network cross validation. Knowledge and Information Systems **30 (1)** (2012), 31–55

[34] Perlich, C., Provost, F.: Distribution-based aggregation for relational learning with identifier attributes. Machine Learning **62(1–2)** SPEC. ISS. (2006), 65–105

[35] Perovšek, M., Vavpetič, A., Kranjc, J., Cestnik, B., Lavrač, N.: Wordification: propositionalization by unfolding relational data into bags of words. Expert Systems with Applications **42(17–18)** (2015), 6442–6456

[36] Richardson, M., Domingos, P.: Markov logic networks. Machine Learning **62 (1–2)** SPEC. ISS. (February 2006), 107–136

[37] Schulte, O., Bina, B., Crawford, B., Bingham, D., Xiong, Y.: A hierarchy of independence assumptions for multi-relational Bayes net classifiers. Proceedings of the 2013 IEEE Symposium on Computational Intelligence and Data Mining, CIDM 2013 – 2013 IEEE Symposium Series on Computational Intelligence, SSCI 2013, 2013, 150–159

[38] Srinivasan, A., King, R. D., Muggleton, S. H., Sternberg, M. J. E.: Carcinogenesis predictions using ILP. Inductive Logic Programming **1297** (1997) 273–287

[39] Taskar, B., Abbeel, P., Koller, D.: Discriminative probabilistic models for relational data. UAI'02 Proceedings of the Eighteenth Conference on Uncertainty in Artificial Intelligence, 2002, 485–492

[40] Vens, C., Van Assche, A., Blockeel, H., Džeroski, S.: First order random forests with complex aggregates. Lecture Notes in Computer Science **3194** (2004), 323–340

[41] Wolpert, D.: The existence of a priori distinctions between learning algorithms. Neural Computation **8(7)** (1996), 1391–1420

[42] Yin, X., Han, J., Yang, J., Yu, P. S.: CrossMine: efficient classification across multiple database relations. Lecture Notes in Computer Science (including subseries Lecture Notes in Artificial Intelligence and Lecture Notes in Bioinformatics) **3848** LNAI(6) (2006), 172–195

[43] Zou, M., Wang, T., Li, H., Yang, D.: A general multi-relational classification approach using feature generation and selection. In: 6th International Conference, ADMA, 21–33, Springer-Verlag, 2010

J. Yaghob (Ed.): ITAT 2015 pp. 179–185
Charles University in Prague, Prague, 2015

Feature Extraction for Terrain Classification with Crawling Robots

Jakub Mrva and Jan Faigl

Czech Technical University in Prague, Technická 2, 166 27 Prague, Czech Republic
jakub.mrva|faiglj@fel.cvut.cz

Abstract: In this paper, we address the problem of terrain classification using a technically blind hexapod walking robot. The proposed approach is built on top of the existing method based on analysis of the feedback from the robot's actuators and the desired trajectory. The formed method uses features for the Support Vector Machine classification method that assumes a regular time-invariant gait to control the robot. However, such a gait does not allow the robot to traverse rough terrains, and therefore, it is necessary to consider adaptive motion gait to deal with small obstacles, which is, unfortunately, not a regular gait with some fixed predefined period. Therefore, we propose to alter the features extraction process to utilize the terrain classification method also for an adaptive motion gait, which enables the robot to traverse rough terrains. The proposed method has been experimentally verified on several terrains that are not traversable by a default regular gait. The achieved results not only confirmed the high accuracy of the terrain classification as the existing approach, but also expanded the area of operation of a hexapod walking robot into more challenging terrains.

1 Introduction

Crawling robots can operate in a much greater scope in terms of terrain diversity than classical wheeled robots. The control complexity is, however, much greater due to the high number of degrees of freedom (DOF). One way to handle a high DOF is to generate a walking pattern—a gait [1]. A simple regular gait gives the robot predefined trajectories for all legs, which are therefore alternating in their support and transfer phases.

In order to increase the robot's perception of the environment—for example to classify the terrain the robot is traversing—one can employ the robot with a variety of sensors. There can be found two complementary approaches based on exteroceptive and interoceptive sensors. In the case of exteroceptive sensing, we can utilize camera [2, 3] or laser-based range measurements [4] for terrain classification.

However, if the robot is technically blind and dependent solely on interoceptive sensing, we can use force, torque [5, 6], or other tactile sensors to gather data about the interaction of the robot with the terrain. Moreover, we can utilize the robot's actuators themselves and develop a classifier based on the differences between the expected and real trajectories of the robot servo drives [7] without the need of any additional sensor.

Figure 1: Used hexapod walking robot for the terrain classification

The existing method [7] uses a default robot motion gait with regular and periodic phases of the leg movements, and therefore, it is suitable only for flat terrains without significant obstacles. Based on this method, we consider several terrains with obstacles or stairs and their classification using an adaptive motion gait [8] that allows a smooth transition while reducing the workload of the servos and thus avoiding overheating. Such a gait does not preserve the predefined trajectories of each leg as a default motion gait does. Hence, the existing method of feature extraction proposed in [7] is not directly applicable because it assumes a regular time-invariant gait with fixed trajectories. Therefore, in this paper, we propose a modification of the feature extraction process of [7] to enable the terrain classification based only on the servo drive feedback also in crawling rough terrain using the adaptive motion gait. Therefore, the method can be used in more challenging terrains up to the structural limits of the hexapod walking robot.

The paper is organized as follows. A brief overview of the adaptive motion gaits for rough terrains and terrain classification methods is provided in the next section. A description of the considered robotic platform and definition of the problem is presented in Section 3. The utilized adaptive motion gait is briefly described in Section 4. The proposed feature extraction method is presented in Section 5 and experimental results in Section 6. The concluding remarks are in Section 7.

2 Related Work

Adaptive motion of a walking robot to traverse a rough terrain has been addressed by many researchers and several approaches can be found in literature. A complex control architecture of quadruped walking robot to traverse challenging terrains has been presented in [9] using several sensors attached to the robot and a precise map created off-line. The off-line scanning can be avoided by using an elevation map created from an on-board laser scanner and used further to alter the gait according to the terrain structure [10].

Another existing direction of the adaptive motion gaits are based on approaches that do not utilize a terrain map. They are based on a tactile information from force sensors [11] (or torque-based estimation of the force [12]) utilized to adapt the gait according to the terrain and to ensure the leg reaches the foothold. A passive actuator to measure the ground reaction force has been proposed in [13] to substitute direct force or torque sensors, which is a suitable approach for the deployment of cheap robotic platforms. In [8], we proposed a similar approach that is even more minimalist since it does not need additional servos and thus it is solely based on the robot's actuators.

The problem of terrain classification is widely investigated also regarding on-board processing. A camera can be used to estimate the terrain class based on extracted features [3] that can be further used to select an energy efficient motion gait [2]. Authors of [14] used a laser range finder for distinguishing between twelve terrains and achieved promising results; however, under specific laboratory conditions only.

Focusing on a structural point of view, an off-line scan of the terrain from a precise external laser-scanning system was used in [15] to generate a database of terrain templates that are used for proper foothold planning. Authors of [4] proposed an approach to avoid building a large database of templates. Their idea is based on a creation of a set of several templates that define good and poor footholds based on local concavity and sloppiness, which are useful attributes for predicting slipperiness of the terrain.

Beside exteroceptive sensing, tactile sensors are used to classify the terrain based on the direct measurements of the robot interaction with the terrain. In [5], authors used features extracted from the measurements of force sensors placed at the tip of the leg that are combined with the measurements of the motor current of the knee joint of a single vibrating robot leg detached from the body. A 6-DOF torque-force sensor was used in [16] (under the same laboratory conditions as in [14]) for a discriminant analysis between six types of terrain. However, all of these approaches are based on additional sensors, and therefore, they increase a complexity of the robot.

A slightly different approach that utilizes only interoceptive sensors built within the actuators was presented in [7]. The actuators consist of position controllers that can send both the desired and the current position of the servo. The difference in these positions is then analyzed in time and frequency domains to extract a 660-dimensional feature vector from the two front legs during each gait cycle. This method is limited by using a periodic time-invariant gait and thus it is applicable almost exclusively on flat terrains without obstacles.

In this paper, we proposed a combination of the terrain classification method [7] with the adaptive motion gait [8]. Both these approaches are based solely on interoceptive sensing using active actuators and we propose a new feature extraction procedure to overcome limitations of [7] and enable on-line classification of rough terrains.

3 Problem Statement

The main problem being addressed in this paper is to extend the existing terrain classification approach [7] to the adaptive motion gait [8] and thus generalize terrain classification also for traversing rough terrains. A new feature extraction method is needed to deal with rough terrains because the adaptive motion gait does not preserve the required condition on the motion gait of [7], i.e., a time-invariant motion gait. In the proposed approach, we consider a relatively cheap and easy-to-use platform PhantomX Hexapod Mark II with Dynamixel AX-12A actuators, see Fig. 1, which is further described in the next section. An overview of the terrain classification method [7] is provided in Section 3.2 to provide a background for the proposed approach.

3.1 Hexapod Structure

The used hexapod platform has six legs each with three joints formed from the Dynamixel actuators. The schema of the leg and the description of its parts is depicted in Fig. 2. All joints (θ_C, θ_F, and θ_T) are controlled with a position controller that provides every 33 ms the following information:

- Desired position θ^{des};
- Current position θ^{cur};
- Error in position $e = |\theta^{des} - \theta^{cur}|$.

Using the adaptive motion gait [8], the robot can traverse small obstacles up to the limits of the robot structure.

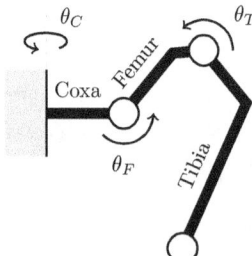

Figure 2: Schema of the leg consisting of three parts (links)—Coxa, Femur, and Tibia. The three joints (θ_C, θ_F, and θ_T) are indexed according to the next respective link. The joint θ_C is fixed to the body with a vertical rotation axis while the other two joints have a horizontal axis.

We consider the robot is operating in an environment that satisfies the robot's structural limits and there is not a large obstacle that the robot cannot traverse. Hence, we are not addressing obstacle avoidance and other high-level navigation problem in this paper. Thus we are strictly focused on the problem of terrain classification and its practical validation in real experiments.

3.2 Terrain Classification

The method of the terrain classification [7] has been proposed for the same hexapod platform as we are using; however, a regular time-invariant gait is utilized for robot motion. The very general idea of the terrain classification is based on the small errors in position control (e) of all servo drives of the front legs (i.e., six servos) that are measured in the time domain at a non-uniform sample rate of approximately 20 Hz.

In order to obtain a more dense data and to get a uniform sample rate for the FFT used in the feature extraction, the signal is interpolated using a cubic Hermite spline interpolation method that creates a continuous function with a continuous first derivative. The interpolated function is then resampled at the frequency of 100 Hz. After that, a feature vector for the classification is created from the sampled data and computed characteristics of the signal.

Having the default regular gait, the data is windowed using a uniform window that contains the last three full gait cycles; so, each terrain-class prediction is based on the past three gait cycles worth of data. Given the motivation that different terrain surfaces induce a specific behavior in different sections of the gait cycle, the data are divided into 16 equally wide segments within a gait cycle to form a gait-phase domain. Respective segments from the last three gait cycles are joined together and basic statistics of all data samples that fall within are computed yielding in 5 values (features) for each segment (i.e., minimum, maximum, mean, median, and standard deviation). Repeated for each servo, we obtain a total of 480 gait-phase features, i.e., 2 (legs) × 3 (servos per leg) × 5 (features) × 16 (segments).

Additional 180 features are calculated in the frequency domain. A Hamming window and discrete Fourier transform are applied on the same resampled position error signal to obtain a frequency spectrum of 25 bins (0–12 Hz). All amplitude values of frequency bins are used alone, giving another 25 features (for each servo), supplemented by 4 features obtained from the shape of the spectrum (i.e., centroid, standard deviation, skewness, and kurtosis) and finalized with the energy of the spectrum. The overall 660-dimensional feature vector consists of:

- Statistics of each segment for each servo ($5 \times 16 \times 6 = 480$);
- Bins of the frequency spectrum ($25 \times 6 = 150$);
- Shape of the frequency spectrum ($4 \times 6 = 24$);
- Energy of the frequency spectrum (6).

Such 660-dimensional feature vectors for particular type of the terrain and several trials of traversing the terrain are used to train a multi-class linear Support Vector Machine (SVM) classifier and authors of [7] report 95% accuracy in distinguishing between 3 terrain classes (concrete, grass, and rocks/mulch).

In our approach presented in this paper, we follow the same idea of the terrain classifier based on the SVM, but we propose a new feature extraction process to address the absence of regularity in the adaptive motion gait for crawling rough terrains.

4 Adaptive Motion Gait

The adaptive gait is originally based on a regular tripod gait in terms of predefined trajectories for each particular leg. However, the trajectories can be changed and the gait-cycle is divided into separate phases of the leg and body motion.

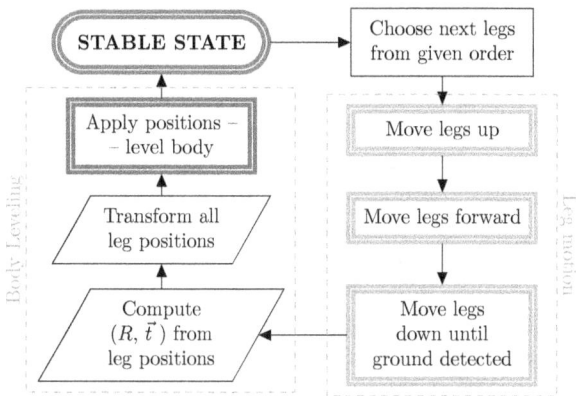

Figure 3: Diagram of a gait cycle. Firstly, the legs in the transfer phase move to find new footholds. Secondly, the body is leveled to adapt the new footholds. Finally, another legs are chosen for the next transfer phase. Orange color highlights the motion of legs in the transfer phase only, while red color highlights the motion of all legs.

The gait diagram is shown in Fig. 3, where it can be seen that legs in the transfer phase move through predefined checkpoints (up and forward) and begin approaching another predefined checkpoint, which is situated far below the current ground level (but still reachable). The ground sensing is done via observing the position error e of the joint θ_F during lowering the leg with respect to a certain threshold value. Although all legs in the transfer phase are moving simultaneously, each ground contact stops only the particular corresponding leg.

After the legs found their new footholds (the rest legs stay motionless), a new body posture is found given the feet positions in order to adapt to the terrain the robot is traversing. The body motion itself is provided by moving all legs according to a transformation of the feet positions.

In summary, the leg motion phase consists of 3 steps (up, forward, and down) and is followed by a body leveling step. Given the tripod gait in which the legs are grouped into two triplets that are alternating, we repeat the same steps for the other triplet to obtain a total of 8 discrete steps per one full gait cycle. Notice, that as a consequence of the adaptive-gait model, a leg is moving only in the transfer phase (3 steps) and in both body leveling steps (4th and 8th steps), where all legs are needed to move the body. A more detail description can be found in [8].

5   Feature Extraction

The proposed feature extraction process is based on the terrain classification originally developed for a regular gait [7], which has to be altered to deal with a different behavior of the adaptive gait [8]. The key difference between these two gaits is in their regularity and in the fact that the regular gait is synchronized by a time signal and a leg never stops moving, whereas the adaptive gait splits the leg and body motion according to the gait phases.

A leg trajectory during the phases of the gait is depicted in Fig. 4. The regular gait is periodic and the robot is able to traverse a flat terrain at the constant speed. Although the robot can pass very small obstacles at the cost of a high servo load, the robot is incapable to traverse a rough terrain using this regular default gait [8].

On the other hand, the adaptive motion gait utilizes a tactile information to detect the ground-contact point and thus it is able to decrease the servo load and adjust the robot to the terrain. However, ground-contact points along the vertical line (during moving the leg down) are not known. Hence the time the leg spends in the ground-approaching phase is also not known. Moreover, the trajectory of the particular foot in the support phase is also influenced by the contacts of the other legs with the grounds, and therefore, the trajectory is not regular during crawling rough terrain and it may vary significantly. These variances have to be considered in the analysis of the servo position signal in the feature extraction process to avoid possible misinterpretation of the data.

Due to the variances of the gait phases, which depends on the roughness of the terrain the robot is traversing, we cannot rely on a uniform partitioning of the gait phases into 16 segments for the feature extraction as in [7]. On the other hand, we can utilize the gait phases of the adaptive gait, as it is shown in the diagram in Fig. 3, and the data from one gait cycle can be therefore divided into 8 segments according to the gait phases.

Authors of [7] extended the feature vector by features extracted from a frequency analysis. However, such analysis requires a condition of periodicity that is not fulfilled in the adaptive gait. Nevertheless, during a practical experimenting, it has been observed that the absence of frequency-based features did not prevent the classifier to achieve accurate classification, which is shown in Section 6. Notice that the authors also did not consider features selection to reduce the 660-dimensional feature vector; so, the frequency analysis may be expendable.

6   Experimental Results

Since the proposed method extends an existing approach by adding more rough terrains where the robot can operate, we focused the experimental evaluation of the proposed method solely on those challenging terrains. Nevertheless, we also used datasets from simple outdoor terrains for completeness.

The proposed multi-class SVM classifier (with linear kernel) was trained for feature vectors collected from 7 different classes:

- Wooden stairs
- Wooden blocks of different height
- Office floor with small obstacles
- Office floor
- Asphalt
- Grass
- Dirt

(a) Small obstacles   (b) Wooden blocks   (c) Wooden stairs

Figure 5: Terrains traversable by the adaptive gait.

The outdoor terrains (grass, dirt and asphalt) are very flat and easily traversable by a default regular gait. However, the rough terrains shown in Fig. 5 are traversable only by the used adaptive gait. Default gait is able to traverse only small obstacles (cf. 5a) and this simple terrain type is used to fill a gap between the flat terrains (outdoor and office floor) and the rough terrains (blocks and stairs).

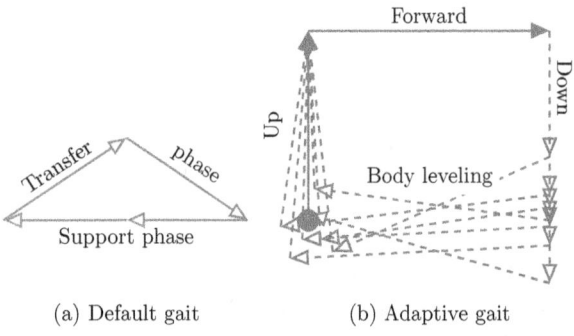

(a) Default gait   (b) Adaptive gait

Figure 4: Comparison of the leg trajectory using a regular default gait and an adaptive gait.

Terrain	Dirt	Asphalt	Grass	Office	Obstacles	Blocks	Stairs
Dirt	62	0	0	0	0	0	0
Asphalt	1	79	0	0	0	0	0
Grass	0	0	75	0	0	0	0
Office	0	0	0	89	0	0	0
Obstacles	0	0	0	0	69	1	0
Blocks	0	0	0	0	0	21	0
Stairs	0	0	0	0	0	1	90

Table 1: Confusion matrix of 2-fold cross-validation with overall accuracy 99.4%

Each terrain class was trained from several trials with 3–7 minutes worth of data; the exact number of feature vectors extracted from the data can be read from columns of Table 1.

The evaluation strategy is based on the verification of the distinguishability of the terrain using the features. Then, we evaluate online detection of the terrain in a separate scenario, where particular types of the terrains are altered and the robot is requested to traverse them and continuously detect the terrain. These two evaluation scenarios are described in the following sections.

6.1 Distinguishability of the Terrains

Two-fold cross-validation with all datasets involved has been used to validate whether the classifier is able to distinguish between the considered 7 terrain classes. As can be seen from the confusion matrix in Table 1, the overall accuracy of 99.4% is very high even for only 2-fold cross-validation (with more folds, we can easily get 100%).

However, notice this test is based on using always data from the same single experiment in both training and testing partition, and therefore, the data are more likely referring to themselves than to a generalized model of particular terrain class. In [7], authors achieved the same high accuracy when evaluating on the same datasets that were used for training.

6.2 Terrain Classification

A more realistic practical scenario is based on evaluation of the classification for traversing rough terrains in a single run, where undefined terrains at the overlap of the particular terrain types are provided. The scenario setup is shown in Fig. 6 and it consists of a sequence of rough terrains used for the learning. The robot starts from a defined position and crosses progressively few small obstacles, a pool of wooden blocks, and wooden stairs. This scenario was repeated five times and the extracted feature vectors were evaluated against the model previously learned from a single-pass of the individual terrains.

The predicted terrain labels from each of five runs are shown in Fig. 7. The class prediction is made once per

Figure 6: Testing scenario consisting of small obstacles (bottom right corner) on the office floor, followed by a pool of wooden blocks and ending with climbing the stairs.

each gait cycle and is computed from the last three gait cycles worth of data. Therefore (with respect to the robot's length) there are long transition areas of the overlapping terrains.

The transition between the office floor and the pool of wooden blocks is mostly characterized as the stairs, which corresponds with the entry side of the pool with increasing height of the blocks.

The other transition between the blocks and the stairs is undefinable and can be predicted as either terrain class, or a similar class (obstacles) based on the actual footholds in the area during the experiment. We can also see that there is some confusion between the dirt and the office floor terrain with obstacles which are both relatively flat and slippery for the robot.

Despite not analyzed, if another terrain class of stairs being traversed down was trained, it is highly probable that a transition from the blocks to the office floor would have been classified as this terrain (for the same reason as the opposite transition mentioned above).

The main aspect of the challenging terrain traversing, which cannot be seen on the simple flat terrains, is the

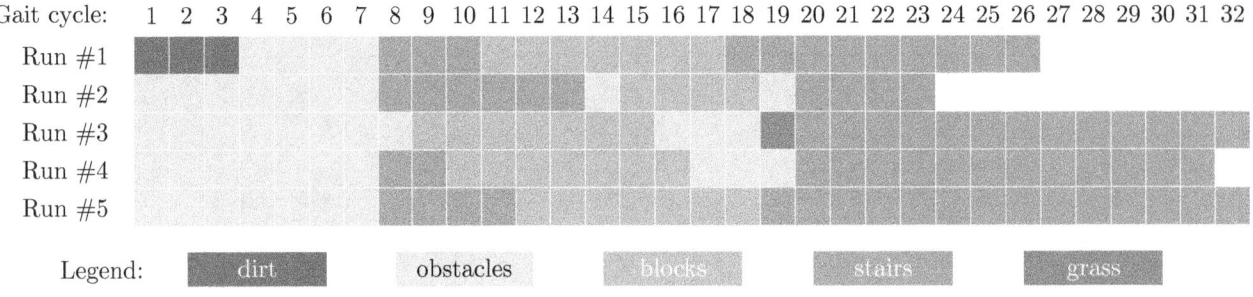

Figure 7: Successive predictions of terrain labels in the testing scenario.

occurrence of a foot slippage on the edge of an obstacle (stair) that yields in a sudden fall to a lower level and impacts all legs. More slippages in a short time can lead in a confusion in the prediction, as can be seen in Fig. 7 where the transition between the blocks and the stairs was once predicted as a grass terrain.

The unequal length of all runs is purely dependent on the event when the robot steps over the side edge of the stairs and thus stops the experiment. This happens due to the fact that the robot cannot steer and is strictly going straight ahead.

Notice, it is not possible to show the ground truth for the predictions in Fig. 7 because the robot can spend different number of gait cycles to get to the same point in the scenario in particular runs. Therefore, the only measure we can get is to compare the results from Fig. 7 to the overview of the testing scenario shown in Fig. 6.

7 Conclusion

We proposed an alternative method to extract features from servo drives to classify terrains for a technically blind robot traversing rough terrains. Although the proposed method simplifies the original feature extraction process, the results indicate it is sufficient to distinguish evaluated terrain classes. Moreover, the results also indicate we can employ the learned classifier in the on-line terrain classification in scenarios with rough terrains.

The classifier is based on the features of the robot motion and interaction with the terrain. However, the features of the terrain itself (e.g., slopiness, slipperiness, convexity) are not analyzed directly—they may be hidden inside the SVM layer and could be addressed in the future work.

Acknowledgments

The presented work has been supported by the Czech Science Foundation (GAČR) under research project No. 15-09600Y.

References

[1] Dudek, G., Jenkin, M.: Computational principles of mobile robotics. New York, NY, USA, Cambridge University Press, 2000

[2] Zenker, S., Aksoy, E., Goldschmidt, D., Worgotter, F., Manoonpong, P.: Visual terrain classification for selecting energy efficient gaits of a hexapod robot. In: IEEE/ASME International Conference on Advanced Intelligent Mechatronics (AIM), 2013, 577–584

[3] Filitchkin, P., Byl, K.: Feature-based terrain classification for littledog. In: IROS, 2012, 1387–1392

[4] Belter, D., Skrzypczyński, P.: Rough terrain mapping and classification for foothold selection in a walking robot. Journal of Field Robotics **28(4)** (2011), 497–528

[5] Hoepflinger, M. A., Remy, C. D., Hutter, M., Spinello, L., Siegwart, R.: Haptic terrain classification for legged robots. In: ICRA, 2010, 2828–2833

[6] Schmidt, A., Walas, K.: The classification of the terrain by a hexapod robot. In: Proceedings of the 8th International Conference on Computer Recognition Systems (CORES), 2013, 825–833

[7] Best, G., Moghadam, P., Kottege, N., Kleeman, L.: Terrain classification using a hexapod robot. In: Proceedings of the Australasian Conference on Robotics and Automation (ACRA), 2013

[8] Mrva, J., Faigl, J.: Tactile sensing with servo drives feedback only for blind hexapod walking robot. In: Proceedings of the 10th International Workshop on Robot Motion and Control (RoMoCo), 2015

[9] Kalakrishnan, M., Buchli, J., Pastor, P., Mistry, M., Schaal, S.: Learning, planning, and control for quadruped locomotion over challenging terrain. The International Journal of Robotics Research **30 (2)** (2011), 236–258

[10] Belter, D.: Gait modification strategy for a six-legged robot walking on rough terrain. In: Proceedings of the 15th International Conference on Climbing and Walking Robots, Adaptive Mobile Robotics, World Scientific, A. Azad et al. (Eds.), Singapore, 2012, 367–374

[11] Winkler, A., Havoutis, I., Bazeille, S., Ortiz, J., Focchi, M., Dillmann, R., Caldwell, D., Semini, C.: Path planning with force-based foothold adaptation and virtual model control for torque controlled quadruped robots. In: ICRA, 2014, 6476–6482

[12] Walas, K., Belter, D.: Supporting locomotive functions of a six-legged walking robot. International Journal of Applied Mathematics and Computer Science **21 (2)** (2011)

[13] Palmer, L., Palankar, M.: Blind hexapod walking over uneven terrain using only local feedback. In: IEEE International Conference on Robotics and Biomimetics (ROBIO), 2011, 1603–1608

[14] Walas, K., Nowicki, M.: Terrain classification using laser range finder. In: IROS, 2014, 5003–5009

[15] Kalakrishnan, M., Buchli, J., Pastor, P., Schaal, S.: Learning locomotion over rough terrain using terrain templates. In: IROS, 2009, 167–172

[16] Walas, K.: Tactile sensing for ground classification. Journal of Automation, Mobile Robotics & Intelligent Systems **7 (2)** (2013), 18–23

J. Yaghob (Ed.): ITAT 2015 pp. 186–193
Charles University in Prague, Prague, 2015

Comparing SVM, Gaussian Process and Random Forest Surrogate Models for the CMA-ES

Zbyněk Pitra[1,2], Lukáš Bajer[3,4], and Martin Holeňa[3]

[1] National Institute of Mental Health
Topolová 748, 250 67 Klecany, Czech Republic
z.pitra@gmail.com

[2] Faculty of Nuclear Sciences and Physical Engineering, Czech Technical University in Prague
Břehová 7, 115 19 Prague 1, Czech Republic

[3] Institute of Computer Science, Academy of Sciences of the Czech Republic
Pod Vodárenskou věží 2, 182 07 Prague 8, Czech Republic
{bajer,holena}@cs.cas.cz

[4] Faculty of Mathematics and Physics, Charles University in Prague
Malostranské nám. 25, 118 00 Prague 1, Czech Republic

Abstract: In practical optimization tasks, it is more and more frequent that the objective function is black-box which means that it cannot be described mathematically. Such functions can be evaluated only empirically, usually through some costly or time-consuming measurement, numerical simulation or experimental testing. Therefore, an important direction of research is the approximation of these objective functions with a suitable regression model, also called surrogate model of the objective functions. This paper evaluates two different approaches to the continuous black-box optimization which both integrates surrogate models with the state-of-the-art optimizer CMA-ES. The first Ranking SVM surrogate model estimates the ordering of the sampled points as the CMA-ES utilizes only the ranking of the fitness values. However, we show that continuous Gaussian processes model provides in the early states of the optimization comparable results.

1 Introduction

Optimization of an expensive objective or fitness function plays an important role in many engineering and research tasks. For such functions, it is sometimes difficult to find an exact analytical formula, or to obtain any derivatives or information about smoothness. Instead, values for a given input are possible to be obtained only through expensive and time-consuming measurements and experiments. Those functions are called black-box, and because of the evaluation costs, the primary criterion for assessment of the black-box optimizers is the number of fitness function evaluations necessary to achieve the optimal value.

The Covariance Matrix Adaptation Evolution Strategy (CMA-ES) [5] is considered to be the state-of-the-art of the black-box continuous optimization. The important property of the CMA-ES is that it advances through the search space only according to the ordering of the function values in current population. Hence, the search of the algorithm is rather local which predisposes it to premature convergence in local optima if not used with sufficiently large

population size. This issue resulted in development of several restart strategies [12], such as IPOP-CMA-ES [1] and BIPOP-CMA-ES [6] performing restarts with population size successively increased, or aCMA-ES [9] using also unsuccessful individuals for covariance matrix adaptation.

Furthermore, the CMA-ES often requires more fitness function evaluations to find the optimum than many real-world experiments can offer. In order to decrease the number of evaluations in evolutionary algorithms, it is convenient to periodically train a surrogate model of the fitness function and use it for evaluation of new points instead of the original function. The second option is to use the model for selection of the most promising points to be evaluated by the original fitness.

Loshchilov's surrogate-model-based algorithm [s]*ACM-ES [13] utilizes the former approach: it estimates the ordering of the fitness values required by the CMA-ES using Ranking Support Vector Machines (SVM) as an ordinal regression model. Moreover, it has been shown [13] that model parameters (hyperparametres) used to construct Ranking SVM model can be optimized during the search by the pure CMA-ES algorithm. Later proposed [s]*ACM-ES extensions, referred to as [s]*ACM-ES-k [15] and BIPOP-[s]*ACM-ES-k [14], use a more intensive exploitation of the surrogate model by increasing population size in generations evaluated by the model.

More recently, a similar algorithm based on regression surrogate model called S-CMA-ES [3] has been presented. As opposed to the former algorithm, S-CMA-ES is performing continuous regression by Gaussian processes (GP) [17] and random forests (RF) [4].

This paper compares the two mentioned surrogate CMA-ES algorithms, [s]*ACM-ES-k and S-CMA-ES, and the original CMA-ES itself. We benchmark these algorithms on the BBOB/COCO testing set [7, 8] not only in their one population IPOP-CMA-ES version, but also in combination with the two-population-size BIPOP-CMA-ES.

The remainder of the paper is organized as follows. The next chapter briefly describes tested algorithms: the CMA-ES, the BIPOP-CMA-ES, the ^{s*}ACM-ES-k, and the S-CMA-ES. Section 3 contains experimental setup and results, and Section 4 concludes the paper and suggests further research directions.

2 Algorithms

2.1 The CMA-ES

In each generation g, the CMA-ES [5] generates λ new candidate solutions $\mathbf{x}_k \in \mathbb{R}^D$, where $k = 1, \ldots, \lambda$, from a multivariate normal distribution $N(\mathbf{m}_{(g)}, \sigma^2_{(g)} \mathbf{C}_{(g)})$, where $\mathbf{m}_{(g)}$ is the mean interpretable as the current best estimate of the optimum, $\sigma^2_{(g)}$ the step size, representing the overall standard deviation, and $\mathbf{C}_{(g)}$ the $D \times D$ covariance matrix. The algorithm selects the μ points with the lowest function value from λ generated candidates to adjust distribution parameters for the next generation.

The CMA-ES uses restart strategies to deal with multimodal fitness landscapes and to avoid being trapped in local optima. A multi-start strategy where the population size is doubled in each restart is referred to as IPOP-CMA-ES [1].

2.2 BIPOP-CMA-ES

The BIPOP-CMA-ES [6], unlike IPOP-CMA-ES, considers two different restart strategies. In the first one, corresponding to the IPOP-CMA-ES, the population size is doubled in each restart i_{restart} using a constant initial step-size $\sigma^0_{\text{large}} = \sigma^0_{\text{default}}$:

$$\lambda_{\text{large}} = 2^{i_{\text{restart}}} \lambda_{\text{default}} . \tag{1}$$

In the second one, the smaller population size λ_{small} is computed as

$$\lambda_{\text{small}} = \left\lfloor \lambda_{\text{default}} \left(\frac{1}{2} \frac{\lambda_{\text{large}}}{\lambda_{\text{default}}} \right)^{U[0,1]^2} \right\rfloor , \tag{2}$$

where $U[0,1]$ denotes the uniform distribution in $[0,1]$. The initial step-size is also randomly drawn as

$$\sigma^0_{\text{small}} = \sigma^0_{\text{default}} \times 10^{-2U[0,1]} . \tag{3}$$

The BIPOP-CMA-ES performs the first run using the default population size λ_{default} and the initial step-size $\sigma^0_{\text{default}}$. In the following restarts, the strategy with less function evaluations summed over all algorithm runs is selected.

2.3 ^{s*}ACM-ES-k

Loshchilov's version of the CMA-ES using the ordinal regression by Ranking SVM as surrogate model in specific generations instead of the original function is referred to as ^{s*}ACM-ES [13], and its extension using a more intensive exploitation is called ^{s*}ACM-ES-k [15].

Before the main loop starts, the ^{s*}ACM-ES-k evaluates g_{start} generations by the original function, then it repeats the following steps: First, the surrogate model is constructed using hyperparameters θ, and the original function-evaluated points from previous generations. Second, the surrogate model is optimized by the CMA-ES for g_m generations with population size $\lambda = k_\lambda \lambda_{\text{default}}$ and the number of best points $\mu = k_\mu \mu_{\text{default}}$, where $k_\lambda, k_\mu \geq 1$. Third, the following generation is evaluated by the original function using $\lambda = \lambda_{\text{default}}$ and $\mu = \mu_{\text{default}}$. To avoid a potential divergence when g_m fluctuate between 0 and 1, $k_\lambda > 1$ is used only in the case of $g_m \geq g_{m\lambda}$, where $g_{m\lambda}$ denotes the number of generations suitable for effective exploitation using the model. Then the model error is calculated according to the comparison of ranking between the original and model evaluation of the last generation. After that, the g_m is adjusted in accordance with the model error. As the last step, the ^{s*}ACM-ES-k searches a hyperparameter space by one generation of the CMA-ES minimizing the model error to find the most suitable hyperparameter settings θ_{new} for the next model-evaluated generations.

The ^{s*}ACM-ES-k version using BIPOP-CMA-ES proposed in [14] is called BIPOP-^{s*}ACM-ES-k.

2.4 S-CMA-ES

As opposed to the former algorithms, a different approach to surrogate model usage is incorporated in the S-CMA-ES [3]. The algorithm is a modification of CMA-ES where the original evaluating and sampling phases are substituted by the Algorithm 1 at the beginning of each CMA-ES generation.

In order to avoid the false convergence of the algorithm in the BBOB benchmarking toolbox, the model-predicted values are adapted to never be lower then the so far minimum of the original function (see the step 17 in the pseudocode).

The main difference between the S-CMA-ES and the ^{s*}ACM-ES-k is in the manner how the CMA-ES is utilized. Considering S-CMA-ES, the model prediction or training is performed within each generation of the CMA-ES. On the contrary in the ^{s*}ACM-ES-k, individual generations of the CMA-ES are started to optimize either original fitness, surrogate fitness, or model itself.

3 Experimental Evaluation

The core of this paper lies in a systematic comparison of the two mentioned approaches to using surrogate models with the CMA-ES and the original CMA-ES algorithm itself. The first group of surrogate-based algorithms is formed by the S-CMA-ES algorithms using Gaussian processes and random forests models, and the other group is formed by the ^{s*}ACM-ES algorithm. These four algorithms (CMA-ES, GP-CMA-ES, RF-CMA-ES, ^{s*}ACM-ES) are tested in their IPOP version (based on IPOP-CMA-ES) [1] and in the bi-population restart strategy version (based on BIPOP-CMA-ES and its derivatives) [6].

3.1 Experimental Setup

The experimental evaluation is performed through the noiseless part of the COCO/BBOB framework (COmparing Continuous Optimizers / Black-Box Optimization Benchmarking) [7, 8]. It is a collection of 24 benchmark functions with different degree of smoothness, uni-/multimodality, separability, conditionality etc. Each function is

---

Algorithm 1 Surrogate CMA-ES Algorithm [3]

Input: g (generation), g_m (number of model generations),
  σ, λ, $\mathbf{m}$, $\mathbf{C}$ (CMA-ES internal variables),
  r (maximal distance between training points and $\mathbf{m}$),
  n_{REQ} (minimal number of points for model training),
  n_{MAX} (maximal number of points for model training),
  A (archive), f_{M} (model), f (original fitness function)
1: $\mathbf{x}_k \sim \mathsf{N}\left(\mathbf{m}, \sigma^2 \mathbf{C}\right)$ $k = 1, \ldots, \lambda$ {*CMA-ES sampling*}
2: **if** g is original-evaluated **then**
3:    $y_k \leftarrow f(\mathbf{x}_k)$ $k = 1, \ldots, \lambda$ {*fitness evaluation*}
4:    $\mathsf{A} = \mathsf{A} \cup \left\{(\mathbf{x}_k, y_k)\right\}_{k=1}^{\lambda}$
5:    $(\mathbf{X}_{\text{tr}}, \mathbf{y}_{\text{tr}}) \leftarrow \left\{(\mathbf{x}, y) \in \mathsf{A} \,|\, (\mathbf{m} - \mathbf{x})^\top \sigma \mathbf{C}^{-1/2}(\mathbf{m} - \mathbf{x}) \leq r\right\}$
6:    **if** $|\mathbf{X}_{\text{tr}}| \geq n_{\text{REQ}}$ **then**
7:      $(\mathbf{X}_{\text{tr}}, \mathbf{y}_{\text{tr}}) \leftarrow$ choose n_{MAX} points if $|\mathbf{X}_{\text{tr}}| > n_{\text{MAX}}$
8:      {*transformation to the eigenvector basis:*}
        $\mathbf{X}_{\text{tr}} \leftarrow \left\{(\sigma \mathbf{C}^{-1/2})^\top \mathbf{x}_{\text{tr}} \text{ for each } \mathbf{x}_{\text{tr}} \in \mathbf{X}_{\text{tr}}\right\}$
9:      $f_{\mathsf{M}} \leftarrow \text{trainModel}(\mathbf{X}_{\text{tr}}, \mathbf{y}_{\text{tr}})$
10:     mark $(g+1)$ as model-evaluated
11:   **else**
12:     mark $(g+1)$ as original-evaluated
13:   **end if**
14: **else**
15:   $\mathbf{x}_k \leftarrow (\sigma \mathbf{C}^{-1/2})^\top \mathbf{x}_k$ $k = 1, \ldots, \lambda$
16:   $y_k \leftarrow f_{\mathsf{M}}(\mathbf{x}_k)$ $k = 1, \ldots, \lambda$ {*model evaluation*}
17:   {*shift y_k values if $(\min y_k) < best\ y\ from\ \mathsf{A}$*}
      $y_k = y_k + \max\left\{0, \min_{\mathsf{A}} y - \min y_k\right\}$ $k = 1, \ldots, \lambda$
18:   **if** g_m model generations passed **then**
19:     mark $(g+1)$ as original-evaluated
20:   **end if**
21: **end if**
Output: f_{M}, A, $(y_k)_{k=1}^{\lambda}$

---

defined for any dimension $D \geq 2$; the dimensions used for our tests are 2, 5, 10, and 20. The set of functions comprises, among others, well-known continuous optimization benchmarks like ellipsoid, Rosenbrock's, Rastrigin's, Schweffel's or Weierstrass' function.

The framework calls the optimizers on 15 different instances for each function and dimension, meaning that 1440 optimization runs were called for each of the eight considered algorithms. The graphs at the end of the paper show detailed results in a per-function and per-group-of-function manner. The following paragraphs summarize the parameters of the algorithms.

The CMA-ES. The original CMA-ES was used in its IPOP-CMA-ES version (Matlab code v. 3.61) with number of restarts = 4, IncPopSize = 2, $\sigma_{start} = \frac{8}{3}$, $\lambda = 4 + \lfloor 3 \log D \rfloor$. The remainder settings were left default.

^{s*}*ACM-ES.* We have used Loshchilov's GECCO 2013 Matlab code `xacmes.m` [14] in its ^{s*}ACM-ES version, setting the parameters CMAactive = 1, newRestartRules = 0 and withSurr = 1, modelType = 1, withModelEnsembles = 0, withModelOptimization = 1, hyper_lambda = 20, λ_{Mult} = 1, μ_{Mult} = 1 and Λ_{minIter} = 4.

S-CMA-ES: GP5-CMA-ES and RF5-CMA-ES. The number after the GP/RF in the names of the algorithms denotes the number of model-evaluated generations g_m, which are evaluated by the model in row. All considered S-CMA-ES versions use the distance $r = 8$ (see algorithm 1). For the GP model, $K_{\text{Matérn}}^{\nu=5/2}$ covariance function with starting values $(\sigma_n^2, l, \sigma_f^2) = \log(0.01, 2, 0.5)$ has been used (see [3] for the details). We have tested RF comprising 100 regression trees, each containing at least two training points in each leaf. The CMA-ES parameters (IPOP version, σ_{start}, λ, IncPopSize etc.) were used the same as in the pure CMA-ES experiments. All S-CMA-ES parameter values were chosen according to preliminary testing on several functions from the COCO/BBOB framework.

BIPOP version of the algorithms. The bi-population versions BIPOP-CMA-ES and BIPOP-^{s*}ACM-ES use the same Loshchilov's Matlab code `xacmes.m` with the parameter BIPOP = 1. The BIPOP-GP5-CMA-ES and BIPOP-RF5-CMA-ES algorithms are constructed in the same manner as the S-CMA-ES was transformed from the CMA-ES – by integration of the Algorithm 1 into every generation of the BIPOP-^{s*}ACM-ES.

3.2 Results

The performance of the algorithms is compared in the graphs placed in Figures 1–3. The graphs in Figure 1 depict the *expected running time (ERT)*, which depends on a given target function value $f_{\text{t}} = f_{\text{opt}} + \Delta f$ – the true optimum f_{opt} of the respective benchmark function raised by a small value Δf. The ERT is computed over all relevant

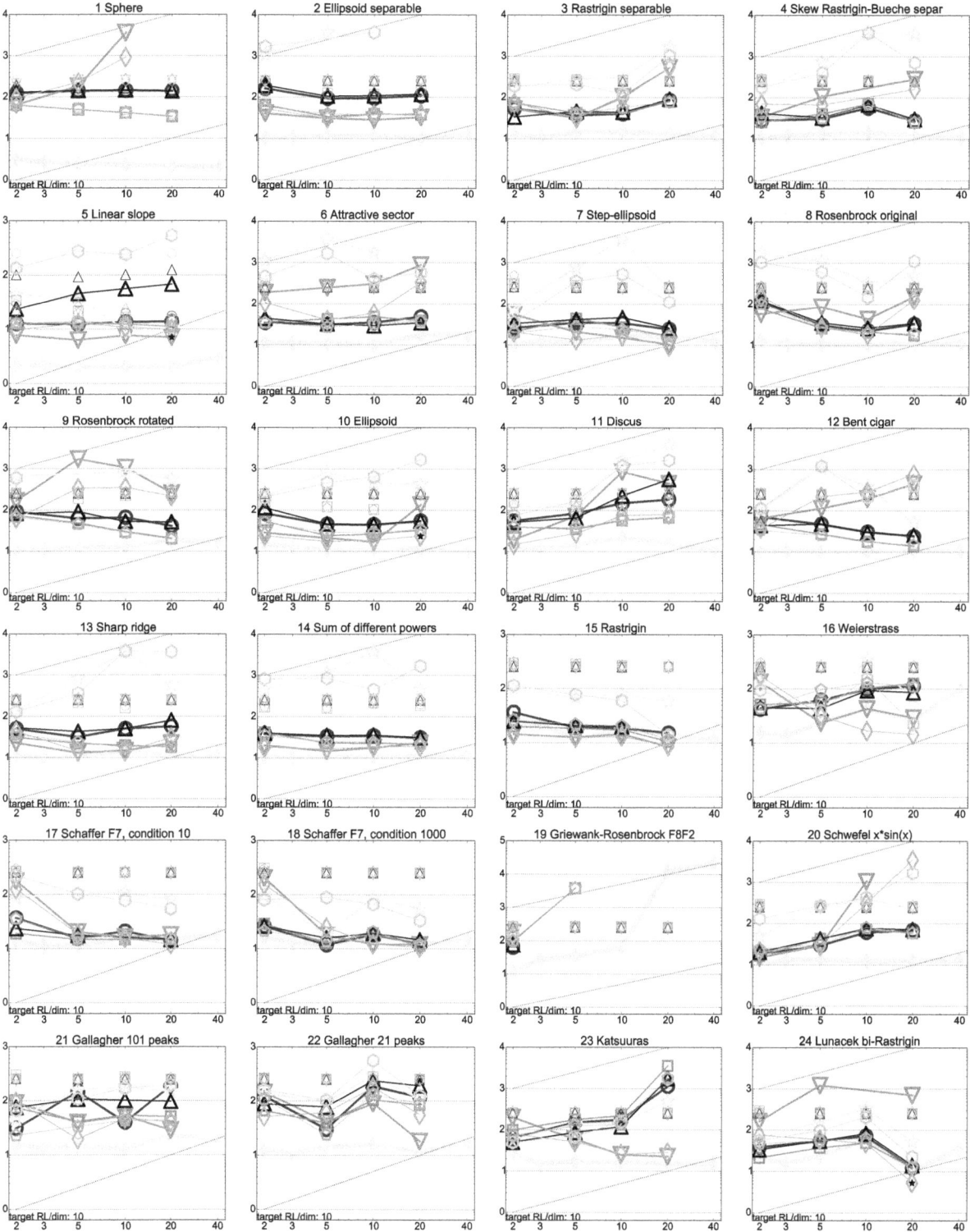

Figure 1: Expected running time (ERT in number of f-evaluations as $\log_{10}$ value) divided by dimension versus dimension. The target function value is chosen such that the bestGECCO2009 artificial algorithm just failed to achieve an ERT of $10 \times \text{DIM}$. Different symbols correspond to different algorithms given in the legend of f_1 and f_{24}. Light symbols give the maximum number of function evaluations from the longest trial divided by dimension. Black stars indicate a statistically better result compared to all other algorithms with $p < 0.01$ and Bonferroni correction number of dimensions (six). Legend: $\circ$:BIPOP-CMAES, $\triangledown$:BIPOP-GP5, :BIPOP-RF5, $\square$:BIPOP-saACMES, $\triangle$:CMA-ES, $\diamond$:GP5-CMAES, $\bigcirc$:RF5-CMAES, $\bigcirc$:saACMES

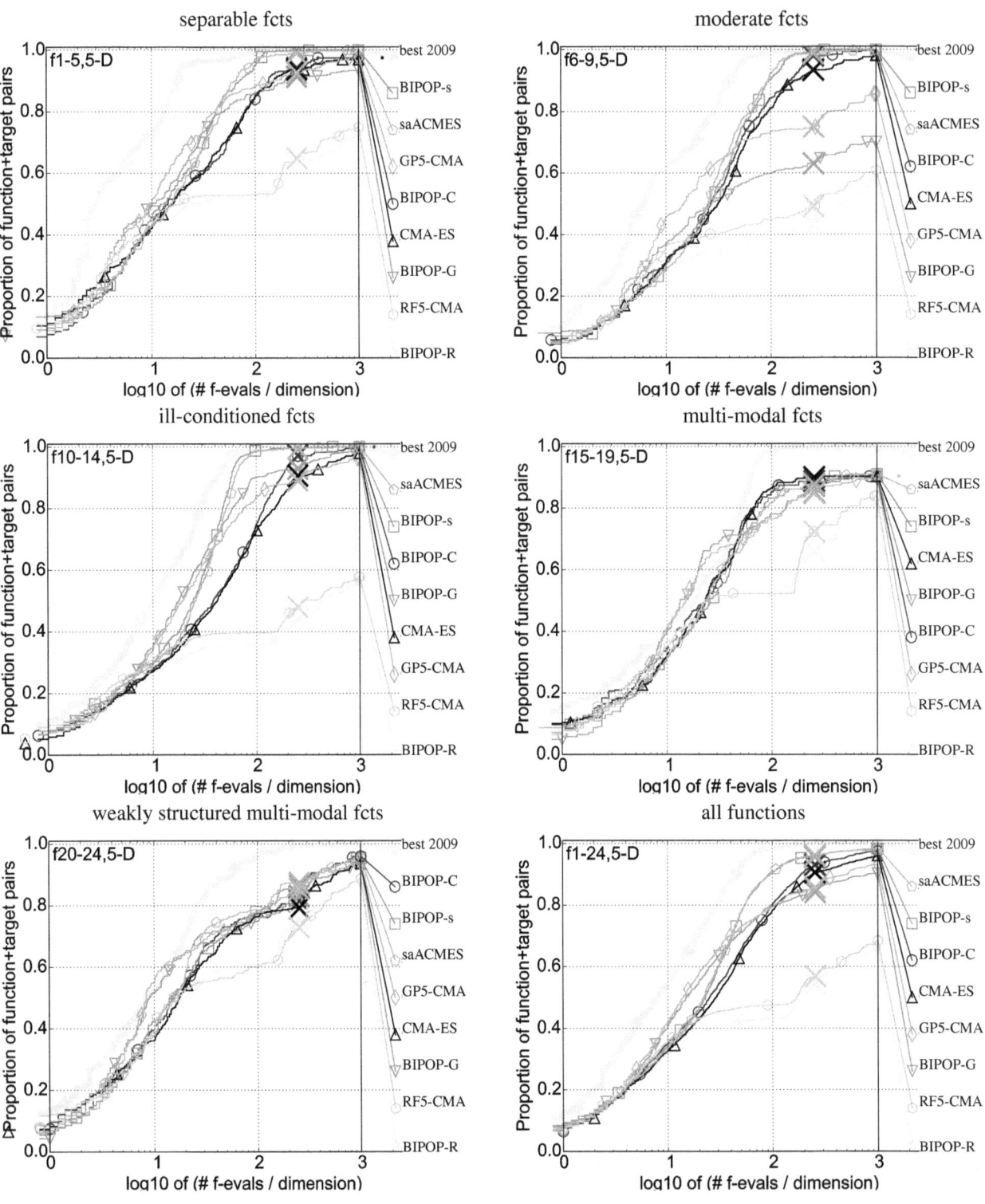

Figure 2: Bootstrapped empirical cumulative distribution of the number of objective function evaluations divided by dimension (FEvals/DIM) for all functions and subgroups in 5-D. The targets are chosen from $10^{[-8..2]}$ such that the bestGECCO2009 artificial algorithm just not reached them within a given budget of $k \times$ DIM, with $k \in \{0.5, 1.2, 3, 10, 50\}$. The "best 2009" line corresponds to the best ERT observed during BBOB 2009 for each selected target.

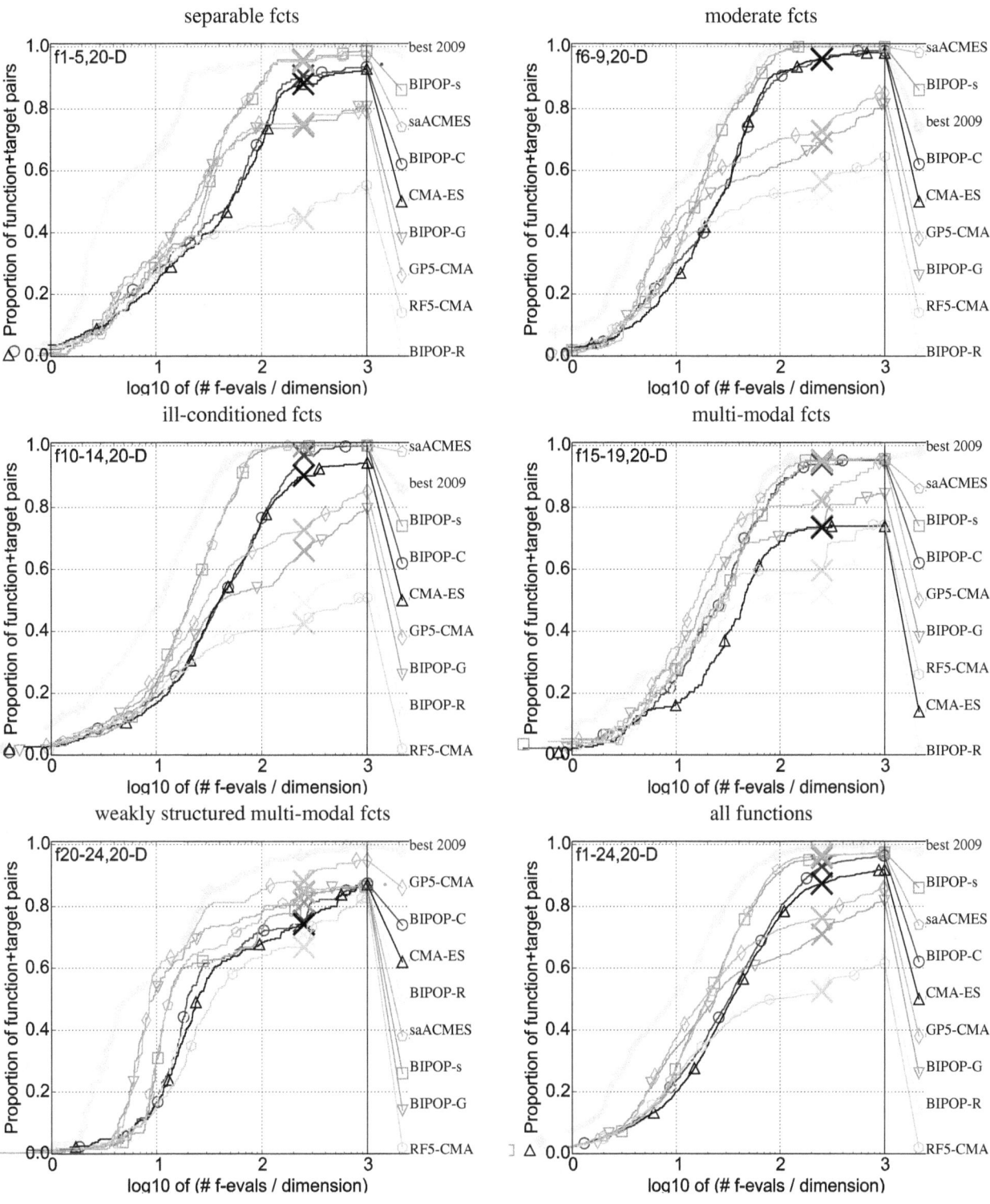

Figure 3:   Bootstrapped empirical cumulative distribution of the number of objective function evaluations divided by dimension (FEvals/DIM) for all functions and subgroups in 20-D. The targets are chosen from $10^{[-8..2]}$ such that the bestGECCO2009 artificial algorithm just not reached them within a given budget of $k \times$ DIM, with $k \in \{0.5, 1.2, 3, 10, 50\}$. The "best 2009" line corresponds to the best ERT observed during BBOB 2009 for each selected target.

trials as the number of the original function evaluations (FEs) executed during each trial until the best function value reached f_t, summed over all trials and divided by the number of trials that actually reached f_t [7].

As we can see in Figure 1, the 24 functions can be roughly divided into two groups according to the algorithm which performed the best (at least in 10D and 20D). The first group of functions where the CMA-ES performed best consists of functions 1, 3, 4, 6, and 20 while on functions 2, 5, 7, 10, 11, 13–16, 18, 21, 23, and 24, GP5-CMA-ES is usually better. The usage of the BIPOP versions generally leads to no improvement or even to performance decrease.

The graphs in Figures 2 and 3 summarize the performance over subgroups of the benchmark functions and show the proportion of algorithm runs that reached the target value $f_t \in 10^{[-8..2]}$ indeed (f_t was actually different for each respective function, see the figures captions). Roughly speaking, the higher the colored line, the better the performance of the algorithm is for the number of the original evaluations given on the horizontal axis.

Thus we can see that our GP5-CMA-ES usually outperforms the other algorithms when we consider the evaluations budget FEs $\leq 10^{1.5}D$, i.e. FEs ≤ 150 for 5D and FEs ≤ 600 for 20D. However, as the number of the considered original evaluations rises, the original CMA-ES or the $^s*$ACM-ES usually performs better. This fact can be summarized that our GP5-CMA-ES is convenient especially for the applications where a very low number of function evaluations is available, such as in [2].

4 Conclusions & Future Work

In this paper, we have compared the surrogate-assisted S-CMA-ES, which uses GP and RF continuous regression models, with $^s*$ACM-ES-k algorithm based on ordinal regression by Ranking SVM, and the original CMA-ES, all in their IPOP and BIPOP versions. The comparison shows that Gaussian process S-CMA-ES usually outperforms the ordinal-based $^s*$ACM-ES-k in early stages of the algorithm search, especially on multimodal functions (BBOB functions 15–24). However, the algorithms and surrogate models should be further analyzed and compared since, for example, the NEWUOA [16] or SMAC [10, 11] algorithms spend a considerably lower number of function evaluations than the CMA-ES in these early optimization phases. The BIPOP versions of the algorithms did not increased performances of appropriate IPOP versions except BBOB function 5.

A natural perspective of improving S-CMA-ES is to make the number of model-evaluated generations self-adaptive. We will additionally investigate different properties of continuous and ordinal regression in view of their applicability as regression models. Different cases and benchmarks where the ordinal regression is clearly superior to continuous regression will be further identified. For example, hybrid surrogate models combining both kinds of regression will be attempted.

Acknowledgements

This work was supported by the Czech Science Foundation (GAČR) grant P103/13-17187S, by the Grant Agency of the Czech Technical University in Prague with its grant No. SGS14/205/OHK4/3T/14, and by the project "National Institute of Mental Health (NIMH-CZ)", grant number CZ.1.05/2.1.00/03.0078 (and the European Regional Development Fund.). Further, access to computing and storage facilities owned by parties and projects contributing to the National Grid Infrastructure MetaCentrum, provided under the programme "Projects of Large Infrastructure for Research, Development, and Innovations" (LM2010005), is greatly appreciated.

References

[1] Auger, A., Hansen, N.: A restart CMA evolution strategy with increasing population size. In: The 2005 IEEE Congress on Evolutionary Computation **2**, 1769–1776, IEEE, Sept. 2005

[2] Baerns, M., Holeňa, M.:. Combinatorial development of solid catalytic materials. Design of high-throughput experiments, data analysis, data mining. Imperial College Press / World Scientific, London, 2009

[3] Bajer, L., Pitra, Z., Holeňa, M.: Benchmarking gaussian processes and random forests surrogate models on the BBOB noiseless testbed. In: Proceedings of the 17th GECCO Conference Companion, Madrid, July 2015, ACM, New York

[4] Breiman, L.: Classification and regression trees. Chapman & Hall/CRC, 1984

[5] Hansen, N.: The CMA evolution strategy: A comparing review. In: J. A. Lozano, P. Larranaga, I. Inza, E. Bengoetxea, (eds), Towards a New Evolutionary Computation, 192 in Studies in Fuzziness and Soft Computing, 75–102, Springer Berlin Heidelberg, Jan. 2006

[6] Hansen, N.: Benchmarking a BI-population CMA-ES on the BBOB-2009 function testbed. In: Proceedings of the 11th Annual GECCO Conference Companion: Late Breaking Papers, GECCO'09, 2389–2396, New York, NY, USA, 2009, ACM

[7] Hansen, N., Auger, A., Finck, S., Ros, R.: Real-parameter black-box optimization benchmarking 2012: Experimental setup. Technical Report, INRIA, 2012

[8] Hansen, N., Finck, S., Ros, R., Auger, A.: Real-parameter black-box optimization benchmarking 2009: Noiseless functions definitions. Technical Report RR-6829, INRIA, 2009, updated February 2010

[9] Hansen, N., Ros, R.: Benchmarking a weighted negative covariance matrix update on the BBOB-2010 noiseless testbed. In: Proceedings of the 12th Annual Conference Companion on Genetic and Evolutionary Computation, GECCO'10, 1673–1680, New York, NY, USA, 2010, ACM

[10] Hutter, F., Hoos, H., Leyton-Brown, K.: Sequential model-based optimization for general algorithm configuration. In: C. Coello, (ed.), Learning and Intelligent Optimization 2011, volume 6683 of Lecture Notes in Computer Science, 507–523, Springer Berlin Heidelberg, 2011

[11] Hutter, F., Hoos, H., Leyton-Brown, K.: An evaluation of sequential model-based optimization for expensive black-box functions. In: Proceedings of the 15th Annual Conference Companion on Genetic and Evolutionary Computation, GECCO'13 Companion, 1209–1216, New York, NY, USA, 2013, ACM

[12] Loshchilov, I., Schoenauer, M., Sebag, M.: Alternative restart strategies for CMA-ES. In: C. A. C. Coello, V. Cutello, K. Deb, S. Forrest, G. Nicosia, and M. Pavone, (eds), PPSN (1), volume 7491 of Lecture Notes in Computer Science, 296–305, Springer, 2012

[13] Loshchilov, I., Schoenauer, M., Sebag, M.: Self-adaptive surrogate-assisted covariance matrix adaptation evolution strategy. In: Proceedings of the 14th GECCO, GECCO '12, 321–328, New York, NY, USA, 2012, ACM

[14] Loshchilov, I., Schoenauer, M., Sebag, M.: BI-population CMA-ES algorithms with surrogate models and line searches. In: Genetic and Evolutionary Computation Conference (GECCO Companion), 1177–1184, ACM Press, July 2013

[15] Loshchilov, I., Schoenauer, M., Sebag, M.: Intensive surrogate model exploitation in self-adaptive surrogate-assisted CMA-ES (saACM-ES). In: Genetic and Evolutionary Computation Conference (GECCO), 439–446, ACM Press, July 2013

[16] Powell, M. J. D.: The NEWUOA software for unconstrained optimization without derivatives. In: G. D. Pillo, M. Roma, (eds), Large-Scale Nonlinear Optimization, number 83 in Nonconvex Optimization and Its Applications, 255–297, Springer US, 2006

[17] Rasmussen, C. E., Williams, C. K. I.: Gaussian processes for machine learning. Adaptative Computation and Machine Learning Series, MIT Press, 2006

J. Yaghob (Ed.): ITAT 2015 pp. 194–199
Charles University in Prague, Prague, 2015

A New Method to Combine Probability Estimates from Pairwise Binary Classifiers

Ondrej Šuch[1], Štefan Beňuš[2], and Andrea Tinajová[3]

[1] University of Žilina and Slovak Academy of Sciences, Slovakia `ondrejs@savbb.sk`,
[2] Constantine the Philosopher University and Slovak Academy of Sciences, Slovakia `sbenus@ukf.sk`
[3] Slovak Academy of Sciences `andrea.tinajova@gmail.com`

Abstract: Estimating class membership probabilities is an important step in many automated speech recognition systems. Since binary classifiers are usually easier to train, one common approach to this problem is to construct pairwise binary classifiers. Pairwise models yield an over-determined system of equations for the class membership probabilities. Motivated by probabilistic arguments we propose a new way for estimating individual class membership probabilities, which reduces to solving a linear system of equations. A solution of this system is obtained by finding the unique non-zero eigenvector of total probability one, corresponding to eigenvalue one of a positive Markov matrix. This is a property shared by another algorithm previously proposed by Wu, Lin, and Weng. We compare properties of these methods in two settings: a theoretical three-way classification problem, and via classification of English monophthongs from TIMIT corpus. **Index Terms**: binary classifiers; multiclass classification; phoneme recognition; English vowels; TIMIT

1 Introduction

Probabilistic approach underlies most current automatic speech recognition (ASR) systems, and very likely also human speech perception. In many ASR systems a common task is to provide estimates of probabilities of a given sample belonging to multiple classes given the observed values of its features. These classes may represent various phonemes, diphones or other kinds of linguistic categories.

In machine learning it is easier to find the boundary between two classes rather than the boundary separating a class from many other classes [1]. Moreover, many discriminative models are naturally suited to pairwise classification, such as logistic regression, LDA or variants of SVM. Thus given k classes C_i, one can readily construct $\binom{k}{2}$ pairwise discriminative models. Let us denote by M_{ij} the model discriminating classes C_i and C_j. Suppose that M_{ij} is able not only to discriminate, but also to compute the pairwise class membership probability r_{ij} of an object X with features **f**:

$$r_{ij} = r_{ij}(X) = p(X \in C_i | \mathbf{f}, X \in C_i \text{ or } X \in C_j). \quad (1)$$

Given the knowledge of $r_{ij}(X)$ the question is then to estimate multi-class probabilities p_i where

$$p_i = p_i(X) = p(X \in C_i | \mathbf{f}). \quad (2)$$

Inspired by Bradley-Terry model, Hastie and Tibshirani suggested [1] to require:

$$\frac{p_i}{p_i + p_j} = r_{ij} \quad (3)$$

$$\sum_i p_i = 1 \quad (4)$$

Note that there are $1 + \binom{k}{2}$ equations for k unknowns, so the system of equations is over-determined for $k \geq 3$ and it may be not possible to solve them.

In the next section we review several approaches which have been suggested to find approximate solution of (3). In Section 3 we will propose a new method to combine pairwise estimates. In Section 4 we will examine its performance with synthetic as well as real world acoustic data. In Conclusion we discuss findings of our experiments.

2 Existing Approaches

One natural requirement for an algorithm which determines probabilities p_i is that if the system (3) has a solution then the algorithm will find them exactly.

Several approaches satisfying this requirement are outlined in the work of Wu, Ling and Wen [2]. They consider the following functionals:

$$\delta_{HT} : \min_{\mathbf{p}} \sum_{i=1}^{k} [\sum_{j:j\neq i}^{k} (r_{ij}\frac{1}{k} - \frac{1}{2}p_i)]^2, \quad (5)$$

$$\delta_1 : \min_{\mathbf{p}} \sum_{i=1}^{k} [\sum_{j:j\neq i}^{k} (r_{ij}p_j - r_{ji}p_i)]^2, \quad (6)$$

$$\delta_2 : \min_{\mathbf{p}} \sum_{i=1}^{k} \sum_{j:j\neq i}^{k} (r_{ij}p_j - r_{ji}p_i)^2, \quad (7)$$

$$\delta_V : \min_{\mathbf{p}} \sum_{i=1}^{k} \sum_{j:j\neq i}^{k} (I_{\{r_{ij}>r_{ji}\}}p_j - I_{\{r_{ji}>r_{ij}\}}p_i)^2, \quad (8)$$

$$(9)$$

where I is the indicator function. Each of the four functionals is nonnegative. When the system (3) does have a solution, each functional is zero at, and only at the solution. One less satisfying feature of these approaches is that they lack probabilistic motivation, unlike the method we propose in the next section.

3 New Method

We will now describe our new algorithm. In general, one has $0 \le r_{ij} \le 1$. To avoid complications arising from degenerate cases we assume sharp inequalities $0 < r_{ij} < 1$, which poses no difficulty in practical applications.

Consider for a moment that an object X belongs to the class C_m. Then for judging its similarity to other classes one may restrict attention to the values r_{mj} (and $r_{jm} = 1 - r_{mj}$), since only classifiers M_{mj} were trained on values from the category C_m. But for those $k - 1$ values equations (3) can be solved exactly, as we will now show.

We have

$$\sum_{j \ne m} \frac{1}{r_{mj}} = \sum_{j \ne m} \frac{p_m + p_j}{p_m} = (k-1) + \frac{1 - p_m}{p_m}. \quad (10)$$

This relation allows us to compute an estimate $p_m^{(m)}$ of p_m explicitly as

$$p_m^{(m)} = \left(\sum_{j \ne m} \frac{1}{r_{mj}} - (k-2) \right)^{-1}, \quad (11)$$

where the upper index indicates that the estimate of p_m is computed by taking into account only values r_{mj}. The remaining probabilities can be then computed by the following formula:

$$p_j^{(m)} = p_m^{(m)} \cdot \left(\frac{1}{r_{mj}} - 1 \right). \quad (12)$$

Now we repeat this argument for $m = 1, 2, \ldots, k$. In general the estimates of p_i thus obtained will be conflicting i.e. in general $p_j^{(m)} \ne p_j^{(n)}$, because given values r_{ij} may not allow for solving (3) consistently. We will now take inspiration from the probability law $p(A) = \sum_i p(A|B_i)p(B_i)$, if B_i is a partition of the probability space. We will require that the estimate $\hat{p}_i$ of p_i should satisfy the following linear system of equations:

$$\hat{p}_j = \sum_m p_j^{(m)} \hat{p}_m, \quad \text{for } j = 1, \ldots, k. \quad (13)$$

These requirements can be interpreted as imposing self-consistency on the estimates $\hat{p}_i$. One readily checks that the matrix of the linear system (13) is Markov and positive, thus (13) has a one-dimensional space of solutions. Imposing an additional condition

$$\sum_m \hat{p}_m = 1 \quad (14)$$

determines a unique estimate $\hat{p}_m$ of p_m.

4 Evaluation of the New Method

First note that our algorithm will yield the correct solution if the system (3) has a solution. In order to see that, one first checks using (10) and (11) that $p_m^{(m)} = p_m$ and $p_j^{(m)} = p_j$. It follows that the vector p_j satisfies equations (13) and (14). Since the solution of (13) and (14) is unique, the method will yield the correct solution. However, this is an ideal, very special situation that will generally not hold for $k \ge 3$.

We have opted to do comparison testing of the proposed method with the method of Wu, Ling and Wen [2] that minimizes functional δ_1 (6). The reason is that that method also involves the construction of a positive Markov matrix whose solution is their estimate of p_m. We conduct two experiments: one is an artificial three-way classification problem, and the other a vowel recognition task.

4.1 Three-Way Classification

The system of equations (3) becomes over-determined for $k = 3$. If one of the classifiers is unreliable then the system (3) will not have a solution. In this section we present the results of a synthetic experiment for three-way classification.

In our experiment we assume that only classifier M_{23} is unreliable. In other words we assume that classifiers M_{12} and M_{13} discriminating respectively categories C_1 versus C_2 and C_1 versus C_3 yield precise estimates of r_{12} and r_{13}. For a fixed value p_1, p_2 we thus set $r_{12} = p_1/(p_1 + p_2)$ and $r_{13} = p_1/(p_1 + p_3) = p_1/(1 - p_2)$. Let $\hat{p}_m$ and p_m^{Wu} denote our and Wu's estimates of p_m. As r_{23} varies in interval $(0,1)$, define the absolute errors

$$\Delta = \sup_{i, r_{23}} |\hat{p}_i - p_i|, \quad (15)$$

$$\Delta_{\mathrm{Wu}} = \sup_{i, r_{23}} |p_i^{\mathrm{Wu}} - p_i|, \quad (16)$$

and the relative error

$$\Delta_{\mathrm{Wu}}^{\mathrm{rel}} = \sup_{i, r_{23}} |p_i^{\mathrm{Wu}} - \hat{p}_i|. \quad (17)$$

The results of our experiment are shown in Table 1. From the table it is clear that sometimes our method gives more precise estimates, but for other values of p_1, p_2, Wu's method will yield more precise results. However, in all cases, the relative error between our results and Wu's results is smaller than the absolute errors, often by an order of one magnitude.

4.2 Vowel Recognition

Unlike consonants, vowels may be perceived non-categorically by listeners [3], making it a good testing ground for multi-class probabilistic estimates. We opted for English language, because it has a large variety of vowels and because there are large corpora of annotated speech available. We worked with TIMIT, a phonetically segmented corpus of American English [4]. Our categories consisted of 15 monophthongs as shown in Table 2. For

p_1	p_2	Δ	Δ_{Wu}	Δ_{Wu}^{rel}
0.05	0.05	0.66	0.7	0.09
0.1	0.1	0.57	0.61	0.09
0.85	0.1	0.07	0.05	0.05
0.85	0.05	0.07	0.05	0.05
0.05	0.85	0.66	0.70	0.1
0.1	0.85	0.58	0.61	0.06
0.33	0.33	0.21	0.22	0.05

Table 1: Errors of estimation for various values of p_1 and p_2

vowel	sample word	sample word's transcription
iy	beet	bcl b IY tcl t
ih	bit	bcl b IH tcl t
eh	bet	bcl b EH tcl t
ae	bat	bcl b AE tcl t
aa	bott	bcl b AA tcl t
ah	but	bcl b AH tcl t
ao	bought	bcl b AO tcl t
uh	book	bcl b UH kcl k
uw	boot	bcl b UW tcl t
ux	toot	tcl t UX tcl t
er	bird	bcl b ER dcl d
ax	about	AX bcl b aw tcl t
ix	debit	dcl d eh bcl b IX tcl t
axr	butter	bcl b ah dx AXR
ax-h	suspect	s AX-H s pcl p eh kcl k tcl t

Table 2: Sample words containing 15 different monophong sounds of American English as segmented in TIMIT corpus

vowel	success rate	Wu's success rate	agreement
iy	48 %	48 %	96.6%
ih	21 %	21 %	94.8 %
eh	22 %	23 %	95.4 %
ae	60 %	60 %	94.4 %
aa	48 %	48 %	96.2 %
ah	20 %	21 %	94.6 %
ao	60 %	61 %	97.2 %
uh	18 %	18 %	95 %
uw	40 %	39 %	96.4 %
ux	40 %	40 %	97.4 %
er	34 %	35 %	95.6 %
ax	31 %	31 %	96.4 %
ix	16 %	18 %	94.4 %
axr	48%	46 %	96.2 %
ax-h	81 %	81 %	98.8 %

Table 3: Evaluation of our and Wu's [2] methods on individual monophthongs from the test data from TIMIT corpus. The first column indicates agreement between classification by our method and TIMIT annotation, the second column the statistics for method of Wu et al, and the third column indicates how often our method and Wu's method agreed on the most-likely classified class.

each of the categories we randomly chose their realizations from the set of male speakers in the corpus. Each realization was analyzed with a window 512 samples wide (at 16kHz sampling rate its length was 32ms). If the center of the window was less than 256 samples away from the next phoneme, it was proportionally less likely to be selected into our dataset. We have trained pairwise classifiers using linear discriminant analysis (LDA). The feature set was log-periodogram, where the analysis window was weighted with Hanning window before computing FFT.

We have performed comparison testing of our and Wu's method by selecting 500 random samples from the test subset. Per phone results are shown in Table 3. The key statistics is that overall there was 96% agreement between most-likely classifications by our method and Wu's method.

The overall success rate was slightly below 40% for both our and Wu's method. Due to the limitations of the features (no F0, no vowel duration, no dynamic information, no multiframe data), suboptimal performance may be expected. For instance without intensity baseline, it is nearly impossible to correctly distinguish some accented vowels.

We decided to do a more detailed case study. From the test subset we have chosen sentence SA1 spoken by speaker MREB0 and examined each monophthong at two points in time. The first was 5 milliseconds after the onset, and the other one approximately near the vowel's center. The results are shown in Table 4.

Likelihoods of most likely estimates of our and Wu's method are again quite close. There are two differences between onset and center predictions. The first one is misprediction of /er/ at the beginning of the word 'greasy', which is quite understandable, since the vowel is preceded by /r/. To gain an insight into the other mispredictions as well as deeper insight into dynamical behavior of the resulting multiclass classifier we present time plots in Fig. 1. In Fig. 1a the mis-classification of /iy/ instead of TIMIT's /ix/ in the word 'in' is shown. We speculate that the problem might be attributed to greater weight put on F2, that is relatively high and within the region for /iy/, compared to F1 that is quite high and definitely within the region for /ix/. In other words, the vowel might be a bit fronter than canonical /ix/. In Fig. 1b, the first vowel of 'greasy' is mis-classified as /ux/ instead of TIMIT's /iy/.

This problem might be attributed to coarticulation from the flanking consonants. The first vowel does have lower F2, which is plausibly responsible for /ux/ prediction, but it is preceded by /r/, which is commonly associated with lip protrusion, which lowers F2. In Fig. 1c in the vowel of word 'wash', we see that it is only in the beginning in the word 'wash' that the classifier gives more weight to /ao/, and then it increasingly agrees that the vowel is /aa/.

offset	TIMIT label	Wu's method		our method	
3831	iy	**iy**	80.1 %	**iy**	79.9 %
6053	ae	**ae**	79.7 %	**ae**	79.6 %
9187	axr	**axr**	62.2 %	**axr**	61.6 %
11780	aa	**aa**	32.9 %	**aa**	32.6 %
19677	ux	**ux**	60.3 %	**ux**	58.2 %
25544	ix	iy	66.4 %	iy	64.9 %
28905	iy	er	41.8 %	er	40.3 %
31328	iy	**iy**	53.4 %	**iy**	53.3 %
34210	aa	ao	76.3 %	ao	75.8 %
39080	ao	aa	77.1 %	aa	76.9 %
40680	er	axr	56.8 %	axr	56.3 %
42512	ao	**ao**	87.2 %	**ao**	87.1 %
46827	ih	iy	58.3 %	iy	57.9 %
48248	axr	**axr**	52.1 %	**axr**	52.4 %

(a) 5ms after vowel's start

offset	TIMIT label	Wu's method		our method	
4200	iy	**iy**	83.2 %	**iy**	82.9 %
6800	ae	**ae**	83.7 %	**ae**	83.6 %
9600	axr	**axr**	50.6 %	**axr**	50.7 %
12500	aa	**aa**	80.7 %	**aa**	79.5 %
21000	ux	**ux**	67.2 %	**ux**	66.4 %
25800	ix	iy	55.5 %	iy	53.3 %
29000	iy	ux	23.5 %	ux	22.8 %
31800	iy	**iy**	72.8 %	**iy**	72.7 %
35000	aa	**aa**	57.6 %	**aa**	57.7 %
39600	ao	aa	78.8 %	aa	78.5 %
41500	er	axr	66.4 %	axr	66.4 %
43500	ao	**ao**	86.3 %	**ao**	86.3 %
47500	ih	ux	37.9 %	ux	37 %
49000	axr	**axr**	71.1 %	**axr**	71.1 %

(b) near the center of the vowel

Table 4: Results of monophthong classification using spectral information in 32ms window centered at the offset indicated in the first column. Vowels were extracted from sentence SA1 spoken by speaker MREB0 from region 1 (New England). Most likely classes are shown computed by Wu's method and our method together with multi-class likelihoods.

In this particular case, we conclude that our classification is closer to the phonetic realization than TIMIT's. The beginning of the vowel is influenced by the preceding /w/ with lip rounding similar to /ao/. The rest of the vowel sounds like an /aa/ to phonetically trained listeners, and the formant values correspond to this perception. Finally, Fig. 1d shows the preference for /aa/ as the first vowel of 'water' in our model over /ao/ in TIMIT's. Similarly to Fig. 1c, this vowel sounds more, and its formant values correspond to our model more, than to TIMIT's. It should be noted, however, that /ao/ and /aa/ have merged in several American dialects and more tokens would be needed for a more thorough analysis.

A common way to improve the performance in automatic speech recognition is to tune the parameters of the system for a particular speaker. To that end we carried one more experiment. We extracted formants for TIMIT vowels spoken by speaker MREB0 using package *phonTools* in R [5]. Next we performed pairwise LDA training as previously but this time used values F1 and F2 for features rather than the log-periodogram. These first two formants are key perceptual features of vowels [6, 7, 8, 9]. Finally, we performed multiclass classification on the first vowel in the word 'water'. The formants contours for this vowel are shown in Fig. 2.

The somewhat suprising results are shown in Fig. 3. One would expect that it would have little problem with classification of the vowel. As seen in Fig. 3, except for a brief start, the classifier overwhelmingly believes that the phoneme is much closer to /aa/ than TIMIT annotated /ao/. However, compared to Fig. 1d the likelihood of /aa/ is markedly smaller near the vowel's boundaries.

5 Conclusions

We have described a new method for combining probability estimates from pairwise classifiers. It is quite general and for its application needs only pairwise classifiers that provide posterior likelihoods. We believe that since the rationale for our method is probabilistically motivated, it has the potential to edge out other methods in practice. In particular by its construction it avoids the problem of 'pairwise coupling' approaches pointed out by G. Hinton [1, pg. 467]. Another important feature is that the resulting probabilities are computed as the dominant eigenvector of a Markov matrix, allowing for efficient computation via iterations when the matrix of binary likelihoods varies slowly in time. Finally, since the method is not hierarchical, it avoids compounding of errors common in hierarchical approaches.

In presented synthetic and phonetic experiments its performance was very close to a method previously suggested by Wu [2]. The classification of English vowels was suboptimal, but that may not be indicative of performance in real world scenarios for several reasons.

- We have used all TIMIT vowel categories, some of which are in previously published performance benchmark tests fused because they are extremely hard to discriminate.

- Other pairwise classifiers, for instance logistic regression or SVM may yield better results.

- Based on the last experiment presented, we question whether TIMIT annotation is consistent throughout the corpus even for individual speakers.

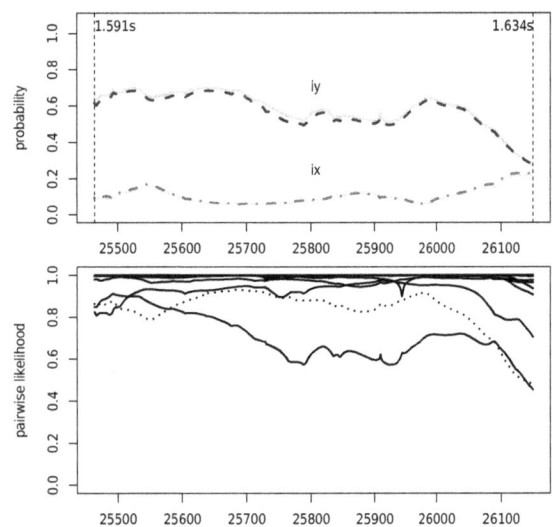

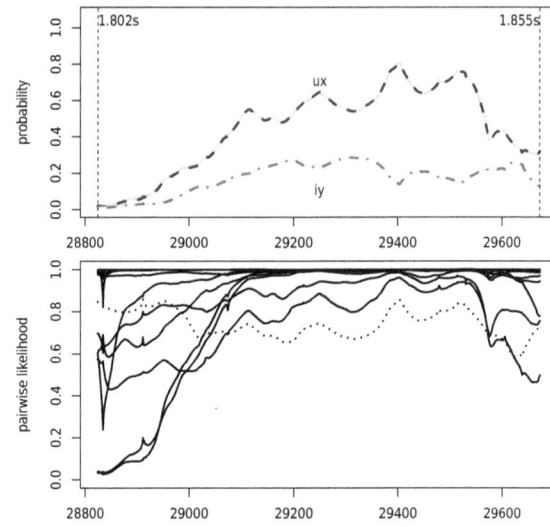

(a) TIMIT annotation is /ix/ in the word 'in'. We considered an alternative classification that the vowel is /iy/.

(b) TIMIT annotation is /iy/ for the first vowel in the word 'greasy'. We considered an alternative classification that the vowel is /ux/.

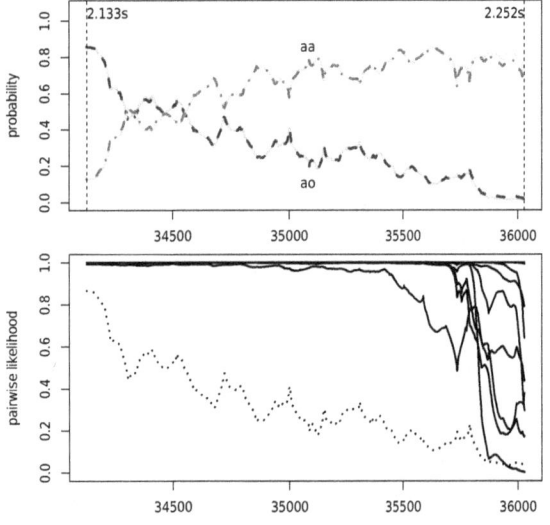

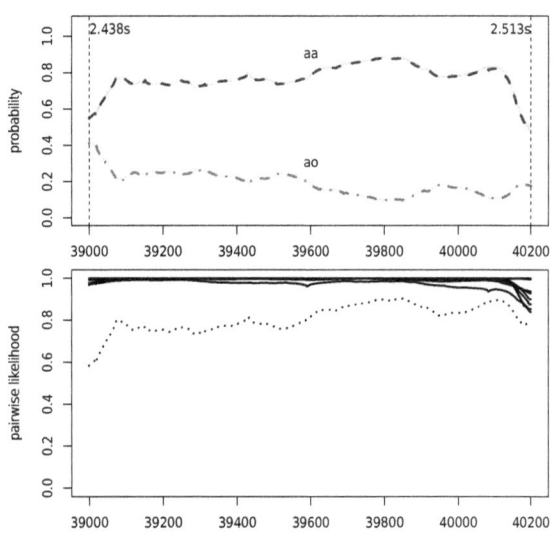

(c) TIMIT annotation is /aa/ in the word 'wash'. We considered an alternative classification that the vowel is /ao/.

(d) TIMIT annotation is /ao/ for the first vowel in the word 'water'. We considered an alternative classification that the vowel is /aa/.

Figure 1: Time series plots of multiclass and pairwise classification likelihoods for four vowels in sentence SA1 spoken by MREB0. The top plot in each subfigure shows multiclass likelihoods, and the bottom plot shows binary classification likelihoods r_{ij}. In multiclass plots, dashed dark curve indicates the likelihood of the alternative hypothesis and dark dash-dotted curve that of TIMIT annotation computed by our method (i.e. $\hat{p}_i$). Solid curves in multiclass plots indicate corresponding but visually nearly indistinguishable estimates obtained via Wu's method. In binary plots we plot likelihoods of the alternative hypothesis against all other classes. The dotted curve in each binary plot indicates likelihood of the alternative hypothesis compared to the TIMIT annotation.

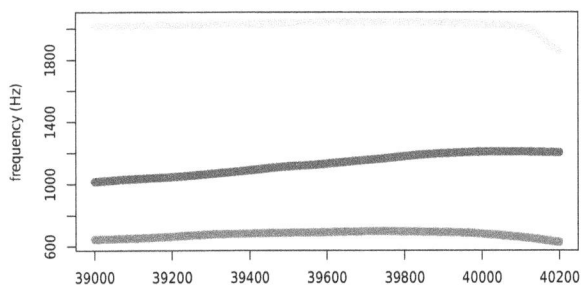

Figure 2: Formant contours F1-F3 for the first vowel of word 'water' in sentence SA1 spoken by MREB0.

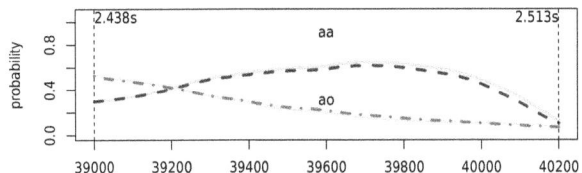

Figure 3: Time series plots of multiclass likelihoods for the first vowel in the word 'water' spoken in sentence SA1 by speaker MREB0. Dark dashed curve indicates likelihood of /aa/, whereas dot-dashed curve indicates likelihood of /ao/. Solid curves, as in Fig. 1, indicate estimate by Wu's method.

Further experiments with a complete ASR system may shed more light on the applicability of the proposed algorithm.

Acknowledgements

Our research was supported by the project University Science Park ITMS 26220220184 and grants APVV-0219-12, APVV-14-0560 and VEGA 2/0197/15. The authors are thankful to Paul Foulkes, K. Bachratá, and Martin Klimo for helpful discussion.

References

[1] Hastie, T. H., Tibshirani, R.: Classification by pairwise coupling. Annals of Statistics **26 (2)** (1998), 451–471

[2] Wu, T.-F., Lin, C.-J., Weng, R.: Probability estimates for multi-class classification by pairwise coupling. Journal of Machine Learning Research **5** (2004), 975–1005

[3] Fry, D., Abramson, A., Eimas, P., Liberman, A.: The identification and discrimination of synthetic vowels. Language and Speech **5** (1962), 171–189

[4] Garofolo, J., Lamel, L., Fisher, W., Fiscus, J., Pallett, D., Dahlgren, N., Zue, V.: TIMIT acoustic-phonetic continuous speech corpus, [Online], 1993.

[5] Barreda, S.: phonTools: functions for phonetics in R, R package version 0.2-2.0, 2014

[6] Potter, R., Steinberg, J.: Toward the specification of speech. J. Acoust. Soc. Amer. **22 (6)** (1950), 807–820

[7] Peterson, G., Barney, H.: Control methods used in a study of vowels. J. Acoust. Soc. Amer. **24 (2)** (1952), 175–184

[8] Turner, R., Patterson, R.: An analysis of the size information in classical formant data: Peterson and Barney (1952) revisited. J. Acoust. Soc. Jpn. **33** (2003)

[9] Kiefte, M., Nearey, T., Assmann, P.: Vowel perception in normal speakers. In: Handbook of vowels and vowel disorders, M. Ball and F. Gibbon, (Eds.) Psychology Press, 2012, ch. 6, 160–185